Air Pollution—Physiological Effects

Research Topics in Physiology

Charles D. Barnes, *Editor*
Department of Physiology
Texas Tech University School of Medicine
Lubbock, Texas

1. Donald G. Davies and Charles D. Barnes (Editors). Regulation of Ventilation and Gas Exchange, 1978

2. Maysie J. Hughes and Charles D. Barnes (Editors). Neural Control of Circulation, 1980

3. John Orem and Charles D. Barnes (Editors). Physiology in Sleep, 1981

4. M. F. Crass, III and C. D. Barnes (Editors). Vascular Smooth Muscle: Metabolic, Ionic, and Contractile Mechanisms, 1982

5. James J. McGrath and Charles D. Barnes (Editors). Air Pollution—Physiological Effects, 1982

Air Pollution— Physiological Effects

Edited by

JAMES J. McGRATH

CHARLES D. BARNES

Department of Physiology
Texas Tech University Health Sciences Center
School of Medicine
Lubbock, Texas

1982

ACADEMIC PRESS
A Subsidiary of Harcourt Brace Jovanovich, Publishers

New York   London
Paris   San Diego   San Francisco   São Paulo   Sydney   Tokyo   Toronto

ACADEMIC PRESS, INC.
111 Fifth Avenue, New York, New York 10003

United Kingdom Edition published by
ACADEMIC PRESS, INC. (LONDON) LTD.
24/28 Oval Road, London NW1 7DX

Library of Congress Cataloging in Publication Data
Main entry under title:

Air pollution--physiological effects.

(Research topics in physiology)
Includes index.
1. Air--Pollution--Physiological effect.
I. McGrath, James J. II. Barnes, Charles Dee.
III. Series.
QP82.2.A3A38   615.9'1   82-6650
ISBN 0-12-483880-4   AACR2

PRINTED IN THE UNITED STATES OF AMERICA

82 83 84 85   9 8 7 6 5 4 3 2 1

Contents

1. The Biochemistry of Cytotoxicity

Kathleen E. Everse and Johannes Everse

2. Effects of Gases and Airborne Particles on Lung Infections

Donald E. Gardner

7. Respiratory Airway Deposition of Aerosols

Richard M. Schreck

8. Mechanisms of Silica and Diesel Dust Injury to the Lung

Milos Chvapil

9. Physiological Effects of Cotton Dusts: Byssinosis

Richard L. Ziprin, Stephen R. Fowler, and Gerald A. Greenblatt

10. Physiological Effects of Lead Dusts

Samarendra N. Baksi

11. Work at High Altitude in Dusty Environments

Robert F. Grover

Contributors

Numbers in parentheses indicate the pages on which the authors' contributions begin.

William M. Abraham (107), Division of Pulmonary Disease, Mount Sinai Medical Center, Miami Beach, Florida 33140

Samarendra N. Baksi (281), Department of Physiology, Texas Tech University Health Sciences Center, School of Medicine, Lubbock, Texas 79430

Milos Chvapil (223), Department of Surgery, University of Arizona Health Sciences Center, Tucson, Arizona 85724

Johannes Everse (1), Department of Biochemistry, Texas Tech University Health Sciences Center, School of Medicine, Lubbock, Texas 79430

Kathleen E. Everse (1), Department of Anatomy, Texas Tech University Health Sciences Center, School of Medicine, Lubbock, Texas 79430

Stephen R. Fowler[1] (243), Veterinary Toxicology and Entomology Research Laboratory, Agricultural Research Service, U.S. Department of Agriculture, College Station, Texas 77841

Donald E. Gardner (47), Inhalation Toxicology Division, Health Effects Research Laboratory, U.S. Environmental Protection Agency, Research Triangle Park, North Carolina 27711

Gerald A. Greenblatt (243), Department of Plant Sciences, Texas Agricultural Experiment Station, Texas A & M University, College Station, Texas 77843

[1]Present address: Department of Pathology, The University of Texas Health Science Center at San Antonio, San Antonio, Texas 78284.

Robert F. Grover (311), Cardiovascular Pulmonary Research Laboratory, Division of Cardiology, Department of Medicine, University of Colorado Health Sciences Center, Denver, Colorado 80262

Guillermo Gutierrez (127), Department of Internal Medicine, The University of Michigan, Ann Arbor, Michigan 48103

Marc J. Jaeger[2] (81), Department of Physiology, College of Medicine, University of Florida, Gainesville, Florida 32610

James J. McGrath (147), Department of Physiology, Texas Tech University Health Sciences Center, School of Medicine, Lubbock, Texas 79430

Richard M. Schreck (183), Biomedical Science Department, General Motors Research Laboratories, Warren, Michigan 48090

Richard L. Ziprin (243), Veterinary Toxicology and Entomology Research Laboratory, Agricultural Research Service, U.S. Department of Agriculture, College Station, Texas 77841

[2]Present address: Meakins Christie Laboratories, McGill University, Montreal, Quebec, Canada.

Preface

This volume, *Air Pollution—Physiological Effects*, is the fifth in a series prepared under the auspices of the Department of Physiology, Texas Tech University Health Sciences Center, School of Medicine. Each year, we examine an area in which research is progressing rapidly, but which has not been discussed recently in an advanced, comprehensive review. Previous volumes surveyed current work in the regulation of ventilation and gas exchange, neural control of circulation, the physiology of sleep, and the physiology of vascular smooth muscle.

In the present work, eminent investigators from industry, government, and academe review their studies of physiological responses to air pollutants. Each author presents the historical basis and theory from which his interest evolved, the current status of his specialized area, and directions of research. Furthermore, each contributor places special emphasis on critical evaluation of the experimental data in his respective area.

The contributions are organized in three sections. The first comprises a chapter exploring cellular injury, with emphasis on lung tissue. The second (five chapters) is concerned primarily with the physiological responses to the potentially toxic gases (e.g., oxidant gases, sulfur dioxide, and carbon monoxide) that are inherently part of a technologically advanced society. The third (five chapters) discusses both particulate (e.g., silica, diesel, cotton, and lead dusts) pollution, an ever increasing problem affecting an ever broader spectrum of lives, and the special physiological problems posed by working at high altitudes in dusty environments.

We expect that this volume will be useful to not only environmental health scientists but also students and researchers in areas peripheral to environmental physiology. Furthermore, we believe that the book is provocative and likely to stimulate productive research in environmental physiology.

James J. McGrath
Charles D. Barnes

Air Pollution—Physiological Effects

1

The Biochemistry of Cytotoxicity

Kathleen E. Everse and Johannes Everse

I. INTRODUCTION

Air pollutants drawn into the respiratory system may produce various effects on the alveolar and bronchial tissues. Some pollutants are chemical compounds that react with the tissue constituents and cause alterations in the chemical structure of these constituents. Examples of such pollutants are HCN, NO_2, H_2S, CO, and a large number of other inorganic and organic compounds. Other pollutants are chemically rather inert and do not promote a chemical reaction in the bodily tissues. Instead, such pollutants accumulate in the respiratory system. Examples of

1

these pollutants are coal dust, asbestos, and chalk dust. Such pollutants, usually present in the form of small particles, are normally ingested by a type of phagocytic cell, the alveolar macrophages. Macrophages containing ingested material are subsequently excreted. Phagocytic cells are also able to release destructive enzymes. They do so in a futile attempt to destroy particles too large to be engulfed. These enzymes can cause damage and even death to adjacent normal lung tissue. Processes or events that cause damage to a cell and, if sustained, will ultimately result in the demise of the cell are cytotoxic processes. Many organic and inorganic compounds possess the ability to kill cells and are therefore considered to be cytotoxic.

Another class of air pollutants (although not always thought of in this manner) are living organisms such as bacteria and viruses. These pollutants, if not properly controlled by the host, can not only invade the host's tissues, but cause severe pathological conditions. The host controls this type of pollutant by using specific cells that kill the invading organisms. This process is generally referred to as cell cytotoxicity.

In this chapter we aim to discuss the cytotoxic activity displayed by various cell types that are present in higher organisms. The cytotoxic killing of cells by other cells (cell cytotoxicity) is an important and common event that occurs in higher organisms during many diseases as well as under healthy conditions.

Cell death is normally viewed as a terminal phenomenon. It should be recognized, however, that cell death is also an essential process in the normal development of multicellular organisms. The reabsorption of the amphibian tail and gills during metamorphosis and the formation of interdigital spaces during the development of many vertebrates are typical examples. Glücksman (1965) has provided a classical review listing 74 examples of cell death that occur and are necessary during normal vertebrate development. An understanding of the regulation of cell death during development is a major area of concern in the field of teratology. There are many examples where a failure of cells to die that were destined for removal or, conversely, the premature death of cells that were destined to multiply results in congenital anomalies (Menkes *et al.*, 1970).

Cytotoxic events are also part of the normal maintenance and repair processes that occur in higher organisms. Several examples of such cytotoxic processes are the removal of extravascular blood, the shedding of a snake's skin, normal cellular turnover such as the shedding of the endometrium or the removal of aged blood cells, and the process of wound healing.

Cell cytotoxicity is obviously important in the defense system of an organism; the killing of foreign invaders such as microbes and parasites and the removal of endogenous hazards such as tumor cells are typical examples. A failure in the cytotoxic activity of the defense system may result in pathological occurrences such as bacterial, viral, and parasitic infections, as well as uncontrolled tumor invasion.

In addition to the normal processes already mentioned, there are many pathological processes that have as a result unnecessary or undesired cell death. Among these pathological processes are toxic responses to environmental hazards such as oxygen toxicity, metal poisoning, chemical injuries, radiation damage, and bacterially induced toxic shock. Also to be included among the pathological occurrences that have cell death as a component are inflammation, autoimmune diseases such as arthritis, and other diseases such as Parkinson's disease, muscular dystrophy, and ulcers. Indeed, many of the diseases affecting the animal kingdom result in some unnecessary or undesired cell death. An understanding of the mechanisms involved in cytotoxicity and the regulation of cytotoxic events will undoubtedly contribute to our understanding of these phenomena and, hopefully, to the eventual control of many of the pathological states that affect mankind.

Cellular cytotoxicity is the damage to one cell caused by another cell, a killer cell. Killer-cell activity has been implicated in many (but not all) of the examples of cell death mentioned above, both normal and pathological. There are at least two, if not more, basic types of cellular cytotoxicity: (1) an attack on the target cell membrane constituents by lysosomal-type hydrolytic enzymes, and (2) an attack on the target cell membrane constituents by active oxygen species. Many cytotoxic cells have the potential to kill by either mechanism, but the oxidative mechanism is more efficient and faster. Cell death that results from environmental hazards and in which cellular cytotoxicity has not been implicated could well be promoted by oxidative mechanisms similar to those used by cytotoxic cells.

In this chapter we hope to provide the reader with an overview of cellular cytotoxicity and of the recent advances in our understanding of the biochemical basis of oxidative cell killing.

II. CELLULAR CYTOXICITY

A variety of cell types have the capacity to be cytotoxic toward other cells. Among these are the circulating cells of the immune system such as the neutrophil, the eosinophil, various lymphocytes including the natural killer (NK) cell, and the blood monocyte. Other phagocytic cells reside in various tissues. These include the brain microglia, the Schwann

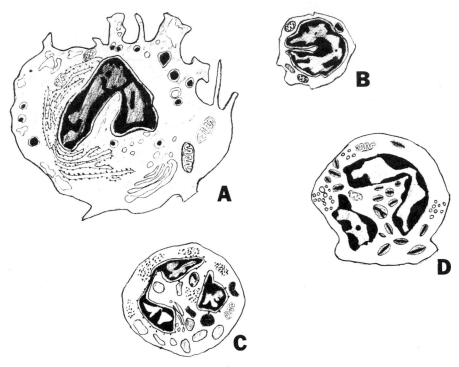

Fig. 1. Morphology and relative sizes of various leukocytes: (A) Macrophage. (B) Lymphocyte. (C) Neutrophil. (D) Eosinophil.

cells of the peripheral nervous system, and the various tissue macrophages which include the Kupffer cells of the liver, the histiocytes of connective tissue, the dendritic cells of the lymph nodes, arfd the alveolar and peritoneal macrophages.

The various types of killer cells of the immune system are identified and grouped according to their morphology and coloration with histochemical stains. The morphology and relative size of the cytotoxic cells are depicted in Fig. 1. Several of these cell types are further subclassified by their function, by their location, or by the presence of various antigens on their plasma membranes.

A. The Neutrophil

Neutrophils constitute the primary defense against microbial infection in mammals. They have been studied more extensively than other cytotoxic cells of the immune system, perhaps because they comprise the largest proportion of the leukocytes. They comprise 50–70% of the white cells of the peripheral blood. Neutrophils are found scattered throughout many tissues, particularly in areas of acute inflammation.

Neutrophils belong to a group of cells called either poly-morphonuclear leukocytes or granulocytes. Granulocytes are characterized by a multilobulated nucleus and many prominent cytoplasmic granules. After being stained with hematoxylin and eosin, the neutrophils show only a pale pink coloration (neutral-staining), because their cytoplasmic granules neither absorb the acidic (red) nor the basic (blue) stain. The cytoplasm of the neutrophil is rich in glycogen particles but it contains scant other cytoplasmic organelles such as mitochondria or endoplasmic reticulum.

At least two types of granules are found in the cytoplasm of the neutrophil: the azurophil (or primary) and the specific (or secondary) granules (Scott and Horn, 1970). They can be distinguished by cytochemical as well as isolation techniques. The ratio of specific granules to azurophil granules found in the mature neutrophil is approximately 2–3:1.

The neutrophil granules play an important role in the destruction of ingested bacteria. Table I lists the various components associated with the human neutrophil granules. Some differences in the composition of these granules are found between species. Note that most of the enzymes associated with the granules are hydrolytic enzymes specifically geared to the destruction and digestion of bacteria. It is also noteworthy that one of the enzymes found in the azurophil granule, myeloperoxidase, is present in amounts greater than 5% of the total dry weight of the cell (Schultz and Kaminker, 1962; Rohrer et al., 1966).

A bacterial infection or an acute inflammatory response involves the mobilization of neutrophils from the bone marrow and blood vessels to the site of injury. For the first 6–24 hours, neutrophils are the predominant leukocyte at the site of tissue injury. They are relatively short-lived and disappear during the next 24 hours. Their mean half-life in the circulation is approximately 7 hours.

The main function of the neutrophil at the inflammatory site is the destruction of foreign particles or cells. The first step in the cytotoxic action of the neutrophils is their attraction to the location of the invading matter. This is accomplished by a process called chemotaxis, in which the cells move toward increasing concentrations of an attractant. The most significant chemotactic factors that induce this response in the neutrophil are bacterial products and components of the complement system. Chemotaxis proceeds until contact between the cytotoxic cell and the target cells is established.

The establishment of contact between the killer cell and the target cell initiates the process of phagocytosis. During phagocytosis, foreign

TABLE I

Human Neutrophil Granule Components[a]

Azurophil (primary)	Specific (secondary)	Membrane fraction
Acid hydrolases	Lysozyme	Alkaline phosphatase[b]
Acid β-	Lactoferrin	Acid p-
glycerophosphatase	Collagenase[b]	nitrophenylphosphatase
β-Glucuronidase	Alkaline phosphatase[b]	
N-Acetyl-β-glucosamini-	Vitamin B_{12}-binding	
dase	proteins	
α-Mannosidase		
Arylsulfatase		
β-Galactosidase		
5'-Nucleotidase		
α-Fucosidase		
Acid protease (cathepsin)		
Neutral proteases		
Cathepsin G		
Elastase		
Collagenase[b]		
Cationic proteins		
Myeloperoxidase		
Lysozyme		
Acid mucopolysaccharide		

[a] From Klebanoff and Clark (1978).
[b] There are conflicting data on the location of these components.

bodies are taken up by the phagocytic cell (endocytosis), packaged into a vacuole (phagosome), and subsequently destroyed. A brief description of the processes associated with phagocytosis may be useful.

Adherence of foreign matter to receptor sites of the phagocyte is followed very rapidly by ingestion. Pseudopodia of the neutrophil surround and engulf the particle by a process of differential adherence called the zipper mechanism (Fig. 2) (Griffin *et al.*, 1975, 1976). The net result of this mechanism is that part of the outer surface of the plasma membrane becomes the inner membrane of the phagosome.

Stossel and Hartwig (1975, 1976) have proposed a biochemical mechanism for phagocytosis. They suggest that the adherence of a foreign particle to the phagocyte activates the actin-binding protein. This activation causes a gelation of the actin at the outer edge of the plasma membrane, followed by a myosin-induced contraction of the gel toward the particle. This gelation—contraction process is initiated in the adjacent cytoplasm to form pseudopods that gradually engulf the particle (see

Fig. 2). This process requires ATP and Ca^{2+} similar to the process in muscle contraction (Stossel and Hartwig, 1975, 1976; Hartwig and Stossel, 1976; Stossel, 1977).

The ATP that supplies the energy for phagocytosis is mostly derived from glycolysis. Inhibitors of glycolysis such as deoxyglucose, fluoride, arsenite, and iodoacetate inhibit the ingestion process (Sbarra and Karnovsky, 1959; Kvarstein, 1969; Boxer *et al.*, 1977). Because there are few mitochondria in neutrophils, energy derived from oxidative phosphorylation probably does not play a major role in the phagocytic process. This is supported by the fact that phagocytosis is not inhibited by dinitrophenol (Sbarra and Karnovsky, 1959; Allison *et al.*, 1963).

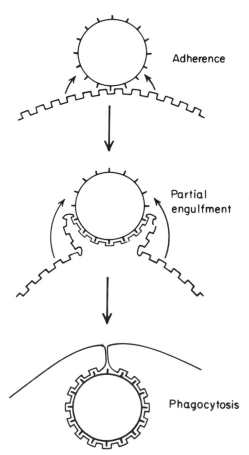

Adherence

Partial engulfment

Phagocytosis

Fig. 2. The zipper model of phagocytosis. (Reproduced with permission from Klebanoff and Clark, 1978.)

The next step in the phagocytic process after the compartmentation of the microorganism into a phagosome is degranulation. Both the specific and the azurophil granules adjacent to the phagosome move toward and fuse with the phagosomal membrane (Poste and Allison, 1973). After fusion, the granules rupture and discharge their contents into the phagosome, which initiates the process of destruction of the ingested foreign body.

Many of the constituents of these granules are capable of interacting with and causing injury to the neutrophil itself as well as other host cells. During the process of phagocytosis, the enzymes of these granules may be regurgitated to the extracellular fluid, thus endangering other host cells (Weissmann *et al.,* 1971a,b; Bainton, 1973; Henson, 1971). Damage to the neutrophil itself can be caused by the ingestion of materials with sharp edges. For example, the ingestion of urate crystals by neutrophils causes damage to the phagosomal membrane and the leakage of the enzyme contents into the cytoplasm (Spilberg, 1975).

B. The Eosinophil

The eosinophil, like the neutrophil, is a polymorphonuclear leukocyte. The morphology of the cell is illustrated in Fig. 1. The eosinophil has a bilobed nucleus with no nucleolus. In contrast to the neutrophil, there are many cytoplasmic organelles such as numerous mitochondria, Golgi apparatus, ribosomes, and rough endoplasmic reticulum. Also scattered throughout the cytoplasm is an abundance of glycogen particles and granules. The name eosinophil is derived from the high affinity that the basic protein of the granules has for the red dye eosin.

Eosinophils reside chiefly in the body tissues. They are localized particularly in areas exposed to the external environment such as the epithelium of the skin, bronchi, gastrointestinal tract, and vagina. The ratio of eosinophils in the circulating blood to those residing in the tissue is about 1:500 (Hirsch, 1965).

Eosinophils are formed in the bone marrow. They remain in the circulation with a half-life of 3–8 hours and then migrate into the tissues (Parwaresch *et al.,* 1976). The normal concentration range of the circulating eosinophil in man is 1–4% of the blood leukocytes.

One of the most characteristic morphological aspects of the eosinophil is the presence in the cytoplasm of numerous granules, both large and small. Three distinct types of granules can be identified in the eosinophil cytoplasm. The most outstanding are the "specific" granules. These are ovoid with a matrix and a central electron-dense crystalloid region called the internum (Miller *et al.,* 1966). The precursors of the specific granules are the large, round, electrondense "primary" granules. Furthermore there are two small granules: one spherical, containing arylsulfatase and acid phosphatase (Parmley and Spicer, 1974), and the other called specific microgranules, having a cup or ring shape (Schaefer et al., 1973).

It may be instructive to compare the enzymatic content of the eosinophil granules with that of the neutrophil granules. Both granules contain β-glucuronidase, ribonuclease, cathepsin, β-galactosidase, mannosidase, collagenase, histaminase, and acid and alkaline phosphatase. In contrast to the neutrophil, the eosinophil granules contain no lysozyme or phagocytin, and the activity of alkaline phosphatase is low (Makita and Sanborn, 1970; West et al., 1975). Several enzymes are present in the eosinophil granules in larger amounts than in the neutrophil granules. They include arylsulfatase B, lysophospholipase, phospholipase D, and phospholipid exchange protein, as well as NADPH oxidase (Weller et al., 1980a; DeChatelet et al., 1977). Differences vary from 3-fold, for the NADPH oxidase, to as high as 25-fold for the phospholipid exchange protein. It is tempting to speculate that the differences in enzyme content between the eosinophil and neutrophil granules reflect a difference in the specificity of these cytotoxic cells toward their target cells. It is particularly noteworthy in this respect that the peroxidase, which is intimately associated with the killing process (see page 24), is a different enzyme in the two phagocytes (Klebanoff et al., 1980).

Two proteins comprise the major portion of the specific granule. Gleich et al. (1976, 1980) have characterized an 11,000 dalton protein of the crystalloid internum called the "major basic protein," which constitutes approximately 50% of the total granule protein. The "eosinophil cationic protein" has been isolated from specific granules of the human eosinophil. This 21,000 dalton protein constitutes about 30% of the specific granule protein and contains 2.5 moles Zn/mole protein (Venge et al., 1980). Although these two proteins reportedly constitue 80% of the total specific granule protein, their biological function(s) has thus far not been identified.

Characteristic pale green, slender, hexagonal crystals form after eosinophil death in vitro. Such crystals maybe found in areas of eosinophilic inflammation in vivo. These Charcot–Leyden crystals contain an appreciable quantity of Zn, 188 parts/million (Archer and Blackwood, 1965). Recently, Weller et al. (1980b) identified the protein of these crystals as lysophospholipase. Whether or not this protein is related to the zinc-containing eosinophil cationic protein remains to be established.

Eosinophils accumulate at a variety of immunological and inflammatory sites throughout the body. Eosinophils are especially abundant at the sites of parasitic infections; such infections may also be associated with marked peripheral blood eosinophilia. In a study by Vadas *et al.*, (1979), the eosinophil was the only cell that was capable of causing irreversible damage to schistosomula. This clearly indicates that the eosinophil function as a cytotoxic cell, perhaps with specificity for parasites.

The accumulation of eosinophils at tissue sites as well as peripheral blood eosinophilia is a characteristic response to immune reactions mediated by IgE, such as the immediate-type hypersensitivity reaction. Phagocytosis of IgE complexes by eosinophils has been reported to be more efficient than the phagocytosis of other immunoglobulin complexes (Greco *et al.*, 1973).

Eosinophils may also be found localized in the uterine tissue. This localization is cyclical and dependent on estrogen levels. Tchernitchin (1967) reported that labeled estradiol given to female rats accumulated in the uterus and that most of the radiolabeled material was bound to the surface of eosinophils. In later studies Tchernitchin *et al.* (1974, 1975) described a dramatic increase in the eosinophil population of the uterus after the administration of estrogen. Again, the estrogen was found to be bound tightly to the eosinophil membranes. This could be a method of regulating the cytotoxic activity of the eosinophil toward the cells in the uterine lining.

In a manner similar to that of the neutrophil, the eosinophil migrates toward areas of an inflammatory response. This migration can be quite rapid as demonstrated by the retest reaction. When an antigen is reinjected into the site of a former antigenic response, the eosinophil is the predominant cell at the site within 2 hours after reinjection (Arnason and Waksman, 1963). Obviously, a potent and specific attractant is necessary for such a rapid response.

Many factors have been identified that can serve as attractants for the eosinophil migratory response, either *in vitro* or *in vivo*. These substances may be grouped into the following categories: (1) lipid cell products, (2) lymphokines, (3) substances released during an anaphylactic response, and (4) products of complement activation. Thus far, most of the substances identified to be chemotactic for the eosinophil are nonspecific and attract other cells equally well. Chemotactic effectors of eosinophils that provoke the greatest response are (1) lipoxygenation products of arachidonic acid, (2) the C5a fragment of complement activation, (3) synthetic formyl methionyl peptides, and to a lesser degree, (4) histamine, and (5) the eosinophil chemotactic factor of anaphylaxis (Goetzl and Austen, 1975; Goetzl *et al.*, 1980; Goetzl, 1980).

Eosinophils, like neutrophils, are capable of bacterial phagocytosis. In a recent study, Sher and Wadee (1980) observed two types of phagocytosis in the eosinophil: (1) the formation of pseudopodia that enveloped the bacteria, and (2) the formation of craters in the plasma membrane through which bacteria entered. As with the neutrophils, phagocytosis is followed by a fusion of the granules with the phagosome, degranulation, and death of the ingested organisms (Mickenberg *et al.*, 1972). Phagocytosis by eosinophils, however, occurs at a much lower rate than that of neutrophils (Cline *et al.*, 1968).

In addition to endocytosis, the eosinophil can accomplish its killing by extracellular degranulation or exocytosis. Glauert *et al.* (1980) observed that eosinophils make a much more intimate contact with a nonendocytosable target than do the neutrophils and then secrete enzymes by a direct fusion of the granules with the other plasma membrane. Caulfield *et al.* (1980a) observed eosinophils to discharge their granules onto the surface of schistosomula. After disruption of the cells by pipetting, the eosinophil plasma membranes adhered to the electron-dense material on the parasite (Caulfield *et al.*, 1980b), again demonstrating the intimate contact between the eosinophil and the target cell.

Finally, eosinophils have been implicated in the pathological destruction of normal tissue (Beeson and Bass, 1977). Several types of cellular damage to the nervous system exist, which may be caused by eosinophil cytotoxicity (Seiler *et al.*, 1969; Chusid *et al.*, 1975). Moreover, the continued presence of high numbers of eosinophils in the circulation appears to cause a clinical entity called Löffler's eosinophilic myocarditis, which is associated with destruction of cardiac tissue (Yam *et al.*, 1972; Spry and Tai, 1976).

The information available thus far supports the concept that the cytotoxic activity of eosinophils is directed preferentially, but not absolutely, toward eukaryotic cells, whereas that of the neutrophil appears to be more geared to the killing of prokaryotes. Such a difference in function would be consistent with the differences in granule enzymes of the two granulocytes, as we discussed earlier.

C. The Macrophage

The macrophages or mononuclear phagocytes comprise a group of large cells which are highly specialized in their endocytic and digestive functions. Macrophages are found throughout the body at a variety of sites, and depending on their tissue location, they may be known by other names. Macrophages may be either free-moving cells (e.g., the blood monocyte, the alveolar macrophage, and the peritoneal macrophage) or remain fixed within the tissue (e.g., the spleen macrophage, the connective tissue histiocyte, the lymph node dendritic cell, and the liver Kupffer cell). The morphological and biochemical similarities of these various cells, as well as the probable similarity of their origin, have been the primary reasons to consider them as a group of cells called the macrophages. For the same reasons we suspect that these various macrophages share some similar or common functions. Nonetheless, it should be recognized that in spite of the common grouping, these various macrophages do exhibit differences in their properties and may not even appear similar, because they have acquired specialized functions to suit their particular tissue location. (For a comprehensive treatment of the various macrophages, see Van Furth, 1975; Carr and Daems, 1980; Sbarra and Strauss, 1980.)

For many years the macrophage has been viewed as a debris-eating scavenger; however, research during the past 10 years has provided a wider perspective on the role of this cell. It is now apparent that the macrophage may perform a striking multiplicity of functions in addition to housecleaning. The macrophage is probably involved in most types of immune response. It participates in interactions among the various immune cells either directly or by activating or inhibiting; it may even regulate these interactions (Rosenthal, 1980). Furthermore, the many secretory products of the macrophage (see Table II) suggest that it functions as an effector cell. The products of macrophage secretion effect not only the immune response, but also the inflammatory response, wound healing, and iron metabolism (Nathan et al., 1980). The cytotoxic activity of the macrophage, about which we are concerned in this chapter, may suggest a major role for the macrophage in the maintenance of the organism as well as in its defense system. Macrophage cytotoxicity may serve in the protection against viruses (Silverstein, 1975), intracellular organisms (Draper and D'Arcy Hart, 1975), senescent cells (Simon, 1980), and transformed cells (Fink, 1976).

The mature macrophage is a large cell (10–15 μm) that contains a large round or slightly irregular-shaped nucleus as well as a nucleolus (see Fig. 1). The cytoplasm contains many organelles, including mitochondria, rough endoplasmic reticulum, and a well-developed Golgi region that encompasses numerous coated vesicles. There may be many large and small vacuoles as well as debris-containing inclusions within the cytoplasm. The number of the vacuoles, storage granules, and other cytoplasmic inclusions appears to vary with the age and activity of the cell. Cytoplasmic filaments are a prominent feature and may occur in bundles (Nichols and Bainton, 1975).

Generally, the tissue macrophage is thought to evolve from stem cells

TABLE II

Secretory Products of Mononuclear Phagocytes[a]

Enzymes	Reactive metabolites of oxygen
Lysozyme	Superoxide
Neutral proteases	Hydrogen peroxide
Plasminogen activator	Hydroxyl radical
Collagenase	Singlet oxygen (?)
Elastase	
Angiotensin-convertase	Bioactive lipids
Acid hydrolases	Arachidonate metabolites
Proteases	Prostaglandin E_2
Lipases	6-Ketoprostaglandin $F_1\alpha$ (from
(Deoxy)ribonucleases	prostacyclin)
Phosphatases	Thromboxane
Glycosidases	Leukotriene
Sulfatases	Hydroxyeicosatetraenoic acids
Arginase	Platelet-activating factors
Complement components	Factor chemotactic for neutrophils
C1	
C2	Factor regulating synthesis of proteins
C3	by other cells
C4	Hepatocytes
C5	Serum amyloid A
Factor B	Haptoglobin
Factor D	Synovial-lining cells
Properdin	Collagenase
C3b inactivator	
β1H	Factors promoting replication of:
	Lymphocytes (lymphocyte-activating
Enzyme inhibitors	factors)
Plasmin inhibitors	Myeloid precursors (colony-stimulat-
α_2-Macroglobulin	ing factors)
	Erythroid precursors
Binding proteins	Fibroblasts
Transferrin	Microvasculature
Transcobalamin II	
Fibronectin	Factors inhibiting replication of:
	Lymphocytes
Nucleosides and metabolites	Tumor cells
Thymidine	Viruses (interferon)
Uracil	*Listeria monocytogenes*
Uric acid	
	Endogenous pyrogens

[a] Taken from Nathan *et al.* (1980) with permission from *The New England Journal of Medicine.*

present in the bone marrow. When the macrophage precursor cells leave the bone marrow and enter the circulation, they become the blood monocytes. The blood monocyte, which accounts for about 5% of the circulating white blood cells, remains in the circulation for an average period of 1½–5 days. After leaving the circulation, the monocyte matures into the tissue macrophage. The tissue macrophage may be distinguished from the monocyte by the presence of fewer storage granules and more vacuoles in the cytoplasm. The peritoneal macrophage is capable of reentering the circulation and migrating to other sites (Whitelaw and Batho, 1975). In addition to their capability to mobilize by migration, tissue macrophages can divide readily. The importance of cell division of preexisting macrophages, in terms of the total numbers of a local population, is a matter of controversy; cell division is thought, however, to be significant to the population of the Kupffer cells and the alveolar macrophages (Yoffey, 1978; Volkman, 1976).

Surprisingly little is known about the enzymatic contents of the lysosomes and granules of the macrophage. The reasons that such information has not yet been obtained are not immediately clear. It is obvious that it is difficult to obtain a pure population of fixed tissue macrophages; however, populations of either the alveolar or peritoneal macrophage may be obtained readily. The homogeneity of a population of macrophages in terms of either their state of maturation or their state of armament for phagocytosis is open to question. An additional problem may be posed by the ready ingestion of other phagocytic cell types by the macrophage; this adds enzymes not of macrophage origin to the macrophage's own enzyme population (Heifets et al., 1980).

The macrophage is identified cytochemically because of its characteristic acid phosphatase and esterase content (Ornstein et al., 1976). Peroxidase is present in the granules but its presence in the mature macrophage is the subject of some dispute (Nichols and Bainton, 1975; Daems et al., 1975). A list of the secretory contents of the macrophage is given in Table II. In addition, the following enzymes have been identified within the macrophage: catalase (Van Berkel, 1974), aminopeptidase (Sannes et al., 1977), NADPH oxidase (Bellavite et al., 1981), glutathione peroxidase (Rossi et al., 1978), and xanthine oxidase (Tubaro et al., 1980).

Macrophages are distributed throughout the body at a wide variety of sites such as the spleen, liver, lymph nodes, and alveoli, as well as within the peritoneal cavity. In many organs macrophages are found within or near streams of moving fluid, i.e., blood, lymph, or air. Their location makes it attractive to speculate that one of the macrophage functions is to clear these streams of toxic materials. An example illustrating the loss of this type of function is provided by the "Kupffer cell failure." The

Kupffer cell is in a strategic position to remove toxic products coming from the intestinal area via the portal vein and thus to prevent bacteria and bacterial endotoxin from entering the general circulation. In the diseased or alcoholic liver Kupffer cell failure leads to endotoxemia and bacterial invasions (Liehr and Grün, 1977; Ruggiero *et al.*, 1977).

Macrophages accumulate at sites of inflammation or invasion. The macrophages are known to arrive at such sites later than the other leukocytes and to engulf all sorts of debris as well as other cells, including lymphocytes, erythrocytes, and neutrophils. Chronic granulomas present at invasive sites, as well as tumors, include large populations of macrophages (Carr, 1980; Gauci, 1976).

The initial step in the cytotoxic process is chemotaxis. In the case of the macrophage the chemotactic attraction is probably mediated by the products of earlier events. Cells that respond early to an inflammatory site produce factors that attract macrophages. Neutrophils release a peptide that is selectively chemotactic for macrophages (Ward, 1968) and lymphocytes produce several macrophage attractants (Ward *et al.*, 1969; Meltzer *et al.*, 1977). Postlethwaite and Snyderman (1975) have isolated a factor of 12,500 daltons from peritoneal fluid that attracts macrophages. This factor appears in the peritoneal cavity during an inflammatory response concomitant with the influx of macrophages, suggesting that the factor may be a specific attractant.

Chemotactic agents have been identified and studied in *in vitro* systems. Macrophages respond to several agents that are also chemotactic for other phagocytic cells. These include components C5a and C3a of the complement cascade as well as the synthetic peptide *N*-formylmethionylleucylphenylalanine (Pike and Snyderman, 1980). Bacteria as well as neoplastic cells have been reported to produce factors that mobilize the army of macrophages (Russel *et al.*, 1976; Meltzer *et al.*, 1977). Some neoplastic cells, however, are capable of defending themselves against macrophage invasion by producing a factor that inhibits macrophage chemotaxis (Stevenson and Meltzer, 1976; Meltzer and Stevenson, 1978). The agents that cause the macrophage migration *in vivo* have not been identified yet. Little is known about the mechanism of chemotaxis, except that an *S*-adenosylmethionine-mediated methylation appears to be required (Pike and Snyderman, 1980).

The process of phagocytosis by the macrophage is similar to phagocytosis by the neutrophil. The macrophage surface membranes contain receptors that allow them to bind particles destined for phagocytosis. Cell surface recognition structures have been identified that enable the macrophage to bind the Fc portion of immunoglobulin, antigen–antibody complexes, and the third component of complement (Scott and Rosenthal, 1977; Kaplan *et al.*, 1975). After the target binds to the macrophage, the target is enclosed in a vesicle formed by the plasma membrane. For antibody-mediated phagocytosis, the macrophage appears to use the zipper mechanism (Fig. 2). On the other hand, macrophage phagocytosis that is not mediated by antibody appears to involve crater formation (Jones et al., 1977). The uptake of antibody-coated organisms proceeds at a greater rate than the uptake of nonopsonized organisms (Pesanti, 1979).

The macrophage mechanism of phagocytosis also appears to resemble that of the neutrophil. The energy required for macrophage phagocytosis is derived from anaerobic glycolysis (Karnovsky et al., 1970). As previously described for the neutrophil, actin-binding protein and myosin are the major components of the macrophage structural unit for endocytosis (Stendahl et al., 1980).

After the particle-filled vesicle is internalized it moves toward the nucleus. Eventually, the phagosome fuses with a lysosome. The rate of lysosomal fusion is not affected by the rate of phagocytosis (Kielian and Cohn, 1980).

Besides being capable of destroying unwanted cells by a phagocytic mechanism, the macrophage, like the eosinophil, is also capable of secreting toxic substances. Hanna (1978) has observed the exocytosis of cytoplasmic organelles by macrophages onto the surface of a tumor.

The capacity of the macrophage to be phagocytic and cytotoxic exhibits a striking variability. Resting macrophages, when exposed to a variety of agents, will undergo a range of changes in their cell membranes and their enzymatic contents. These progressive changes are expressed as changes in their ability to synthesize, secrete, engulf, and kill. Table III lists some of the characteristics of macrophages undergoing changes. These altered macrophages are described variously as activated, stimulated, elicited, or armed. Because these terms are used inconsistently, depending often on the appearance of one or another particular biochemical or functional parameter, they often create confusion. Hibbs et al. (1978) proposed a classification of the various states of activation of the macrophage. They identified four different functional states that have progressively lower thresholds for the expression of nonspecific tumor cell killing. These activation states are described as representing a continuum in the differentiation from a resting cell toward a killing cell.

Some of the agents that can elicit an activation response in the macrophage are the following: irritants such as thioglycolate and various oils; large polymers such as glycogen, double-stranded RNA, dextran, and pyran; immunopotentiators such as interferon, macrophage activation factor, endotoxin, and the killed vaccine of Bacillus Calmette Guérin (BCG); and products of chronic infections.

TABLE III

Some Characteristics of Activated Macrophages[a]

Increased size, enhanced spreading, and membrane ruffling
Increased rate of glucose oxidation
Enhanced protein synthesis
Secretion and enhanced synthesis of lysosomal enzymes
Synthesis and secretion of neutral proteinases
Enhanced production of lymphocyte activating factor
Increased number of Fc receptors
Altered function of complement receptor: mediation of phagocytosis
Enhanced (occasionally decreased) phagocytosis of unopsonized particles
Expression of new surface antigen
Increased responsiveness to chemotactic stimulation
Ability to kill or inhibit the growth of intracellular pathogens
Cytotoxic effects against malignant and other rapidly dividing cells

[a] Taken from Ögmundsdottir and Weir (1980) with permission.

The activation of macrophages toward killing has been a subject of considerable research and interest in recent years. Macrophages activated toward killing have been shown to destroy tumor cells both *in vitro* and *in vivo* (Hibbs *et al.*, 1978; Fidler *et al.*, 1978). Macrophage-activating agents are being used in the experimental therapy of cancer in man and have produced some promising results (Samak *et al.*, 1978; Schultz, 1981; Old, 1981). A more detailed discussion of the antitumor effects will be presented later in this chapter (page 30). The activation of macrophages may also have potential in the treatment of diseases such as leprosy (Delville and Jacques, 1978) and malaria (Gillet *et al.*, 1978), among others.

An oxidative mechanism similar to those present in the neutrophil and the eosinophil appears to be operating in the cell killing by activated macrophages (Nathan and Cohn, 1980; Buchmüller and Mauel, 1981). The details of the biochemical mechanism of cytotoxicity will be addressed in Section III of this chapter.

D. Other Cytotoxic Cells

Several other cell types have been shown to express either phagocytic or cytotoxic activity. These types include the lymphocytes, the natural killer cells, and the phagocytic cells of the nervous system. The cytotoxic activity of these cells is not characterized as well as that of the neutrophil, the eosinophil, or the macrophage. Nevertheless, a brief discussion of the properties of these cells appears warranted.

1. The Lymphocytes

The lymphocytes are a group of cells that comprise approximately 30% of the circulating blood leukocytes. Although lymphocytes may be found scattered throughout many body tissues, they are found predominantly within the lymphoid tissue, including the tonsils, appendix, lymph nodes, spleen, and thymus. The general morphology of the lymphocyte is illustrated in Fig. 1. The lymphocyte may be characterized as a small round cell, slightly larger than an erythrocyte, with a very large round nucleus, and therefore containing very little cytoplasm.

The population of lymphocytes, when classified according to function, consists of a number of different cell types. These different cell types appear indistinguishable in electron micrographs unless their characteristic cell-surface antigens have been labeled with electron-dense markers. Certain lymphocytes may be classified, according to their antigenic markers, into two distinct groups called the B and the T cells. Lymphocytes with neither the characteristic B cell nor T cell markers are termed null cells.

The B cells derive from precursor cells formed in the fetal liver or the bone marrow. The presence of immunoglobulin on the cell surface is the distinguishing marker of the B cell. The immune response of the B cell causes it to differentiate into the plasma cell, which then produces antibodies. B cells do not possess cytotoxic properties.

The T or thymus-derived lymphocytes develop in the thymus from precursor cells that arise in the bone marrow. The surface of the T cell does not contain immunoglobulin, but is characterized by the presence of an antigen, the T antigen, which is not present on B lymphocytes. Functions of various groups of T cells, which have been observed *in vitro,* are used to distinguish the various subpopulations such as the T suppressor cell (T_S), the T helper cell (T_H), the T delayed hypersensitivity mediator cell (T_D), and the T killer (T_K) cell. We will limit our discussion of the T lymphocytes to the T killer cells because they form the only subpopulation that exhibits a cytotoxic function.

The cytotoxicity of the T killer lymphocytes is poorly understood. The addition of certain mitogens to a population of T cells *in vitro* will stimulate cytotoxic activity. This may result from an increase in the percentage of killers in the total population or from an increased activity of the killer-cell population. The former may result from a mitogen-induced differentiation of the T cells into killer cells or from the induction of a preferential mitosis of the killer cells. Which of these possibilities is operative is unknown (Waterfield *et al.,* 1976). Both foreign cells and autologous cells infected by virus can be targets for the T killer *in vitro.* In order to kill a virus-infected cell, the T_K cell must be able to distinguish a virus-infected cell from a noninfected cell (Zinkernagel, 1976). How the cell accomplishes this distinction is totally obscure. To be able to kill, the T_K lymphocyte must make close contact with its target cell (Biberfield and Holm, 1970). Studies on the cytotoxic action of the T killer lymphocyte have been hampered, because it is difficult to separate the T_K cells from the other T cells.

Another lymphocyte, called the killer or K lymphocyte, is also involved in cytotoxic actions. The K cell appears to be a null lymphocyte, i.e., a lymphocyte with no B or T cell markers (Perlman and Holm, 1969). The K lymphocyte is involved in a type of cell killing termed antibody-dependent cell-mediated cytotoxicity (ADCC). In the antibody-dependent cytotoxic interaction the target cell is linked by antibodies to the killer cell. The immunoglobulin first binds to the target cell through its antigen binding site. Then the K cell, which has a binding site for the Fc portion of immunoglobulin, binds to the antibody-coated target cell (Parillo and Fauci, 1977). Details of the mechanism of K lymphocyte killing are as yet unknown; however, the tumoricidal function of the null cells appears to be induced by interferon (Ortaldo *et al*, 1980).

2. *The Natural Killer (NK) Cell*

For many years, researchers performing *in vitro* cytotoxic assays had overlooked a spontaneous background lysis. Kiessling *et al.* (1976) attributed this spontaneous lysis to a non-B, non-T, nonmacrophage cell type called the natural killer cell. Herberman (1980a) provides a thorough review of our current understanding of the natural killer cell.

Neither the natural killer cell nor its killing activity is well defined; therefore, the literature on the natural killer cell abounds with controversy. The NK cell has been described as a small resting lymphocyte that has scanty cytoplasm and is neither a mature B nor T cell (Kiessling *et al.*, 1980). On the other hand, Saksela and Timonen (1980) describe the NK cell as a large lymphocyte that has a comparatively large cytoplasmic volume rich in electron-dense and azurophil-staining granules. Apparently, most investigators agree that the natural killer cell belongs to the group of null lymphocytes; however, Lohmann-Matthes and Domzig (1980) classify the NK cell as a promonocyte according to both its morphology and its identified receptors.

Because both the NK cell and the K cell may be null lymphocytes, the possibility exists that they are the same cell expressing two different killing activities. Other similarities, such as the presence of receptors for the Fc portion of the immunoglobulin molecule and an interferon response have been found in both cell types. Interferon augments NK cell killing (Zarling, 1980), perhaps by causing noncytotoxic pre-NK cells to acquire killer activity (Saksela *et al.,* 1980). The NK cell has also been shown to have Fc receptors that, in contrast to the Fc receptors of the K cell, have a low affinity for immunoglobulin (Roder, 1980). Gidlund *et al.* (1980) have demonstrated that natural killer cells can also express ADCC activity. Whether or not these two cytotoxic cells, the NK and the K cell, are the same cell therefore remains to be established.

It does seem clear, however, that the natural killer cells are a heterogeneous population of cells capable of tumoricidal activity (Jensen and Koren, 1980; Herberman, 1980b; Burton, 1980). Certainly, further research into natural killer cell activity is needed to clarify its biological function and the mechanism of killing.

3. The Phagocytic Cells of the Nervous System

Several cell types within the nervous system are known to possess phagocytic activity, i.e., they are capable of removing debris. These phagocytic cells are the pericytes, the astrocytes, and the microglia of the central nervous system and the Schwann cell of the peripheral nervous system. The degree of importance of their phagocytic activity both in homeostasis and under pathological conditions, as well as their mechanism of phagocytosis, have not yet been established. Most phagocytic cells capable of ingesting debris also possess the capacity to kill. It is, therefore, tempting to speculate that this highly specialized function of the phagocytic cells of the nervous system may be implicated in the neuronal death associated with certain disease states of the nervous system, e.g., multiple sclerosis and Parkinson's disease.

The pericytes and the microglia are capable of migration and mobilization as well as activation and local proliferation (Baldwin, 1973; Boya, 1975). Both the pericyte and the microglia are thought to have a monocyte origin and, therefore, to be brain macrophages (Oehmichen and Torvik, 1976; Imamoto and Leblond, 1977; Baldwin, 1980). The pericyte, like other macrophages, has been implicated in tumoricidal activity (Torack, 1961).

The Schwann cell of the peripheral nervous system is another cell that is capable of migration and undergoes a proliferative response to local injury (Raine *et al.,* 1969). The migratory and proliferative response, as well as the phagocytic nature of the Schwann cell, may suggest that it, too, is a macrophage.

The astrocyte may develop from the microglia (Skoff, 1975). Astrocytes may be the cells responsible for the removal of cellular debris resulting from the massive neuronal death during early development (Baldwin, 1980). It does not seem unreasonable to suggest that the astrocyte may be the cell responsible for the death of these neurons.

An example of the physiological effects of cytotoxic activity within the nervous system will be presented later in this chapter.

III. BIOCHEMICAL MECHANISM OF CYTOTOXICITY

The biochemical mechanism by which phagocytic cells kill target cells is rather complex. A somewhat simplified schematic illustration of this mechanism is presented in Fig. 3. The mechanism consists of two parts: (1) the biochemical reactions involved in the production of the cytotoxic compounds, and (2) the biochemical reactions involved in the neutralization of excess toxic material. The second part is necessary to prevent self-destruction of the phagocytic cell.

Although a variety of cells possess the capacity to kill other cells (as

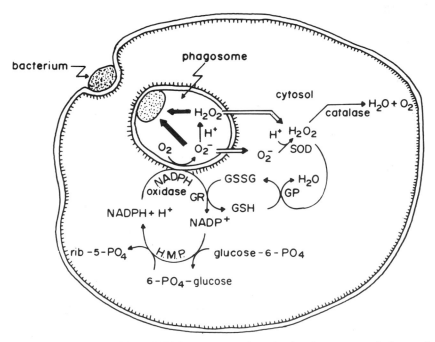

Fig. 3.   Formation of bactericidal oxygen products in the phagosomes of phagocytic leukocytes and protection of the cytosol against oxidative injury. The NADPH oxidase in the phagosome membrane generates superoxide into the phagosome. The cytosol of the leukocyte is protected by superoxide dismutase, catalase, and the glutathione redox system. →, reaction; ➡, attack; ⇒, diffusion; GR, glutathione reductase; GP, glutathione peroxidase; SOD, superoxide dismutase; HMP, hexose-monophosphate shunt. (Reproduced with permission from Roos and Weening, 1979.)

outlined in the previous section), much evidence suggests that the biochemical reactions involved in the cytotoxic activity are very similar, if
not identical, in all phagocytic cells. This may seem surprising, especially
because the various phagocytes appear to have some degree of specificity for their target cells. It is well established, however, that the toxic
material produced by the various phagocytes is not specific for any given
type of target cell, but is toxic to all cells, including endogenous cells.
Thus, the specificity for the target cell must be brought about by a
recognition mechanism that is distinct from the mechanism involving
the production of toxic materials.

A.   Production of Reduced Oxygen Species

Within seconds after contact of the phagocyte with a target cell, one
observes a large increase in the oxygen consumption of the phagocyte.
This phenomenon is generally referred to as the metabolic or oxidative
burst. The metabolic burst involves a large increase in glucose metabolism via the hexose monophosphate shunt, which results in the production of large amounts of NADPH. The overall reaction of the hexose
monophosphate shunt is:

$$3 \text{ glucose-6-P} + 6 \text{ NADP}^+ \rightarrow 2 \text{ fructose-6-P} + \text{glyceraldehyde-3-P} + 6 \text{ NADPH} + 6 \text{ H}^+ + 3 \text{ CO}_2$$

The fructose-6-P is readily reconverted to glucose-6-P by an isomerase,
which simplifies the overall reaction to:

$$\text{glucose-6-P} + 6 \text{ NADP}^+ \rightarrow \text{glyceraldehyde-3-P} + 6 \text{ NADPH} + 6 \text{ H}^+ + 3 \text{ CO}_2$$

Hence, 6 moles of NADPH are produced per mole of glucose that is
oxidized.
    The NADPH, in turn, reduces molecular oxygen to various reduced
oxygen species, including superoxide and hydrogen peroxide:

$$\text{H}^+ + \text{NADPH} + 2 \text{ O}_2 \rightarrow \text{NADP}^+ + 2 \text{ O}_2^-$$
$$2 \text{ O}_2^- + 2 \text{ H}^+ \rightarrow \text{H}_2\text{O}_2 + \text{O}_2$$

These reactions are catalyzed by NADPH oxidase and superoxide dismutase, respectively. The overall reaction taking place during the metabolic burst is therefore:

$$\text{glucose-6-P}^+ + 12 \text{ O}_2 \rightarrow \text{glyceraldehyde-3-P} + 3 \text{ CO}_2 + 12 \text{ O}_2^-$$

This explains the large increase in oxygen uptake by the activated phagocyte. The NADPH oxidase is located in the plasma membrane (Dewald *et al.*, 1979; Cohen *et al.*, 1980). Cytochemical evidence has shown that the superoxide and/or hydrogen peroxide is produced at the outside of the membrane, i.e., in the phagosomes or outside the cell (Briggs *et al.*, 1975; Root and Metcalf, 1977). Thus, the reduced oxygen compounds are produced close to the target cell, where they can readily exert their cytotoxic activity, as shown in Fig. 3.

In vitro studies have shown that although hydrogen peroxide is cytotoxic to bacteria as well as to eukaryotic cells, it takes fairly high concentrations of H_2O_2 to be effective (Bayliss and Waites, 1976). Furthermore, some evidence suggests that superoxide may be less toxic than hydrogen peroxide (Klebanoff, 1974). The cytotoxicity of hydrogen peroxide is dramatically enhanced, however, in the presence of certain compounds. Thus, the toxic effect of H_2O_2 may be increased as much as 3000-fold by the addition of 0.1 mM Cu^{2+} (Bayliss and Waites, 1976). Ascorbate as well as iodide also increase the antibacterial activity of H_2O_2 (Drath and Karnovsky, 1974).

Evidence has also been presented suggesting the formation of hydroxyl free radicals. The inhibition of the superoxide production of PMN homogenates by scavengers of hydroxyl radicals such as benzoate, ethanol, and mannitol has been taken as presumptive evidence for the formation of hydroxyl radicals (Beauchamp and Fridovich, 1970; Misra and Fridovich, 1976). These radicals are presumably formed by the Haber–Weiss reaction:

$$O_2^{\cdot} + H_2O_2 \rightarrow O_2 + OH^- + OH\cdot$$

Some serious questions as to the occurrence of this reaction *in vitro* have been raised, however (Halliwell, 1976; McClune and Fee, 1976).

Finally, the occurrence of chemiluminescence during phagocytosis suggests the presence of electronically excited species that may be toxic to the organism. One such species that could be formed from the dismutation of superoxide is singlet oxygen. Although the presence of singlet oxygen in intact PMNs has not yet been shown, circumstantial evidence suggesting the involvement of singlet oxygen in the phagocytic process has been presented. Krinsky (1974) found that the carotenoid-containing *Sarcina lutea* is killed poorly by human PMNs, in contrast to a pigmentless mutant. Since carotenoids are potent singlet oxygen quenchers, the involvement of singlet oxygen in the phagocytic process was suggested. A quenching of the PMN chemiluminescence by a singlet oxygen quencher that parallels an inhibition of the cytotoxic activity, however, has yet to be shown.

B. Role of Peroxidases

Peroxidases are abundantly present in neutrophils, eosinophils, and monocytes; in fact, 5% of the total protein in PMNs may be peroxidase (Schultz and Kaminker, 1962; Rohrer *et al.*, 1966), with lesser amounts being present in monocytes and macrophages. The enzyme present in neutrophils is myeloperoxidase. In purified form myeloperoxidase is a green protein that contains two heme-liganded ferric ions per molecule of about 150,000 daltons.

Myeloperoxidase is located almost exclusively in the azurophil granules, together with a number of hydrolytic enzymes. After particle ingestion, granules adjacent to the phagocytic vacuole fuse with the vacuole; the common membrane then ruptures and the granule contents are discharged into the vacuolar space, thus converting the phagosomes into a phagolysosome (degranulation). The number of granules moving into the phagosomes depends on the number of ingested particles, and when this number is large, almost complete degranulation of the phagocyte may occur. It is interesting to note that granule formation does not occur beyond the myelocyte stage of neutrophil development; hence, during active phagocytosis, the mature neutrophil can be irreversibly depleted of its granules.

These observations suggest that myeloperoxidase may be involved in the killing and/or subsequent degradation of the ingested particles. The question to be answered is: In what manner? Klebanoff and colleagues showed many years ago that myeloperoxidase in conjunction with hydrogen peroxide and a halide ion forms a powerful cytotoxic species (Klebanoff, 1967; Klebanoff and Smith, 1970), its cytotoxic activity being many times more powerful than that of hydrogen peroxide (Klebanoff and Smith, 1970; Clark *et al.*, 1975). The involvement of the myeloperoxidase in the killing of an ingested organism is illustrated schematically in Fig. 4.

The myeloperoxidase–H_2O_2–halide complex has been shown to be cytotoxic to a variety of cells, including bacteria, fungi, tumor cells, and normal eukaryotic cells (Klebanoff and Clark, 1978; Badwey and Karnovsky, 1980). These *in vitro* studies, together with the evidence presented above, strongly suggest a role for myeloperoxidase in the killing of target cells.

Experiments with leukocytes from myeloperoxidase-deficient patients demonstrated, however, that these leukocytes are capable of killing ingested bacteria. The rate of the bactericidal activity of these cells involves a considerable lag time, which is not observed with normal leukocytes (see Fig. 5). These results suggest that the cytotoxic activity of the leukocytes may be facilitated by, but not be totally dependent on, myeloperox-

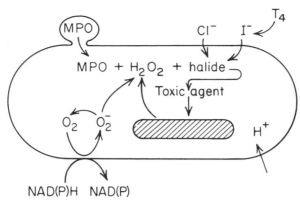

Fig. 4. The myeloperoxidase-mediated antimicrobial system. Myeloperoxidase, re-leased into the phagocytic vacuole during the degranulation process, reacts with H_2O_2, generated either by leukocytic or microbial metabolism, and a halide to form an agent toxic to a variety of microorganisms. (Reproduced with permission from Klebanoff and Clark, 1978.)

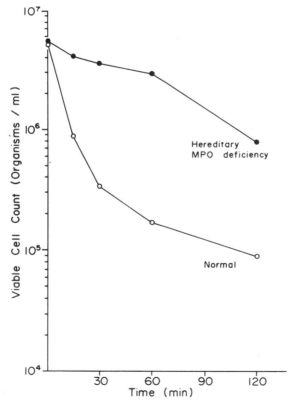

Fig. 5. Staphylocidal activity of normal and myeloperoxidase-deficient leukocytes. (Data taken from Klebanoff and Hamon, 1972.)

idase. This hypothesis is further substantiated because most patients with a hereditary myeloperoxidase deficiency do not suffer from life-threatening infections. It should also be noted that the phagocytosis-dependent increases in oxygen consumption and superoxide formation are greater in myeloperoxidase-deficient cells than in normal cells (Rosen and Klebanoff, 1976). It is therefore possible that in the absence of myeloperoxidase the phagocyte can eventually build up concentrations of hydrogen peroxide high enough to act as the cytotoxic agent.

A complex with cytotoxic properties similar to that of the myeloperoxidase–H_2O_2-halide complex can also be formed by other peroxidases. Thus, horseradish peroxidase forms a potent cytotoxic complex in the presence of H_2O_2 and a halide ion; the cytotoxic activity of this complex is considerably greater in the presence of iodide ions than in the presence of chloride ions (Jacques et. al., 1975). Lactoperoxidase forms a similar complex with iodide ions, but is most active in the presence of thiocyanate ions (Reiter, 1978, 1979). In fact, the existence of such cytotoxic complexes was first discovered with the lactoperoxidase system (Klebanoff and Luebke, 1965).

Studies concerning the structure of the cytotoxic complexes suggest that similar complexes are probably formed readily by most, if not all, peroxidases that contain a heme iron as a functional group. This concept has important implications, because many different peroxidases are found in nature. Thus, the peroxidase found in eosinophils is distinct from the myeloperoxidase (Archer et al., 1965; Desser et al., 1972), which is present in neutrophils and monocytes. Because eosinophils function as a cytotoxic cell it appears reasonable to assume that the eosinophil peroxidase can also form a cytotoxic complex in the presence of H_2O_2 and a halide ion. However, this has not yet been established.

C. Protection against Self-Killing

The production of highly cytotoxic compounds by phagocytic cells immediately raises the question as to how these cells protect themselves from being destroyed by their own products. Although the superoxide and hydrogen peroxide, as well as the cytotoxic form of the peroxidases, are produced at the outside of the cell membrane, cytochemical staining experiments suggest that superoxide and hydrogen peroxide may readily diffuse back into the phagocytic cell. In doing so, these compounds could become lethal to the phagocyte, especially if they could react with the peroxidase present in the endogenous granules.

In addition to this, there is a problem concerning the excess cytotoxic material that is present after a target cell is killed. This material needs to be neutralized before it can attack other cells at random or, when present in phagosomes, can attack the phagocyte itself. The need for a system that can neutralize any toxic material rapidly and efficiently is thus obvious.

Superoxide rapidly dismutates to hydrogen peroxide and oxygen:

$$2\ O_2^- + 2\ H^+ \rightarrow H_2O_2 + O_2$$

At neutral pH the reaction is sufficiently rapid for the superoxide to have a half-life of less than 1 second. Thus, superoxide, which is much less toxic than hydrogen peroxide, does not accumulate to any appreciable extent. Nevertheless, the cytoplasm of PMNs contains an enzyme, superoxide dismutase, that catalyzes the dismutation of superoxide, thus further reducing its already short half-life.

Hydrogen peroxide is quite toxic and its concentration can conceivably build up to rather high levels in the phagosome or in the extracellular space. The cytoplasm of phagocytes contains, however, high concentrations of catalase. This enzyme dismutates hydrogen peroxide to oxygen and water:

$$2H_2O_2 \rightarrow 2H_2O + O_2$$

Catalase has a high K_m as well as a high turnover number. This means that high concentrations of hydrogen peroxide are neutralized rapidly, but when the concentration lowers the enzyme becomes less effective in removing H_2O_2. Low concentrations of H_2O_2 are therefore not removed very efficiently by catalase.

Hydrogen peroxide also reacts readily with lipids containing unsaturated fatty acids to form lipid peroxides. Lipid peroxides are strong oxidants and can promote chain reactions. They are, therefore, quite detrimental to the cell. To prevent damage to itself, the phagocyte possesses an enzyme system that rapidly neutralizes lipid peroxides. This enzyme system, consisting of glutathione peroxidase and glutathione reductase, can also reduce any remaining hydrogen peroxide. The reaction mechanism is as follows:

$$
\begin{array}{ccc}
H_2O_2 & \text{reduced glutathione} & NADP^+ \\
 & \text{(peroxidase)} & \text{(reductase)} \\
H_2O & \text{oxidized glutathione} & NADPH
\end{array}
$$

The NADPH required for this reaction is supplied by the hexose monophosphate shunt. Thus, the phagocyte has two mechanisms for eliminating excess hydrogen peroxide: catalase and the glutathione cycle. This emphasizes further the importance of an effective mechanism for H_2O_2 removal to the viability of the phagocyte.

IV. PHYSIOLOGICAL ASPECTS OF PEROXIDASE ACTIVITY

The information presented in the previous sections amply demonstrates that one of the biological functions, if not the primary biological function, of some peroxidases is that of a cytotoxic agent. This appears to be well established for the peroxidases present in phagocytes, but is less certain for peroxidases present in specific tissues. Thus, the peroxidase present in the thyroid is responsible for the iodination of the tyrosine residues in thyroxine biosynthesis, whereas the glutathione peroxidase in erythrocytes may have just the neutralization of lipid peroxides as its main function. It should be borne in mind that the presence of peroxidase activity in certain tissues may only indicate the presence of phagocytes in such tissues.

There is reason to believe that peroxidases that do not normally have a cytotoxic function may become cytotoxic under certain pathological conditions. This change in function causes tissue damage that would not occur under normal physiological conditions. Such a change in function may be promoted by such abnormalities as the deterioration of a control mechanism, a lack of appropriate substrate, or the overproduction of superoxide or hydrogen peroxide.

In this section we will discuss several examples of physiological and pathological conditions that involve the cytotoxic activity of peroxidases. We hope the examples chosen will illustrate the manner by which peroxidases can and do affect the physiological state of the organism.

A. Endotoxin Shock

During various bacterial infections, bacterial lipopolysaccharides (endotoxins) may be released into the blood circulation, causing a circulatory collapse through humoral influences. Endotoxin shock can be experimentally produced in dogs by the intervenous injection of purified endotoxin. The arterial pressure drops within seconds after the injection (Janssen et al., 1981) and one observes a rapid decrease in leukocyte count: Decreases of up to 80% can be routinely observed in experimental animals. The immediate vascular effect of endotoxin appears to be indirect and mediated by the release of histamine as well as other humoral substances (Hinshaw et al., 1961; Spink et al., 1962; Vick,

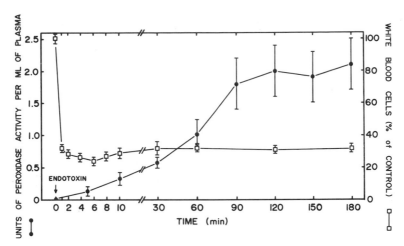

Fig. 6.   Decrease in leukocyte count and increase in circulating peroxidase activity following the injection of endotoxin (Janssen *et al.*, 1981) in a mongrel dog.

1964). Thus, there are striking similarities in the initial responses of the cardiovascular system to endotoxin and to histamine (Gilbert, 1960; Hinshaw *et al.*, 1962); in fact, the injection of endotoxin leads to the release of substantial amounts of histamine (Hinshaw *et al.*, 1961).

The decrease in the leukocyte count caused by endotoxin administration is illustrated in Fig. 6. Because the leukocyte count does not return to normal for long periods of time after the endotoxin injection, the question arises as to the fate of the leukocytes. Either they become sequestered somewhere in the body or they are destroyed. To answer this question, we measured the increase in myeloperoxidase content of the blood, which would indicate destruction of the leukocytes. The results showed a dramatic increase in circulating myeloperoxidase (Fig. 6); however, the increase occurred after a considerable time lag compared to the decrease in the leukocyte count. This may indicate a sequestering of the leukocyte in some tissues, followed by the destruction of at least part of the sequestered cells. Whether the same type of cell killing by endotoxins is responsible for the histamine release is not known.

The release of large amounts of myeloperoxidase in the circulation as a result of endotoxin administration must indicate that other enzymes are also released from the leukocytes. It is thus possible that one or more of the released compounds may be responsible for some of the symptoms associated with endotoxin shock. In this context, it might be of considerable interest to determine what physiological effects, if any, are associated with the presence of large amounts of myeloperoxidase in the circulation. It is well known that the lethal effects of endotoxin toward leukocytes are much less *in vitro* than *in vivo* (Cohn and Morse, 1960; Woods *et al.*, 1961; Proctor, 1979), which suggests that the toxicity of endotoxin *in vivo* must be a secondary effect.

B. Cell Transformation

Cell transformation (carcinogenesis) is promoted by a large number of environmental compounds. This has been a topic of discussion for many years and has been amply reviewed (Fishbein, 1979; Emmelot and Kriek, 1979). In this section we will consider the body's responses to cell transformations.

It appears to be well established that the body possesses one or more defense mechanisms capable of eliminating any transformed cells that may be present. There is, however, a limit to the ability of the defense mechanisms. This is clearly demonstrated by the fact that a large number of tumor cells (usually 10^6 or more) must be administered to an animal in order to have a reasonable degree of certainty that a tumor will develop. Smaller numbers of tumor cells are usually readily eliminated by the animal and do not produce tumors unless they remain concentrated in a local area. It is assumed that endogenous cells that undergo spontaneous or induced transformation are eliminated from the body by the same mechanism as transformed cells that are introduced into the body experimentally.

Much evidence suggests that *in vivo* the recognition of a cell as having been transformed is a function of the immune system. Specific antigens present in the outer membrane of a transformed cell (but not of a normal cell) provide the trigger that eventually leads to the death of the transformed cell. Little is known about the nature of these specific antigens; however, it has been reported that the transformation of normal cells with certain enzymes such as neuraminidase or trypsin results in changes in the cell membrane that are similar or identical to those found in transformed cells.

Several types of cells have been shown to exhibit cytotoxic activity toward tumor cells *in vitro*. These include the neutrophils, the macrophages, and the natural killer cells. As discussed before, the cytotoxicity of none of these cells is specific for tumor cells only; each of them has the ability to kill a variety of foreign cells as well as endogenous cells. Thus, the cytotoxic cell is triggered into action by a signal, ultimately received from the target cell, which activates the cytotoxic cell into eliminating the target cell. No such signal is received from normal cells. It is conceivable that the presence or absence of specific antigens or antigen–antibody complexes on the surface of the target cell would provide the signal.

The cytotoxic action of neutrophils *in vitro* toward tumor cells was first demonstrated by Clark and Klebanoff (1975). The cytotoxic action in the form of exocytosis closely resembles the mechanisms described in earlier sections of this chapter. No evidence that neutrophils play a major role in the elimination of tumor cells *in vivo* has been obtained as yet.

There is a considerable body of evidence suggesting macrophages are involved in the tumoricidal activity of the body. Tumors growing *in vivo* are commonly infiltrated with macrophages (Evans, 1972; Eccles and Alexander, 1974; Russell *et al.*, 1976). The degree of infiltration is such that macrophages may constitute as much as 60% of the cellular content of the tumor (Evans, 1972). Activation of the macrophages by a chronic BCG infection causes resistance to tumor growth in mice (Zbar *et al.*, 1971), and macrophages isolated from regressing tumors have a higher cytotoxic activity for tumor cells than those isolated from progressing tumors (Russell and McIntosh, 1977). These and other experiments (Weinberg *et al.*, 1978) suggest that macrophages need to be activated before they can exhibit their tumoricidal activity.

A compound that strongly promotes the tumoricidal activity of macrophages is endotoxin. *In vitro* endotoxin at concentrations as low as 0.5 to 1.0 ng/ml is capable of activating the tumoricidal activity of macrophages. Indirect evidence suggests that endotoxin can promote activation of macrophages *in vivo*, also producing tumor necrosis and regression (Nauts *et al.*, 1953; Parr *et al.*, 1973).

Earlier in this chapter, we demonstrated that the injection of endotoxin into dogs results in the release of large amounts of myeloperoxidase into the circulation. This is probably the result of a degranulation of some granulocytes in response to the endotoxin. In this context it is important to point out that myeloperoxidase also possesses antitumor activity. Schultz and colleagues showed many years ago (1955) that injection of myeloperoxidase in conjunction with thiotepa has a tumoristatic if not tumoricidal effect on tumor-bearing rodents. Moreover, recent experiments in our own laboratory have demonstrated that several peroxidases, including horseradish peroxidase and lactoperoxidase, under specified conditions can exert tumoricidal activity *in vivo*. These results suggest that endotoxin assists in the activation of macrophages either directly, by interacting with the macrophage, or indirectly, through the myeloperoxidase it releases.

The involvement of natural killer cells in the regression of tumors *in vivo* is as yet rather speculative. It has been shown that cancer patients may have a deficient natural killer cell activity (Cannon *et al.*, 1977; Takasugi *et al.*, 1977); a deficiency is also observed in animals with transplanted tumors (Becker *et al.*, 1976). Furthermore, transfer of lymphoid cells with natural killer cell activity into irradiated rats (Shellam and Hogg, 1977) and *in vivo* transfer experiments in rats (Harmon *et al.*, 1977) induces a resistance to tumor tare and tumor growth in the host animals. The fact that athymic mice have a diminished incidence of spontaneous tumors and show a longer period of latency upon tumor induction by carcinogens also substantiates an involvement of the killer cells in the host resistance to tumor development and growth. Thus far, however, no experiments proving a direct role for the natural killer cells in the prevention and regression of cancer have been reported.

C. The Role of Peroxidases in Neural Disorders

The peroxidase activity present in extracts of whole brains is very low; however, relatively high concentrations of peroxidase activity are found in specific regions of the brain such as the substantia nigra. The enzyme is virtually uncharacterized, and its biological function has been a subject of speculation.

This peroxidase attracted our attention for several reasons:

1. Ambani *et al.* (1975) hypothesized that the brain peroxidase provides protection against damage induced by H_2O_2 that evolves during the monoamine oxidase inactivation of the catecholamine neurotransmitters. If this hypothesis is correct, the enzyme must function much like the glutathione peroxidase in the phagocytic cells.

2. Wong-Riley (1976) reported that the pigmented cells of the substantia nigra may be histochemically differentiated from surrounding neurons and glial cells by their distinctive elevation in peroxidase activity, dopamine concentration, and melanin content. This suggests that the peroxidase may be associated with dopamine metabolism and melanin production.

3. It is well known that the nigrostriatum of patients with Parkinson's disease shows a selective degeneration; many neuronal bodies have been lost causing distinct lesions in the substantia nigra. Patients with Parkinson's disease also have very low levels of dopamine in the substantia nigra. It occurred to us that the loss of neuronal tissues could be the result of a cytotoxic activity by the nigral peroxidase.

The enzyme was recently purified from dog midbrains and characterized as a heme protein with a molecular weight of 100,000–150,000. The enzyme possesses a potent cytotoxic activity against erythrocytes in the presence of hydrogen peroxide and Cl^-, indicating that under

TABLE IV

Some Effectors of Nigrostriatal Peroxidase Mediated
Cytotoxicity

Conditions	SCI[a]
NADH (1 mM)	—
NSPO (30 µg/ml)	—
NSPO + NADH	14.3
NSPO + NADH + Mn (100 µM)	18.0
NSPO + NADH + kainic acid (100 µM)	58.3
NSPO + NADH + dopamine (100 µM)	−5.3
NSPO + NADH + kainic acid + dopamine	−6.4

[a] The SCI (Specific Cytotoxicity Index) was calculated from the equation: $\frac{1}{C}\left(\frac{A_E}{A_c}-1\right)$, where C represents the amount of nigrostriatal peroxidase (NSPO) in mg. A_E and A_c represent the absorbance at 410 nm for the experimental and control conditions, respectively.

appropriate conditions this peroxidase is capable of killing cells. We also found that the cytotoxic activity of this enzyme can be suppressed completely by 100 µM dopamine. Representative results are given in Table IV.

Disorders similar to Parkinsonism can be induced in experimental animals. Prolonged inhalation or ingestion of manganese results in lesions in the substantia nigra that are almost indistinguishable from those in patients with Parkinson's disease (Chondra et al., 1974). Furthermore, a series of compounds produce the symptoms of Parkinsonism upon injection into experimental animals. Perhaps the best known of these compounds is kainic acid, a glutamic acid derivative (McGeer et al., 1978). Table IV also lists the effects of these compounds on the cytotoxic activity of the brain peroxidase. A small but significant enhancement of the cytotoxic activity is observed upon addition of 100 µM Mn^{2+}, whereas a much more pronounced enhancement is observed with the same concentration of kainic acid. This difference in enhancement parallels the difference in potency in inducing Parkinsonism in experimental animals.

Other results obtained by Dr. Matthew B. Grisham in our laboratory indicate that the midbrain peroxidase has an important function in the oxidation of dopamine to melanin. This is a very important function, because it probably represents the mechanism by which the neurotransmitter is inactivated.

The nigral peroxidase is, therefore, an enzyme with potent cytotoxic activity under appropriate conditions. This activity is normally suppressed by the presence of dopamine, its natural substrate; however, when for some reason (e.g., in Parkinson's disease) the dopamine levels decrease dramatically, the cytotoxic activity of the enzyme becomes apparent and causes the damage to the substantia nigra observed in these patients. The necessary hydrogen peroxide is probably provided by the monoamine oxidase. This hypothesis is consistent with the improvement effected in Parkinsonism patients by treatment with dopa; dopamine severely inhibits the cytotoxicity and thereby further damage.

V. CONCLUDING REMARKS

Phagocytic cells may be visualized as a group of primitive cells. The phagocytic cells are migratory cells and migration is generally a characteristic of undifferentiated cells. Cytotoxic activity of the immune system may be associated predominantly with immature cells. Moreover, the more primitive forms of life, such as amoeba, use phagocytic mechanisms to sustain themselves. On the other hand, phagocytic cells can also be visualized as a highly developed group of cells. They possess a highly developed and sophisticated enzyme system aimed at the killing and digestion of other cells. Several of these enzymes are unique to phagocytic cells. Furthermore, the various cytotoxic cells of the reticuloendothelial system appear to have developed a specificity for certain types of target cells, and in some cases, evidence points toward a certain degree of cooperation among the cytotoxic cells. Clearly, these properties suggest a high degree of development.

Many of the biochemical reactions involved in the cytotoxic activity of the neutrophil have been elucidated, and it appears that other cytotoxic cells use a similar, if not identical, mechanism. These reactions involve, however, only the production of cytotoxic compounds and the mechanisms used by the cell to dispose of any excess. Little is known about the biochemical events in the target cell membrane that eventually lead to its destruction. Some preliminary results obtained in our laboratory suggest that a polymerization reaction may occur in the target cell membrane as a result of its interaction with the cytotoxic products. This polymerization reaction leads to a complete loss of fluidity, and subsequently, the rigid membrane cracks.

Even less is known about how the cytotoxic cell recognizes a target cell from a nontarget cell and what determines the specificity of the various cytotoxic cells for their respective targets. Furthermore, the biochemical bases for such events as manganese toxicity, endotoxin toxicity, and the results of the ingestion of carcinogens are still poorly understood. It is quite interesting, however, that the peroxidases, whose biological functions have thus far been relatively obscure, appear to play a role in all these events.

Much additional work needs to be done to gain a better understanding of the action of peroxidases and of the physiological events with which they are associated. It is the authors' hope that this chapter may encourage the student to experiment further in this exciting and important area of biochemical physiology.

ACKNOWLEDGMENTS

We thank Linda Everse for the drawings presented in Fig. 1. Studies from the authors' laboratory discussed in this chapter were supported by Grant No. RD-94 of the American Cancer Society.

REFERENCES

Allison, F., Lancaster, M. G., and Crosthwaite, J. L. (1963). Studies on the pathogenesis of acute inflammation. V. An assessment of factors that influence *in vitro* the phagocytic and adhesive properties of leukocytes obtained from rabbit peritoneal exudate. *Am. J. Pathol.* **43**, 775–795.

Ambani, L., Van Woert, M., and Murphy, S. (1975). Brain peroxidase and catalase in Parkinson disease. *Arch. Neurol. (Chicago)* **32**, 114–118.

Archer, G. T., and Blackwood, A. (1965). Formation of Charcot-Leyden crystals in human eosinophils and basophils and study of the composition of the isolated crystals. *J. Exp. Med.* **122**, 173–180.

Archer, G. T., Air, G., Jackas, M., and Morell, D. B. (1965). Studies on rat eosinophil peroxidase. *Biochim. Biophys. Acta* **99**, 96–101.

Arnason, B. G., and Waksman, B. H. (1963). The retest reaction in delayed sensitivity. *Lab. Invest.* **12**, 737–747.

Badwey, J. A., and Karnovsky, M. L. (1980). Active oxygen species and the functions of phagocytic leukocytes. *Annu. Rev. Biochem.* **49**, 695–726.

Bainton, D. F. (1973). Sequential degranulation of the two types of polymorphonuclear leukocyte granules during phagocytosis of microorganisms. *J. Cell Biol.* **58**, 249–264.

Baldwin, F. (1973). Ultrastructural characteristics of microglia and response to cortical puncture. *J. Anat.* **114**, 154.

Baldwin, F. (1980). Microglia and brain macrophages. In "The Reticuloendothelial System: A Comprehensive Treatise" (I. Carr and W. T. Daems, eds.), Vol. 1, pp. 635–669. Plenum, New York.

Bayliss, C. E., and Waites, W. M. (1976). Effect of hydrogen peroxide on spores of *Clostridium bifermentans. J. Gen. Microbiol.* **96**, 401–407.

Beauchamp, C., and Fridovich, I. (1970). Mechanism for the production of ethylene from methional. Generation of the hydroxyl radical by xanthine oxidase. *J. Biol. Chem.* **245**, 4641–4646.

Becker, S., Fenyo, E. M., and Klein, E. (1976). The natural killer cell in the mouse does not require H-2 homology and is not directed against type or group-specific antigens of murine C viral proteins. *Eur. J. Immunol.* **6**, 882–885.

Beeson, P. B., and Bass, D. A. (1977). "The Eosinophil," pp. 252–257. Saunders, Philadelphia, Pennsylvania.

Bellavite, P., Berton, G., Dri, P., and Soranzo, M. R. (1981). Enzymatic basis of the respiratory burst of guinea pig resident peritoneal macrophages. *J. Reticuloendothel. Soc.* **29**, 47–60.

Biberfield, P., and Holm, G. (1970). Cytotoxic effects of lymphoid cells *in vitro*. *Exp. Cell Res.* **52**, 672–677.

Boxer, L. A., Baehner, R. L., and Davis, J. (1977). The effect of 2-deoxy-glucose on guinea pig polymorphonuclear leukocyte phagocytosis. *J. Cell. Physiol.* **91**, 89–102.

Boya, J. (1975). Contribution to the ultrastructural study of microglia in the cerebral cortex. *Acta Anat.* **92**, 364–375.

Briggs, R. T., Drath, D. B., Karnovsky, M. L., and Karnovsky, M. J. (1975). Localization of NADH oxidase on the surface of human polymorphonuclear leukocytes by a new cytochemical method. *J. Cell Biol.* **67**, 566–586.

Buchmüller, Y., and Mauel, J. (1981). Studies on the mechanisms of macrophage activation: Possible involvement of oxygen metabolites in killing of *Leishmania enrietti* by activated mouse macrophages. *J. Reticuloendothel. Soc.* **29**, 181–192.

Burton, R. C. (1980). Alloantisera selectively reactive with NK cells: Characterization and use in defining NK cell lines. *In* "Natural Cell-Mediated Immunity against Tumors" (R. B. Herberman, ed.), pp. 19–35. Academic Press, New York.

Cannon, G. B., Bonnard, G. D., Djeu, J., West, W. H., and Herberman, R. B. (1977). Relationship of human natural lymphocyte-mediated cytotoxicity to cytotoxicity of breast-cancer-derived target cells. *Int. J. Cancer* **19**, 487–497.

Carr, I. (1980). Macrophages in connective tissues: The granuloma macrophage. *In* "The Reticuloendothelial System: A Comprehensive Treatise" (I. Carr and W. T. Daems, eds.), Vol. 1, pp. 671–703. Plenum, New York.

Carr, I., and Daems, W. T., eds. (1980). "The Reticuloendothelial System: A Comprehensive Treatise," Vol. 1. Plenum, New York.

Caulfield, J. P., Korman, G., Butterworth, A. E., Hogan, M., and David, J. R. (1980a). The adherence of human neutrophils and eosinophils to schistosomula: Evidence for membrane fusion between cells and parasites. *J. Cell Biol.* **86**, 46–63.

Caulfield, J. P., Korman, G., Butterworth, A. E., Hogan, M., and David, J. R. (1980b). Partial and complete detachment of neutrophils and eosinophils from schistosomula: Evidence for the establishment of continuity between a fused and normal parasite membrane. *J. Cell Biol.* **86**, 64–76.

Chondra, S., Seth, P. K., and Mankeshwar, J. K. (1974). Manganese poisoning: Clinical and biochemical observations. *Brain Res.* **108**, 257–277.

Chusid, M. J., Dale, D. C., West, B. C., and Wolf, S. M. (1975). The hypereosinophilic syndrome: Analysis of fourteen cases with review of the literature. *Medicine (Baltimore)* **54**, 1–27.

Clark, R. A., and Klebanoff, S. J. (1975). Neutrophil-mediated tumor cell cytotoxicity: Role of the peroxidase system. *J. Exp. Med.* **141**, 1442–1447.

Clark, R. A., Klebanoff, S. J., Einstein, A. B., and Fefer, A. (1975). Peroxidase-H_2O_2-halide system: Cytotoxic effect on mammalian tumor cells. *Blood* **45**, 161–170.

Cline, M. J., Hanifin, J., and Lehrer, R. I. (1968). Phagocytosis by human eosinophils. *Blood* **32**, 922–934.

Cohen, H. J., Chovaniec, M. E., and Davies, W. A. (1980). Activation of the guinea pig granulocyte NAD(P)H-dependent superoxide generating enzyme: Localization in a plasma membrane enriched particle and kinetics of activation. *J. Am. Soc. Hematol.* **55,** 355–363.

Cohn, Z. A., and Morse, S. I. (1960). Function and metabolic properties of polymorphonuclear leukocytes. II. The influence of a lipopolysaccharide endotoxin. *J. Exp. Med.* **111,** 689–704.

Daems, W. T., Wisse, E., Brederoo, P., and Emeiss, J. J. (1975). Peroxidatic activity in monocytes and macrophages. *In* "Mononuclear Phagocytes in Immunity, Infection, and Pathology" (R. Van Furth, ed.), pp. 57–82. Blackwell, Oxford.

DeChatelet, L. R., Shirley, P. S., McPhail, L. C., Huntley, C. C., Muss, H. B., and Bass, D. A. (1977). Oxidative metabolism of the human eosinophil. *Blood* **50,** 525–535.

Delville, J., and Jacques, P. J. (1978). Therapeutic effect of intravenously administered yeast glucan in mice locally infected by *Mycobacterium leprae*. *In* "Macrophages and Lymphocytes" (M. R. Escobar and H. Friedman, eds.), Vol. A, pp. 245–253. Plenum, New York.

Desser, R. K., Himmelhoch, S. R., Evans, W. H., Januska, M., Mage, M., and Shelton, E. (1972). Guinea pig heterophil and eosinophil peroxidase. *Arch. Biochem. Biophys.* **148,** 452–465.

Dewald, B., Baggiolini, M., Curnette, J. T., and Babior, B. M. (1979). Subcellular localization of the superoxide-forming enzyming in human neutrophils. *J. Clin. Invest.* **63,** 21–29.

Draper, P., and D'Arcy Hart, P. (1975). Phagosomes, lysosomes, and mycobacteria. *In* "Mononuclear Phagocytes in Immunity, Infection and Pathology" (R. Van Furth, ed.), pp. 575–589. Blackwell, Oxford.

Drath, D. V., and Karnovsky, M. L. (1974). Bactericidal activity of metal-mediated peroxide-ascorbate systems. *Infect. Immun.* **10,** 1077–1083.

Eccles, S. A., and Alexander, P. (1974). Macrophage content of tumours in relation to metastatic spread and host immune reaction. *Nature (London)* **250,** 667–669.

Emmelot, P., and Kriek, E., eds. (1979). "Environmental Carcinogenesis." Elsevier/North Holland Publ., Amsterdam.

Evans, R. (1972). Macrophages in syngeneic animal tumours. *Transplantation* **14,** 468–473.

Fidler, I. J., Fogler, W. E., Barnes, Z., and Fisher, K. (1978). The treatment of established micrometastases with syngeneic macrophages. *In* "Macrophages and Lymphocytes" (M. R. Escobar and H. Friedman, eds.), Vol. B, pp. 399–410. Plenum, New York.

Fink, M. A., ed. (1976). "The Macrophage in Neoplasia." Academic Press, New York.

Fishbein, L. (1979). "Potential Industrial Carcinogens and Mutagens." Elsevier/North-Holland Publ., Amsterdam.

Gauci, C. L. (1976). The significance of the macrophage content of human tumors. *In* "Lymphocytes, Macrophages, and Cancer" (G. Mathé, I. Florentin, and M. C. Simmler, eds.), pp. 122–130. Springer-Verlag, Berlin and New York.

Gidlund, M., Haller, O., Orn, A., Ojo, E., Stern, P., and Wigzell, H. (1980). Characteristics of murine NK cells in relation to T lymphocytes and K cells. *In* "Natural Cell-Mediated Immunity against Tumors" (R. B. Herberman, ed.), pp. 79–88. Academic Press, New York.

Gilbert, R. P. (1960). Mechanisms of the hemodynamic effects of endotoxin. *Physiol. Rev.* **40,** 245–279.

Gillet, J., Jacques, P. J., and Herman, F. (1978). Particulate 1-3 glucan and causal prophylaxis of mouse malaria (*Plasmodium berghei*). *In* "Macrophages and Lymphocytes" (M. R. Escobar and H. Friedman, eds.), Vol. A, pp. 307–313. Plenum, New York.

Glauert, A. M., Oliver, R. C., and Thorne, K. J. I. (1980). The interaction of human eosinophils and neutrophils with non-phagocytosable surface: A model for studying cell-mediated immunity in shistosomiasis. *Parasitology* **80**, 525–537.

Gleich, G. J., Loegering, D. A., Mann, K. G., and Maldonado, J. E. (1976). Comparative properties of the Charcot-Leyden crystal protein and the major basic protein from human eosinophils. *J. Clin. Invest.* **57**, 633–640.

Gleich, G. J., Loegering, D. A., Frigas, E., Wassom, D. L., Solley, G. O., and Mann, K. G. (1980). The major basic protein of the eosinophil granule: Physicochemical properties, localization, and function. *In* "The Eosinophil in Health and Disease" (A. A. F. Mahmoud and K. F. Austen, eds.), pp. 79–94. Grune & Stratton, New York.

Glücksman, A. (1965). Cell death in normal development. *Arch. Biol.* **76**, 419–437.

Goetzl, E. J. (1980). The unique roles of monohydroxyeicosatetraenoic acids (HETEs) in the regulation of human eosinophil function. *In* "The Eosinophil in Health and Disease" (A. A. F. Mahmoud and K. F. Austen, eds.), pp. 167–184. Grune & Stratton, New York.

Goetzl, E. J., and Austen, K. F. (1975). Purification and synthesis of eosinophilotactic tetrapeptides of human lung tissue: Identification as eosinophil chemotactic factor of anaphylaxis. *Proc. Natl. Acad. Sci. U.S.A.* **72**, 4123–4127.

Goetzl, E. J., Weller, P. F., and Sun, F. F. (1980). The regulation of human eosinophil function by endogenous monohydroxyeicosatetraenoic acids (HETEs). *J. Immunol.* **124**, 926–933.

Greco, D. B., Fujita, Y., Ishikawa, T., and Arbesman, C. E. (1973). Antigen-antibody complexes in/on eosinophils in nasal secretions from patients allergic to ragweed. *J. Allergy Clin. Immunol.* **51**, 124–125.

Griffin, F. M., Jr., Griffin, J. A., Leider, J. E., and Silverstein, S. C. (1975). Studies on the mechanism of phagocytosis. I. Requirements for circumferential attachment of particle-bound ligands to specific receptors on the macrophage plasma membrane. *J. Exp. Med.* **142**, 1263–1282.

Griffin, F. M., Jr., Griffin, J. A., and Silverstein, S. C. (1976). Studies on the mechanism of phagocytosis. II. The interaction of macrophages with anti-immunoglobin IgG-coated bone marrow-derived lymphocytes. *J. Exp. Med.* **144**, 788–809.

Halliwell, B. (1976). An attempt to demonstrate a reaction between superoxide and hydrogen peroxide. *FEBS. Lett.* **72**, 8–10.

Hanna, M. G., Jr. (1978). Macrophages in tumor immunity. *In* "Macrophages and Lymphocytes" (Mr. Escobar and H. Friedman, eds.), Vol. B, pp. 353–359. Plenum, New York.

Harmon, R. C., Clark, E. A., Reddy, A. L., Hildemann, U. H., and Mullen, Y. (1977). Immunity to MCA-induced rat sarcomas: Analysis of *in vivo* and *in vitro* results. *Int. J. Cancer* **20**, 748–758.

Hartwig, J. H., and Stossel, T. P. (1976). Interactions of actin, myosin, and an actin-binding protein of rabbit pulmonary macrophages. III. Effects of cytochalasin B. *J. Cell Biol.* **71**, 295–303.

Heifets, L., Imai, K., and Goren, M. B. (1980). Expression of peroxidase-dependent iodination by macrophages ingesting neutrophil debris. *J. Reticuloendothel. Soc.* **28**, 391–404.

Henson, P. M. (1971). The immunologic release of constituents from neutrophil leukocytes. I. The role of antibody and complement on non-phagocytosable surfaces or phagocytosable particles. *J. Immunol.* **107**, 1535–1546.

Herberman, R. B., ed. (1980a). "Natural Cell-Mediated Immunity against Tumors." Academic Press, New York.

Herberman, R. B. (1980b). Summary: Characteristics of NK and related cells. *In* "Natural

Cell-Mediated Immunity against Tumors" (R. B. Herberman, ed.), pp. 277–288. Academic Press, New York.

Hibbs, J. B., Weinberg, J. B., and Chapman, H. A. (1978). Modulation of the tumoricidal function of activated macrophages by bacterial endotoxin and mammalian macrophage activation factor(s). In "Macrophages and Lymphocytes" (M. R. Escobar and H. Friedman, eds.), Vol. B, pp. 433–453. Plenum, New York.

Hinshaw, L. B., Jordan, M. M., and Vick, J. A. (1961). Mechanism of histamine release in endotoxin shock. Am. J. Physiol. 200, 987–989.

Hinshaw, L. B., Emerson, T. E., Jr., Iampietro, P. F., and Brake, C. M. (1962). A comparative study of the hemodynamic actions of histamine and endotoxin. Am. J. Physiol. 203, 600–606.

Hirsch, J. G. (1965). Neutrophil and eosinophil leukocytes. In "The Inflammatory Process" (B. S. Zweifach, L. Grant, and R. T. McCluskey, eds.), pp. 245–280. Academic Press, New York.

Imamoto, K., and Leblond, C. P. (1977). Presence of labeled monocytes, macrophages and microglia in a stab wound of the brain following an injection of bone marrow cells labeled with ^{3}H-uridine into rats. J. Comp. Neurol. 174, 255–280.

Jacques, P. J., Avila, J. L., Pinardi, M. E., and Convit, J. (1975). Germicidal activity of a polyenzyme system on pathogenic protozoa in vitro. Arch. Int. Physiol. Biochim. 83, 976–978.

Janssen, H. F., Lutherer, L. O., and Barnes, C. D. (1981). An observed pressor effect of the cerebellum during endotoxin shock in the dog. Am. J. Physiol. 240, H368–H374.

Jensen, P. J., and Koren, H. S. (1980). Heterogeneity in natural killing. In "Natural Cell-Mediated Immunity against Tumors" (R. B. Herberman, ed.), pp. 277–288. Academic Press, New York.

Jones, T. C., Minick, R., and Yang, L. (1977). Attachment and ingestion of Mycoplasmas by mouse macrophages. Am. J. Pathol. 87, 347–353.

Kaplan, G., Gaudernack, G., and Seljelid, R. (1975). Localization of receptors and early events of phagocytosis in the macrophage. Exp. Cell Res. 95, 365–375.

Karnovsky, M. L., Simmons, S., Glass, E. A., Shafer, A. W., and D'Arcy Hart, P. (1970). Metabolism of macrophages. In "Mononuclear Phagocytes" (R. Van Furth, ed.), pp. 103–117. Blackwell, Oxford.

Kielian, M. C., and Cohn, Z. A. (1980). Phagosome-lysosome fusion: Characterization of intercellular membrane fusion in mouse macrophages. J. Cell Biol. 85, 754–765.

Kiessling, R., Petranyi, G., Kärre, K., Jondal, M., Tracey, D., and Wigzell, H. J. (1976). Killer cells: A function comparison between natural, immune T-cell and antibody-dependent in vitro systems. J. Exp. Med. 143, 772–780.

Kiessling, R., Roder, J. C., and Biberfield, P. (1980). Ultrastructural and cytochemical studies of mouse natural killer (NK) cells. In "Macrophages and Lymphocytes" (M. R. Escobar and H. Friedman, eds.), Part B, pp. 155–163. Plenum, New York.

Klebanoff, S. J. (1967). A peroxidase-mediated anti-microbial system in leukocytes. J. Clin. Invest. 46, 1078.

Klebanoff, S. J. (1974). Role of the superoxide anion in the myeloperoxidase-mediated antimicrobial system. J. Biol. Chem. 249, 3724–3728.

Klebanoff, S. J., and Clark, R. A. (1978). "The Neutrophil: Function and Clinical Disorders." North-Holland Publ., Amsterdam.

Klebanoff, S. J., and Hamon, C. B. (1972). Role of myeloperoxidase-mediated anti-microbial systems in intact leukocytes. RES, J. Reticuloendothel. Soc. 12, 170–196.

Klebanoff, S. J., and Luebke, R. G. (1965). The anti-lactobacillus system of saliva. Role of salivary peroxidase. Proc. Soc. Exp. Biol. Med. 118, 483–486.

Klebanoff, S. J., and Smith, D. C. (1970). Peroxidase-mediated anti-microbial activity of rat uterine fluid. *Gynecol. Invest.* **1**, 21–30.

Klebanoff, S. J., Jong, E. C., and Henderson, W. R., Jr. (1980). The eosinophil peroxidase: Purification and biological properties. *In* "The Eosinophil in Health and Disease" (A. A. F. Mahmoud and K. F. Austen, eds.), pp. 99–111. Grune & Stratton, New York.

Krinsky, N. I. (1974). Singlet excited oxygen as a mediator of the anti-bacterial action of leukocytes. *Science* **186**, 363–365.

Kvarstein, B. (1969). The effect of temperature, metabolic inhibitors, and EDTA on phagocytosis of polystyrene latex particles by human leukocytes. *Scand. J. Clin. Lab. Invest.* **24**, 271–277.

Liehr, H., and Grün, M. (1977). Clinical aspects of Kupffer cell failure in liver diseases. *In* "Kupffer Cells and Other Liver Sinusoidal Cells" (E. Wisse and D. L. Knook, eds.), pp. 427–436. Elsevier, Amsterdam.

Lohmann-Mattes, M. L., and Domzig, W. (1980). Natural cytotoxicity of macrophage precursor cells and of mature macrophages. *In* "Natural Cell-Mediated Immunity against Tumors" (R. B. Herberman, ed.), pp. 117–129. Academic Press, New York.

McClune, G. J., and Fee, J. A. (1976). Stopped flow spectrophotometric observations of superoxide dismutation in aqueous solution. *FEBS Lett.* **67**, 294–298.

McGeer, E., Olney, J., and McGeer, P. (1978). "Kainic Acid as a Tool in Neurobiology." Raven, New York.

Makita, T., and Sanborn, E. B. (1970). The ultrastructural localization of adenine-triphosphatase and alkaline phosphatase activity in eosinophil leukocytes. *Histochemie* **24**, 99–105.

Meltzer, M. S., and Stevenson, M. M. (1978). Macrophage function in tumor-bearing mice. *Cell. Immunol.* **35**, 99–111.

Meltzer, M. S., Stevenson, M. M., and Leonard, E. J. (1977). Characterization of macrophage chemotaxis in tumor cell cultures and comparison with lymphocyte-derived chemotactic factors. *Cancer Res.* **37**, 721–725.

Menkes, B., Sander, S., and Ilies, A. (1970). Cell death in teratogenesis. *Adv. Teratol.* **4**, 169–215.

Mickenberg, I. D., Root, R. K., and Wolff, S. M. (1972). Bacterial and metabolic properties of human eosinophils. *Blood* **39**, 67–80.

Miller, F., DeHarven, E., and Palade, G. E. (1966). The structure of eosinophil leukocyte granules in rodents and man. *J. Cell Biol.* **31**, 349–362.

Misra, H. P., and Fridovich, I. (1976). Superoxide dismutase and the oxygen enhancement of radiation. *Arch. Biochem. Biophys.* **176**, 577–581.

Nathan, C. F., and Cohn, Z. A. (1980). Role of oxygen-dependent mechanisms in antibody-induced lysis of tumor cells by activated macrophages. *J. Exp. Med.* **152**, 198–208.

Nathan, C. F., Murray, H. W., and Cohn, Z. A. (1980). The macrophage as an effector cell. *N. Engl. J. Med.* **303**, 622–626.

Nauts, H. C., Fowler, G. A., and Bogatko, F. H. (1953). Review of the influence of bacterial infection and of bacterial products (Coley's toxins) on malignant tumors in man. *Acta Med. Scand.* **45**, Suppl. 276, 1–103.

Nichols, B. A., and Bainton, D. F. (1975). Ultrastructure and cytochemistry of mononuclear phagocytes. *In* "Mononuclear Phagocytes in Immunity, Infection, and Pathology" (R. Van Furth, ed.), pp. 17–55. Blackwell, Oxford.

Oehmichen, M., and Torvik, A. (1976). The origin of reactive cells in retrograde and Wallerian degeneration. Experiments with intravenous injection of ^{3}H-DFP-labelled macrophage. *Cell Tissue Res.* **173**, 343–348.

Ögmundsdottir, H. M., and Weir, D. M. (1980). Mechanisms of macrophage activation. *Clin. Exp. Immunol.* **40**, 223–234.

Old, L. J. (1981). Cancer immunology: The search for specificity. *Cancer Res.* **41**, 361–375.

Ornstein, L., Ansley, H., and Saunders, A. (1976). Improving manual differential white cell counts with cytochemistry. *Blood Cells* **2**, 557–585.

Ortaldo, J. R., Herberman, R. B., and Djeu, J. Y. (1980). Characteristics of augmentation by interferon of cell-mediated cytotoxicity. *In* "Natural Cell-Mediated Immunity against Tumors" (R. B. Herberman, ed.), pp. 593–607. Academic Press, New York.

Parmley, R. T., and Spicer, S. S. (1974). Cytochemical and ultrastructural identification of a small type granule in human late eosinophils. *Lab. Invest.* **30**, 557–567.

Parr, I., Wheeler, E., and Alexander, P. (1973). Similarities of the antitumor actions of endotoxin, lipid A, and double-stranded RNA. *Br. J. Cancer* **27**, 370–389.

Parrillo, J. E., and Fauci, A. S. (1977). Apparent direct cellular cytotoxicity mediated via cellular antibody. Multiple Fc receptor bearing effector cell populations mediating cytophilic antibody induced cytotoxicity. *Immunology* **33**, 839–850.

Parwaresch, M. R., Walle, A. J., and Arndt, D. (1976). The peripheral kinetics of human radiolabelled eosinophils. *Virchows Arch. B* **21**, 57–66.

Perlman, P. and Holm, G. (1969). Cytotoxic effects of lymphoid cells *in vitro*. *Adv. Immunol.* **11**, 117–193.

Pesanti, E. L. (1979). Kinetics of phagocytosis of *Staphlococcus aureus* by alveolar and peritoneal macrophages. *Infect. Immun.* **26**, 479–486.

Pike, M. C., and Snyderman, R. (1980). Biochemical and biological aspects of leukocyte chemotactic factors. *In* "The Reticuloendothelial System: A Comprehensive Treatise" (I. Carr and W. T. Daems, eds.), Vol. 2, pp. 1–19. Plenum, New York.

Poste, G., and Allison, A. C. (1973). Membrane fusion. *Biochim. Biophys. Acta* **300**, 421–465.

Postlethwaite, A. E., and Snyderman, R. (1975). Characterization of chemotactic activity produced *in vivo* by a cell mediated immune reaction in the guinea pig. *J. Immunol.* **114**, 274–278.

Proctor, R. A. (1979). Endotoxin in vitro interactions with human neutrophils: Depression of chemiluminescence, oxygen consumption, superoxide production, and killing. *Infect. Immun.* **25**, 912–921.

Raine, C. S., Wisniewski, H. H., and Prineas, J. (1969). An ultrastructural study of experimental demyelination and remyelination. *Lab. Invest.* **21**, 316–327.

Reiter, B. (1978). Antimicrobial systems in milk. *J. Dairy Res.* **45**, 131–147.

Reiter, B. (1979). The lactoperoxidase-thiocyanate-hydrogen peroxide antibacterium system. *Ciba Found. Symp.* **65** (new ser.), 285–294.

Roder, J. C. (1980). Phenotypic characteristics of NK cells in the mouse. *In* "Natural Cell-Mediated Immunity against Tumors" (R. B. Herberman, ed.), pp. 161–171. Academic Press, New York.

Rohrer, G. F., von Wartburg, J. P., and Aebi, H. (1966). Myeloperoxidase aus menschlichen Leukocyten. I. Isolierung und Charakterisierung des Enzymes. *Biochem. Z.* **344**, 478–491.

Roos, D., and Weening, R. S. (1979). Defects in the oxidative killing of microorganisms by phagocytic leukocytes. *Ciba Found. Symp.* **65** (new ser.), 225–262.

Root, B. K., and Metcalf, J. A. (1977). H_2O_2 release from human granulocytes during phagocytosis. Relationship to superoxide anion formation and cellular catabolism of H_2O_2. *J. Clin. Invest.* **60**, 1266–1279.

Rosen, H., and Klebanoff, S. J. (1976). Chemiluminescence and superoxide production by myeloperoxidase-deficient leukocytes. *J. Clin. Invest.* **58**, 50–60.

Rosenthal, A. (1980). Regulation of the immune response-role of the macrophage. *N. Engl. J. Med.* **303,** 1153–1156.

Rossi, F., Zabucchi, G., Dri, P., Bellavite, P., and Berton, G. (1978). O_2^- and H_2O_2 production during the respiratory burst in alveolar macrophages. *In* "Macrophages and Lymphocytes" (M. A. Escobar and H. Friedman, eds.), Part A, pp. 53–74. Plenum, New York.

Ruggerio, G., Utili, R., and Andreana, A. (1977). Clearance of viable salmonella strains by the isolated, perfused rat liver: A study of serum and cellular factors involved and of the affect of treatments with carbon tetrachloride or *Salmonella enteritidis* lipopolysaccharide. *J. Reticuloendothel. Soc.* **21,** 79–88.

Russel, R. J. McInroy, R. J., Wilkinson, P. C., and White, R. G. (1976). A lipid chemotactic factor from anaerobic coryneform bacteria including *Corynebacterium parvum* with activity for macrophages and monocytes. *Immunology* **30,** 935–949.

Russell, S. W., and McIntosh, A. T. (1977). Macrophages isolated from regressing Moloney sarcomas are more cytotoxic than those recovered from progressing sarcomas. *Nature (London)* **268,** 69–71.

Russell, S. W., Doe, W. F., and Cochrane, C. G. (1976). Number of macrophages and distribution of mitotic activity in regressing and progressing Moloney sarcomas. *J. Immunol.* **116,** 164–166.

Saksela, E., and Timonen, T. (1980). Morphology and surface properties of human NK cells. *In* "Natural Cell-Mediated Immunity against Tumors" (R. B. Herberman, ed.), pp. 173–185. Academic Press, New York.

Saksela, E., Timonen, T., Virtanen, I., and Cantell, K. (1980). Regulation of human natural killer cell activity by interferon. *In* "Natural Cell-Mediated Immunity against Tumors" (R. B. Herberman, ed.), pp. 645–653. Academic Press, New York.

Samak, R., Israel, L., and Edelstein, R. (1978). Influence of tumor burden, tumor removal, immune stimulation, plasmapheresis on monocyte mobilization in cancer patients. *In* "Macrophages and Lymphocytes" (M. R. Escobar and H. Friedman, eds.), Vol. B, pp. 411–423. Plenum, New York.

Sannes, P. L., McDonald, J. K., and Spicer, S. S. (1977). Dipeptidyl aminopeptidase II in rat peritoneal wash cells. Cytochemical localization and biochemical characterization. *Lab. Invest.* **37,** 243–253.

Sbarra, A. J., and Karnovsky, M. L. (1959). The biochemical basis of phagocytosis. I. Metabolic changes during the ingestion of particles by polymorphonuclear leukocytes. *J. Biol. Chem.* **234,** 1355–1362.

Sbarra, A. J., and Strauss, R. R., eds. (1980). "The Reticuloendothelial System: A Comprehensive Treatise," Vol. 2. Plenum, New York.

Schaefer, H. E., Hubner, G., and Fischer, R. (1973). Spezifische microgranula in eosinophilen. *Acta Haematol.* **50,** 92–104.

Schultz, J., and Kaminker, K. (1962). Myeloperoxidase of the leukocyte of normal human blood. I. Content and localization. *Arch. Biochem. Biophys.* **96,** 465–467.

Schultz, J., Turtle, A., Shay, H., and Gruenstein, M. (1955). Chemistry of experimental chloroma: The effect of antileukemic agents. *Abstr. Pap., 128th Meet., Am. Chem. Soc.* 70C.

Schultz, R. M. (1981). Factors limiting tumoricidal function of interferon-induced effector systems. *Cancer Immunol. Immunother.* **10,** 61–66.

Scott, R. E., and Horn, R. B. (1970). Ultrastructural aspects of neutrophil granulocyte development in humans. *Lab. Invest.* **23,** 202–215.

Scott, R. E., and Rosenthal, A. S. (1977). Isolation of receptor-bearing plasma membrane vesicles from guinea pig macrophages. *J. Immunol.* **119,** 143–148.

Seiler, G., Westerman, R. A., and Wilson, J. A. (1969). The role of specific eosinophil granules in eosinophil-induced experimental encephalitis. *Neurology* **19**, 478–488.

Shellam, G. R., and Hogg, N. (1977). Gross-virus-induced lymphoma in the rat. IV. Cytotoxic cells in normal rats. *Int. J. Cancer.* **19**, 212–224.

Sher, R., and Wadee, Λ. Λ. (1980). A scanning electron microscope study of eosinophil phagocytosis. *J. Reticuloendothel. Soc.* **28**, 179–189.

Silverstein, S. C. (1975). The role of mononuclear phagocytes in viral immunity. In "Mononuclear Phagocytes" (R. Van Furth, ed.), pp. 557–568. Blackwell, Oxford.

Simon, G. T. (1980). Splenic macrophages. In "The Reticuloendothelial System: A Comprehensive Treatise" (I. Carr and W. J. Daems, eds.), Vol. 1, pp. 469–497. Plenum, New York.

Skoff, R. P. (1975). The fine structure of pulse labelled (^{3}H-thymidine cells) in degenerating rat optic nerve. *J. Comp. Neurol.* **161**, 595–612.

Spilberg, I. (1975). Current concepts of the mechanism of acute inflammation in gouty arthritis. *Arthritis Rheum.* **18**, 129–134.

Spink, W. W., Davis, R. B., Potter, R., and Chartrand, S. (1964). The initial stage of canine endotoxin shock as an expression of anaphylactic shock: Studies on complement titers and plasma histamine concentrations. *J. Clin. Invest.* **43**, 696–704.

Spry, C. J. F., and Tai, P. C. (1976). Studies on blood eosinophils. II. Patients with Löffler's cardiomyopathy. *Clin. Exp. Immunol.* **24**, 423–434.

Stendahl, O. I., Hartwig, J. H., Bretschi, E. A., and Stossel, T. P. (1980). Distribution of actin-binding protein and myosin in macrophages during spreading and phagocytosis. *J. Cell Biol.* **84**, 215–224.

Stevenson, M. M., and Meltzer, M. S. (1976). Depressed chemotactic responses *in vitro* of peritoneal macrophages from tumor-bearing mice. *JNCI, J. Natl. Cancer Inst.* **57**, 847–852.

Stossel, T. P. (1977). Contractile proteins in phagocytosis: An example of cell surface-to-cytoplasm communication. *Fed Proc., Fed. Am. Soc. Exp. Biol.* **36**, 2181–2184.

Stossel, T. P., and Hartwig, J. H. (1975). Interactions between actin, myosin and an actin-binding protein from rabbit alveolar macrophages. Alveolar macrophage myosin Mg^+-adenosine triphosphatase requires a cofactor for activation by actin. *J. Biol. Chem.* **250**, 5706–5712.

Stossel, T. P., and Hartwig, J. H. (1976). Interaction of actin, myosin and a new actin-binding protein of rabbit pulmonary macrophages. *J. Cell Biol.* **68**, 602–619.

Takasugi, M., Ramseyer, A., and Takasugi, J. (1977). Decline of natural nonselective cell-mediated cytotoxicity in patients with tumor progression. *Cancer Res.* **37**, 413–418.

Tchernitchin, A. (1967). Autoradiographic study of 6,7-^{3}H-estradiol-17 incorporation into the rat uterus. *Steroids* **10**, 661–668.

Tchernitchin, A., Roorijk, J., Tchernitchin, X., Vandenhende, J., and Garland, P. (1974). Dramatic early increase in uterine eosinophils after oestrogen administration. *Nature (London)* **248**, 142–143.

Tchernitchin, A., Tchernitchin, X., and Galand, P. (1975). Correlation of estrogen-induced uterine eosinophilia with other parameters of estrogen stimulation, produced with estradiol-17 and estriol. *Experientia* **31**, 993–994.

Torack, R. M. (1961). Ultrastructure of capillary reaction to brain tumors. *Arch. Neurol. (Chicago)* **5**, 416–428.

Tubaro, E., Lott, L., Santiangeli, C., and Cavallo, G. (1980). Xanthine oxidase increased in polymorphonuclear leukocytes and macrophages in three pathological situations. *Biochem. Pharmacol.* **29**, 1945–1948.

Vadas, M. A., David, J. R., Butterworth, A. E., Pisani, N. T., and Siongok, T. A. (1979). A new method for the purification of human eosinophils and neutrophils, and a comparison of the ability of these cells to damage schistosomula of *Schistosoma mansoni.* *J. Immunol.* **122,** 1228–1236.

Van Berkel, T. J. C. (1974). Difference spectra, catalase and peroxidase activities of isolated parenchymal and nonparenchymal cells from rat liver. *Biochem. Biophys. Res. Commun.* **61,** 204–209.

Van Furth, R., ed. (1975). "Mononuclear Phagocytes in Immunity, Infection and Pathology." Blackwell, Oxford.

Venge, P., Dahl, R., Hällgren, R., and Olsson, I. (1980). Cationic proteins of human eosinophils and their role in the inflammatory reaction. *In* "The Eosinophil in Health and Disease" (A. A. F. Mahmoud and K. F. Austen, eds.), pp. 131–142. Grune & Stratton, New York.

Vick, J. A. (1964). Trigger mechanisms of endotoxin shock. *Am. J. Physiol.* **206,** 944–946.

Volkman, A. (1976). Disparity in origin of mononuclear phagocyte populations. *J. Reticuloendothel. Soc.* **19,** 249–268.

Ward, P. A. (1968). Chemotaxis of mononuclear cells. *J. Exp. Med.* **128,** 1201–1221.

Ward, P. A., Remold, H. G., and David, J. R. (1969). Leukotactic factor produced by sensitized lymphocytes. *Science* **163,** 1079–1081.

Waterfield, J. D., Waterfield, E. M., Anaclerio, A., and Möller, G. (1976). Lymphocyte-mediated cytotoxicity against tumor cells. Specificity and characterization of concanavalin A-activated cytotoxic effector lymphocytes. *Transplant. Rev.* **29,** 277–310.

Weinberg, J. B., Chapman, H. A., and Hibbs, J. B. (1978). Characterization of the effects of endotoxin on macrophage tumor cell killing. *J. Immunol.* **121,** 72–80.

Weissmann, G., Dukor, P., and Zurier, R. B. (1971a). Effect of cyclic AMP on release of lysosomal enzymes from phagocytes. *Nature (London) New Biol.* **231,** 131–135.

Weissmann, G., Zurier, R. B., Spieler, P. J., and Goldstein, I. M. (1971b). Mechanisms of lysosomal enzyme release from leukocytes exposed to immune complexes and other particles. *J. Exp. Med.* **134,** 149s–165s.

Weller, P. F., Wasserman, S. I., and Austen, K. F. (1980a). Selected enzymes preferentially present in the eosinophil. *In* "The Eosinophil in Health and Disease" (A. A. F. Mahmoud and K. F. Austen, eds.), pp. 115–128. Grune & Stratton, New York.

Weller, P. F., Goetzl, E. J., and Austen, K. F. (1980b). Identification of human eosinophil lysophospholipase as the constituent of Charcot-Leyden crystals. *Proc. Natl. Acad. Sci. U.S.A.* **77,** 7440–7443.

West, B. C., Gelb, N. A., and Rosenthal, A. S. (1975). Isolation and partial characterization of human eosinophil granule. *Am. J. Pathol.* **81,** 575–585.

Whitelaw, D. M., and Batho, H. F. (1975). Kinetics of monocytes. *In* "Mononuclear Phagocytes in Immunity, Infection, and Pathology" (R. Van Furth, ed.), pp. 175–187. Blackwell, Oxford.

Wong-Riley, M. (1976). Endogenous peroxidatic activity in brain stem neurons as demonstrated by their staining with diaminobenzidine in normal squirrel monkeys. *Brain Res.* **108,** 257–277.

Woods, M. W., Landy, M., Whitby, J. L., and Burk, D. (1961). Symposium on bacterial endotoxins. III. Metabolic effects of endotoxin on mammalian cells. *Bacteriol. Rev.* **25,** 447–456.

Yam, L. T., Li, C. Y., Necheles, T. F., and Katayama, I. (1972). Pseudo-eosinophilia, eosinophilic endocarditis and eosinophilic leukemia. *Am. J. Med.* **53,** 193–202.

Yoffey, J. M. (1978). Lymphocytes and macrophages in the lymphomyeloid complex. *In* "Macrophages and Lymphocytes" (M. R. Escobar and H. Friedman, eds.), Vol. B, pp. 127–144. Plenum, New York.

Zarling, J. M. (1980). Augmentation of human natural killer cell activity by purified inter-feron and polyribonucleotides. *In* "Natural Cell-Mediated Immunity against Tumors" (R. B. Herberman, ed.), pp. 687–706. Academic Press, New York.

Zbar, B., Berstein, I. D., and Rapp, H. J. (1971). Suppression of tumor growth at the site of infection with living Bacillus Calmette-Guérin. *JNCI, J. Natl. Cancer Inst.* **46,** 831–839.

Zinkernagel, R. M. (1976). Virus-specific T-cell mediated cytotoxicity across the H-2 barrier to virus-altered alloantigen. *Nature (London)* **261,** 139–141.

2

Effects of Gases and Airborne Particles on Lung Infections*

Donald E. Gardner

*This report has been reviewed by the Health Effects Research Laboratory, U.S. Environmental Protection Agency, and approved for publication. Mention of trade names or commercial products does not constitute endorsement or recommendation for use.

I.   INTRODUCTION

The respiratory system, which daily processes a huge volume of ambient atmosphere, is subjected to a constant assault by potentially hazardous chemicals and microorganisms. Elaborate defense mechanisms have evolved, enabling the lung to protect itself from microbial attack in several fashions. Indeed, these defenses are so effective that in healthy animals the lower respiratory tract is virtually sterile (Pecora and Yegian, 1958; Goldstein et al., 1976). Particles deposited in the tracheobronchial region are transported out of the lung within a few hours by the mucociliary escalator systems (Lauweryns and Baert, 1977). The most effective defense of the deep lung is provided by alveolar macrophages, which are capable of inactivating and killing viable microorganisms within 2–4 hours of invasion (Kass et al., 1966; Green, 1973; Kim et al., 1976).

With the relatively recent increase of fossil fuel consumption and the growth of the synthetic chemical industry, a whole new spectrum of potentially dangerous compounds has been introduced into the atmosphere to be processed by human respiratory systems. Environmental toxicology has traditionally been concerned with describing the effects of single agents on tested individuals or groups. In the case of compounds with directly toxic, caustic, or mutagenic properties (such as carbon monoxide, chlorine gas, or aromatic hydrocarbons) administered in sufficient quantities, the single-agent approach can yield demonstrable causal relationships between specific chemicals and health effects. Nonetheless, effects more subtle and complex, but no less dangerous, attend the attack of chemicals not on lung parenchyma directly but on the macrophagic defenders of the respiratory system. Such undermining of natural defenses against disease can occur at concentrations of toxicants far lower than those eliciting classical toxicological symptoms, but approximating concentrations now encountered by the populations of urban industrial centers.

This review describes an *in vivo* testing methodology, developed and refined over the past two decades, by which the effects of minute concentrations of pollutants on animal respiratory defenses can be evaluated quantitatively. Results from such tests, performed for a variety of gaseous and particulate compounds on several different animal species, are presented, and a discussion of the various respiratory defense mechanisms and their impairment by environmental toxicants is included.

II.   THE *IN VIVO* INFECTIVITY MODEL SYSTEM

Bacterial infection, by itself, in a host respiratory system is not generally lethal to the host; mortality results when unchecked infection produces lesions in respiratory tissue and invades the bloodstream. The suppression of pulmonary defenses by environmental chemicals would be expected to result in prolongation of bacterial viability, during the initial incubation stages, and ultimately, in enhanced disease and mortality rates. The weakened defense against microbes reflects a summation of various deleterious changes within the lung, which may include edema, reduced phagocytic or bactericidal activity of the pulmonary macrophages, altered ciliary activity, inflammation, and immunosuppression.

The net result of these changes can be monitored and quantitated by use of an animal model system such as that depicted schematically in Fig. 1. In this system, animals are exposed in chambers to controlled concentrations of test pollutants, while a similar group of animals is maintained in clean air only. The two groups are then brought together and exposed briefly to an atmosphere containing an aerosol of infectious microorganisms, after which the groups are again separated and monitored for respiratory disease and mortality. The operational details of these experiments, including selection of test animal species, infectious microorganisms, concentration of test pollutant, and duration of exposure periods, are dictated by the objectives of the particular experiment. Procedures that have proved effective in such studies are described in greater detail in the following subsections.

Because studies aiming to assess risk to humans and to establish regulatory guidelines must conform to some specific criteria, both the procedures and the facilities were designed to

1. Ensure the sensitivity of the test animals to a wide variety of gaseous and particulate compounds as well as complex mixtures.

2. Yield reproducible data.

3. Permit correlation among species of test animals.

4. Facilitate direct comparison of data with that from mechanistic studies.

Furthermore, to ensure the efficacy of the model system, the infectious agent itself must

1. Be able to multiply within and infect susceptible tissue.

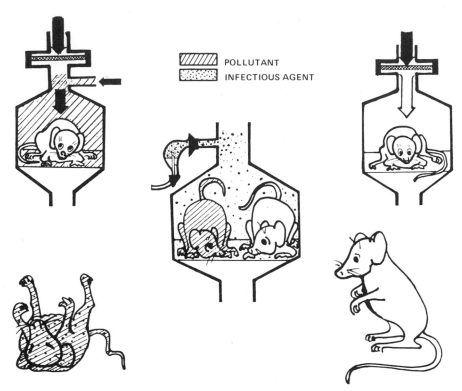

POLLUTANT
INFECTIOUS AGENT

Fig. 1. Schematic representation of the *in vivo* infectivity model system. Animals are exposed to a test pollutant or to clean, filtered air (controls). The two groups are simultaneously challenged with an aerosol of infectious microbes (streptococci), then separated and monitored for different rates of mortality.

2. Produce a low but significant (approximately 10%) mortality in control animals, so that mortality differences after chemical exposure will be statistically significant.

3. Not be affected by small variations in dose or virulence (i.e., changes in mortality should result from altered susceptibility of the host to the infection and not from variations in the initial dose of the microorganism).

4. Be quantifiable at the administered dose.

5. Be able to withstand the rigors of aerosolization (for administration via inhalation).

6. Not be highly pathogenic to humans (e.g., the investigators administering them).

Fig. 2. Animal exposure chambers for administration of gaseous pollutants in infectivity studies.

A. Exposure Chambers

Figure 2 shows the exposure chambers in which the test animals are administered controlled atmospheres. Temperature and relative humidity are controlled by the experimenter, and water and feed can be provided *ad libitum* throughout. Figure 3 depicts schematically the infectivity chamber where both control and test animals are challenged with an infectious microbial aerosol. The aerosol is generated by use of a DeVilbiss No. 40 compressed air nebulizer, which atomizes a liquid (in this case, a suspension of microbes in nutrient broth) by moving a stream of air across a narrow nozzle into which the liquid has been drawn by the

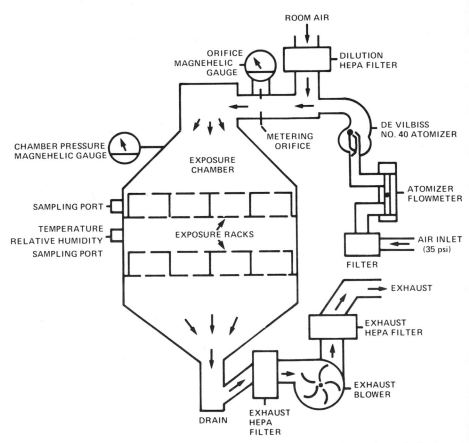

Fig. 3. Schematic diagram of the exposure chamber for administration of infectious microbes. The airflow in the chamber is denoted by arrows.

Bernoulli effect. Large droplets impinge upon the upper wall of the exit tube, whereas smaller droplets, ranging from 1 to 7 μm in diameter, remain aerosolized and are transported by the airflow (arrows) to the animals. The aerosol concentration is generally on the order of a few microliters of fluid per liter of air, and, typically, 5 ml of fluid are atomized over a 15-minute exposure period. To minimize the variation in total dosage per animal, the animals are housed individually and distributed randomly in stainless steel mesh racks, which prevent the animals from huddling in the exposure chamber.

B. Microorganisms

A variety of cultivated disease strains have been used as infectious agents for this model system. The two most commonly used are *Streptococcus pyogenes* and *Klebsiella pneumoniae*, and others, used less frequently, include *Diplococcus pneumoniae, Mycoplasma pneumoniae, Escherichia coli, Salmonella typhimurium*, and influenza virus. For inhalation studies in which the lung is the target organ, it is more appropriate to use respiratory infectious agents than microbes of intestinal origin.

As an essential test of the effect of initial microbe concentration on mortality, cumulative data for 2650 control mice administered *Streptococcus* aerosols in amounts ranging from 200 to 4000 colony-forming units (cfu) per lung are shown in Fig. 4. As is evident from the Fig. 4, there is no correlation between the initial dosages of microbe in this range and overall mortality; thus, one criterion for the infectivity model system, that mortality be attributable to altered host responses to infection rather than minor, random variations in administered doses, is fulfilled (Miller *et al.*, 1978a).

C. Test Animals

The choice of test animal depends on the nature of the information sought and is also subject to constraints of economy, space, and compatibility with other experimental criteria. Mice are the most commonly used subjects in these infectivity studies, although experiments have also been performed with rats, hamsters, and squirrel monkeys (Ehrlich, 1963, 1966; Purvis and Ehrlich, 1963; Henry *et al.*, 1970; Fairchild *et al.*, 1972; Coffin and Gardner, 1972a; Coffin *et al.*, 1976). The natural resistance to respiratory infection varies tremendously from species to species and from microbe to microbe. Although very few comparative

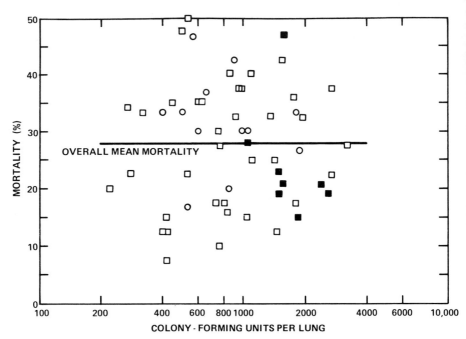

Fig. 4. Average mortality in groups of control animals versus the number of colony-forming units per lung of *Streptococcus pyogenes* initially administered in the infectivity exposure chamber. The wide scatter of points indicates no relation between the initial dosage (in the range from 200 to 3000 cfu/lung) and mortality. Thus, measured mortality differences between control and pollutant-exposed animals reflect differences in resistance to microbial growth.

interspecies studies have been performed, it would appear that susceptibility to respiratory infection could vary inversely with animal size. Such a relation appears reasonable, because the ratio of volume respired to body weight also decreases dramatically with increasing size. Another possibility is that larger animals are more resistant to the effects of inhaled pollutants. It is equally possible, however, that different test species vary in their sensitivity to a particular test microbe. One way to test this possibility is to determine whether there is a significant difference in mortality between the control groups of the two species. For example, the control mortality rates for mice and monkeys exposed to *Klebsiella* were 41 and 0%, respectively (Ehrlich *et al.*, 1975). This illustrates that monkeys are far more resistant to this particular organism than are mice, but one cannot conclude directly from such comparative studies that the monkey is less vulnerable to the test chemical due to its size.

D. Exposure Sequence

The concentration of pollutant, duration of exposure, and length of time between exposure and infectious challenge are all variables that have been observed to affect the infectivity and mortality rates. In general, with materials that elicit any deleterious response in disease defense, increased concentration or duration of exposure results in increased mortality, whereas longer recovery periods between exposure and infectious challenge result in greater viability of the host animals. Animals forced to exercise, and hence to respire more heavily, exhibit greater susceptibility to infection, as do animals exposed to pollutants after infectious challenge. Additional stresses, such as variations in temperature, also act to lower resistance. In the studies described in the subsequent sections, virtually all of these parameters are varied, so comparisons of results from different studies must allow for such variations.

E. Analysis of Data

The primary measure used to monitor resistance to infection is mortality of the host animals; the mortality rate in exposed animals minus the mortality rate in control animals is termed the mortality change or excess mortality. This value, expressed as a percentage of the total number of animals tested, reflects the absolute increase in mortality following exposure to a test atmosphere. For a series of tests to be considered significant in our laboratory, some degree of mortality (usually $10-15\%$) must be observed in the control group. This assures the investigator that the microorganism used is indeed virulent and has the potential to be pathogenic. A second and related measure of mortality frequently used is the relative mean survival time (usually over a 15-day postexposure period), determined from the weighted average number of days survived by animals within a test group. Both mortality measures are treated statistically by analysis of variance (ANOVA); an important criterion in evaluation of excess mortality is the level of statistical significance for which deaths in test animals exceed those in controls. Usually a level of $p < 0.05$ is required to consider differences significant.

In addition to simple mortality, statistical analyses were conducted on a number of other physiological characteristics that also can be observed for alteration due to pollutant exposure. These may include scoring of pulmonary lesions, hematological analysis, histopathology and electron microscopy of respiratory tissues, measurement of bactericidal activity, assays of pulmonary free cells, and measurement of ciliary activity.

These measurements may be compared to mortality data to help demonstrate the significant mechanisms of pulmonary disease resistance.

III.   EFFECTS OF GASEOUS POLLUTANTS ON HOST RESISTANCE TO INFECTION

A major source of air pollution in urban environments is vehicular traffic. One by-product of internal combustion is nitric oxide, from which nitrogen dioxide (NO_2) and ozone (O_3) are produced in the presence of sunlight. In high concentrations, these gases are powerful and dangerous respiratory irritants. Direct monitoring of ambient air in several cities reveals a background NO_2 concentration of approximately 0.1 ppm, interspersed with daily peak levels as high as 0.5 ppm; O_3 is also found at background concentrations of approximately 0.1 ppm (Gardner, 1980a; U.S. Environmental Protection Agency, 1977). In the past decade, many experiments using the *in vivo* infectivity model have been performed to assess the effects of these gases, separately or combined, at varying concentrations and periods of exposure, and accompanied by a variety of external stresses. These studies are summarized below.

A.   Ozone

A number of studies have shown that levels of O_3 as small as 0.1 ppm are sufficient to evoke a statistically significant increase in mortality in mice exposed to pathogenic bacteria (Coffin *et al.*, 1968; Coffin, 1970; Coffin and Blommer, 1970; Coffin and Gardner, 1972a). In these studies, 3-hour exposures to O_3 followed by application of an infectious aerosol resulted in almost complete mortality at levels above 0.5 ppm; in addition, a massive acceleration of bacterial growth rates in the lung and bacterial invasion of the blood were also related directly to O_3 exposure. These results have been reproduced in several studies involving large numbers of animals and multiple replications of experiments. Further studies have shown that mice forced to exercise during O_3 exposure suffered an even greater mortality rate (Illing *et al.*, 1980), presumably on account of an enhanced respiration rate. Another variable strongly affecting mortality is the time between O_3 exposure and infectious challenge. Mice exposed to infectious bacteria during or within 2 hours after O_3 exposure exhibit enhanced mortality, whereas those administered the infectious aerosol 4 to 6 hours later did not (Miller *et al.*, 1978a).

Classical studies on induction of O_3 tolerance indicated that repeated preexposure of animals to low (0.3 ppm or less) levels of O_3 resulted in protection against death from toxic amounts of O_3, presumably because of a significant reduction in edemagenesis (Stokinger and Scheel, 1962; Alpert et al., 1971a; Coffin and Gardner, 1972b). Subsequent studies using the infectivity model showed that the reduction of excess mortality caused by preexposure to O_3 levels from 0.1 to 1.0 ppm was only about one half of the total mortality for mice not exposed (Gardner and Graham, 1977). Thus, prevention of edema does not appear to be closely related to microbial resistance.

B. Nitrogen Dioxide

To evaluate the resistance of different animal species to K. pneumoniae infection, Ehrlich (1975) exposed three different species (mice, hamsters, and squirrel monkeys) to varied concentrations of NO_2 prior to infectious challenge. The resistance of the test animals to the pollutant appeared initially to be related directly to animal size, with mice exhibiting the greatest mortality and monkeys the least. The NO_2 levels required to cause significant mortality in the monkeys were so great that it is unclear whether death resulted from lowered disease resistance or from gross toxicity.

For 3-hour exposures to NO_2 followed by exposure to Streptococcus aerosol, mice exhibited significant increases in mortality at concentrations greater than 2.0 ppm (Ehrlich et al., 1977). Continuous long-term exposures to lower levels of NO_2 also evoked significant mortality increases: 1.5 ppm for as little as 8 hours or 0.5 ppm continuously for 3 months produced significant excess mortality (Ehrlich and Henry, 1968). Figure 5 depicts a summary of experiments in which mice were exposed to a variety of NO_2 concentrations for periods ranging from minutes to months (Gardner et al., 1979). For all concentrations, the mortality rate increases directly with increasing duration of exposure; both the initial rates and the slopes of the dose–response lines increase with increasing NO_2 concentrations.

To simulate further the conditions of an urban environment, mice were exposed continuously to background NO_2 concentrations of 1.0 ppm and were exposed also to either 5.0 ppm NO_2 or 5.0 ppm NO_2 plus 0.1 ppm O_3 daily for 3 hours. Prolonged exposure resulted again in enhanced mortality, and the effects of the two pollutants together were at least additive, i.e., the combined effects of the two gases were at least as much as the sums of the two administered separately (Ehrlich, 1980). As with O_3, exercise during NO_2 exposure exacerbated the effects of the pollutant (as compared to those on resting animals) at concentrations of 3.0 ppm and greater (Illing et al., 1980).

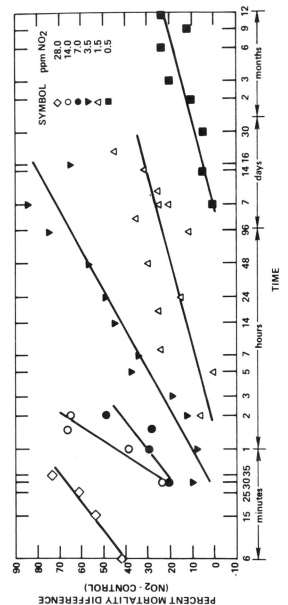

Fig. 5. Mortality enhancement for mice exposed to NO_2 at various concentrations and for various durations prior to challenge with streptococci. At all concentrations prolonged exposure results in enhanced mortality, but the severity of resistance reduction appears more directly related to concentration.

C. Other Gases

Not every gaseous pollutant affects pulmonary disease resistance. One example is sulfur dioxide (SO_2), a by-product of coal combustion. In one series of tests, mice were administered single or multiple 3-hour exposures daily or continuous exposures for 3 months of 5 ppm SO_2 and were challenged with *Streptococcus* aerosol. These exposures elicited no changes in animal mortality or life-span, and produced no changes in physical parameters, including body weight, temperature, development of edema, macrophage viability or activity, or appearance of lungs, trachea, or nasal cavities (Ehrlich, 1980). Similarly, 3-hour exposures of mice to the oxidizing agent peroxyacetyl nitrate (PAN) produced statistically significant enhancement in respiratory pneumonia only at concentrations exceeding 14 mg/m^3 (Thomas *et al.*, 1981).

IV. EFFECTS OF AIRBORNE PARTICLES ON HOST RESISTANCE TO RESPIRATORY INFECTION

In addition to gaseous compounds, air pollution contains a wide variety of metallic, mineral, and organic particulate matter, with sizes ranging from 0.01 to greater than 100 μm. The composition of these particles may be pure or heterogeneous and, in addition, particles may serve as substrates onto which gaseous organic compounds may condense. These particles contribute to a spectrum of respiratory difficulties, including gross toxicity from heavy metals, carcinogenesis from surface organics, direct damage of pulmonary tissue by deposition of granular material, and reduced disease resistance. To evaluate fully the impact of airborne particles on human health is most complex; the subsections that follow describe studies on one facet of the problem, the deterioration of disease defense mechanisms caused by individual compounds.

In these studies, particulate concentrations are measured as mass per unit volume (usually in μg/m^3) rather than in ppm (as are gases). Test animals were primarily female CD-1, COBS strain mice, 4–6 weeks old, as in most of the gaseous pollutant studies. In a few experiments, Syrian golden hamsters or Sprague–Dawley rats were used. The animals were isolated in chambers that allowed only the heads to be exposed to the test aerosol, usually for 2 hours per test. Described below are results from such *in vivo* tests for several different particulate classes.

For gaseous pollutants, it is fairly well established that the duration of reduced disease resistance that follows a 3-hour exposure is relatively short, a matter of hours (see Section II), presumably because the gases are cleared quickly and the impaired alveolar macrophages are replaced. With several chemical classes of particulate pollutants, there is also the possibility that materials deposited in the lungs will reside and exert a toxic influence for a longer time. Thus, studies of the effects of metals on pulmonary systems followed five lines: classic mortality enhancement measurements, measurement of particle deposition and retention in lungs, measurement of macrophage and lymphocyte levels in exposed lungs, comparison of *in vivo* studies with the effects of the metals on isolated alveolar macrophages *in vitro*, and characterization of changes in immunological responses.

A. Mortality Studies

One of the most toxic metals is cadmium, used in electroplating, storage batteries, solders, and, more recently, in photoelectric cells and semiconductors. It is also emitted as a vapor when zinc ores are roasted. Nickel and manganese, like cadmium, are important industrial metals that are emitted as particles, in the size range ($<$ 15 μm) considered inhalable into the lower respiratory tract. Several studies have been performed comparing the effects of these three metals, combined with various anions, on animal mortality, macrophage viability, and ciliary activity (Graham *et al.*, 1975; Adalis *et al.*, 1977; Graham and Gardner, 1977; Adkins *et al.*, 1979).

Both cadmium chloride and cadmium sulfate were administered in an aerosol, at concentrations ranging from 80 to 1600 μg/m^3, to mice for 2 hours prior to streptococcal challenge. The excess mortality for these mice compared to the mortality of the controls was significant at all concentrations, with nearly total lethality at the highest dose rates (Gardner *et al.*, 1977a). For cadmium, the chloride and sulfate forms elicited similar effects per unit mass of metal. By contrast, nickel chloride was more lethal than nickel sulfate, but the chloride of manganese was less toxic than the oxide (Gardner, 1980b). The ranking of toxicity for the three metals was cadmium $>$ nickel $>$ manganese, and they also appeared to exhibit different mechanisms of action (Gardner, 1980c). Cadmium and manganese elicited mortality increases when the animals were exposed to bacteria immediately after metal exposure, whereas nickel-induced mortality was enhanced only when the microbial aerosol was administered 24 hours after the metal exposure (Graham and Gardner, 1977).

To test whether the potentiation of respiratory infection can be attributed to the effect of the metal or just to increased stress on the respiratory system by high concentrations of particulate matter, two par-

TABLE I

Concentrations of Metallic Compounds for Which 3-Hour Exposures Elicit 20% Increase in Mortality in Mice after Streptococcal Challenge

Compound	Concentration (mg/m^3)
$CdSO_4$	0.2
$NiCl_2$	0.5
$NiSO_4$	0.5
$CuSO_4$	0.6
$ZnSO_4$	1.45
$Al_2(SO_4)_3$	2.2
$Zn(NH_4)_2(SO_4)_2$	2.5
$MgSO_4$	3.6

ticulate but generally nontoxic substances, carbon black and iron oxide, were administered at comparatively high concentrations (5.0 and 2.5 mg/m^3, respectively). These substances elicited no significant mortality increases.

Other metal sulfates that significantly enhance mortality in the presence of bacteria are those of copper, zinc, aluminum, and magnesium (see Tables I and II) (Ehrlich, 1980; Ehrlich et al., 1978). Nonmetallic salts (ammonium sulfate, ammonium bisulfate, ammonium nitrate, and

TABLE II

Maximum Tested Concentrations of Sulfates and Nitrates That Did Not Alter the Resistance of Mice to Streptococcal Infection[a]

Compound	Concentration (mg/m^3)
$AlNH_4(SO_4)_2$	2.4
$Fe(NH_4)_2SO_4$	2.5
$Fe_2(SO_4)_3$	2.9
Na_2SO_4	4.0
$(NH_4)_2SO_4$	5.3
$(NH_4)HSO_4$	6.7
$Pb(NO_3)_2$	2.0
$Ca(NO_3)_2$	2.8
$NaNO_3$	3.1
KNO_3	4.3
NH_4NO_3	4.5

[a] From Ehrlich (1980).

potassium nitrate) failed to enhance susceptibility to infection even at concentrations of several mg/m³.

B. Mechanical Clearance of Particulates

One pulmonary defense mechanism is the mucociliary escalator system, involving epithelial cells and the moving mucous blanket of the upper airway, which clears from 10 to 25% of the inhaled bacterial load (Green and Kass, 1964). When this system is destroyed or impaired, both the infectious bacteria and toxic particles can reside in the lung for a longer time and increase the probability of infectious disease. To test the effects of pollutants on ciliary activity, Adalis *et al.* (1977, 1978) exposed hamsters to respirable aerosols of nickel chloride or cadmium chloride for 2 hours, and, after exposure, excised the tracheas, cut them into rings, nourished them in tissue culture, and monitored the ciliary beating frequency with a stroboscopic microscope attachment. The reduction in ciliary beating frequency due to cadmium was very similar to that caused by more than 10 times that (molar) amount of nickel; for both metals, increasing the amount of the metal diminished ciliary activity, which recovered to normal only after 3 days. Seven days were required for recovery from a 1-day *in vivo* exposure to a nickel aerosol and were not sufficient for recovery from a 2-day exposure. By contrast, manganese did not elicit any significant reduction in ciliary activity (Adkins *et al.*, 1980).

C. Free Lung Cell Assays

Besides depressing the activity of the epithelial cells of the upper airway, pollutants also affect the pulmonary cells of the deep lung. The type, numbers, and viability of these cells may be determined by means of pulmonary lavage, whereby free lung cells may be extracted and assayed without artificial stimulation.

For rats exposed to 1500 μg cadmium/m³, the total number of alveolar macrophages decreased significantly immediately after exposure; although the number of these macrophages returned to normal within 24 hours, their viability had not fully recovered in that time (Gardner *et al.*, 1977a). After 24 hours, a marked increase in the total number of free cells in the lung was also evident. This was caused predominantly by an increase in polymorphonuclear leukocytes and lymphocytes; these effects were not observed with exposures of 500 μg cadmium/m³. By contrast, no significant differences were apparent in total cell yield after animals were exposed to 650 μg/m³ of nickel, although the phagocytic capacity of the alveolar macrophages was depressed (Adkins *et al.*, 1979). Exposure to 900 μg/m^3 of manganese elicited an opposite effect: The total number and viability of lavaged free cells decreased markedly, but no decrease was apparent in either the phagocytic ability or in pulmonary cell viability (Adkins *et al.*, 1980).

In addition to chemically pure pollutants, the *in vivo* infectivity model has been used to test the toxicity of complex pollutant samples taken from foundries, smelters, and coal-fired power plants (Aranyi *et al.*, 1981). Mice were exposed to concentrations of 2 mg/m^3 test particles or to filtered air for 3 hours per day for 1, 2, or 4 weeks prior to streptococcal challenge. Dust from a copper smelter and fly ash from a fluidized-bed coal furnace (which had shown very high and low toxicity to cultured rabbit alveolar macrophages, respectively) were administered as the test aerosols.

These *in vivo* tests were performed as a control to evaluate the results of the *in vitro* cytotoxicity tests of these pollutants; it is noteworthy that the copper smelter dust aerosols produced significant enhancement of mortality in mice, whereas the coal fly ash evoked no significant mortality changes. Among the parameters monitored in the *in vivo* tests, total cell counts, differential cell counts, or cell viability from lavaged animals did not change. The copper smelter dust, however, did cause a significant reduction in the bactericidal activity in the animals, as measured by use of radiolabeled *Klebsiella*. The coal fly ash had no effect on this activity.

The copper smelter dust contained high concentrations of the trace elements lead, arsenic, copper, iron, antimony, and zinc; the coal fly ash contained the less toxic trace elements calcium, magnesium, iron, silicon, and sulfur. This relation between high toxicity to alveolar macrophages and overall mortality caused by bacterial infection is again indicative of the main line of disease defense; in this study, the *in vivo* model system proved a useful verifier in the development of a simple and less expensive, if less complete, *in vitro* testing system.

D. Physical Damage

Another real hazard that can result from combustion of high-sulfur coal is the production of sulfuric acid mists, either free or condensed, onto carbon particles. Earlier studies had indicated that inhalation of sulfuric acid aerosols results in greater total respiratory deposition of *Streptococcus* and, in combination with ozone, enhanced mortality (Fairchild *et al.*, 1975; Gardner *et al.*, 1977b). Scanning electron microscopic studies of mouse respiratory tracts have revealed extensive pulmonary tissue damage resulting from exposure to 50–200 mg/m^3 sulfuric acid mist coated onto 5 mg/m^3 carbon particles (Ketels *et al.*, 1977). To evaluate more fully the effects of lower concentrations of sulfuric acid, studies were performed to compare the effects of carbon alone and carbon particles carrying sulfuric acid after exposures lasting up to 20 weeks (Fenters *et al.*, 1979). In addition to observation of mortality, this study involved microscopic examination of tissues and immunological assays of the affected animals.

The infectious agent for the mortality studies was influenza A$_2$/Taiwan virus, administered for 1 hour after either 4 or 20 weeks of pollutant exposure. Statistically significant increases in mortality were observable only after 20 weeks of exposure to the acid–carbon mixture; after only 4 weeks, there was no evident change either in percent mortality, pulmonary consolidation, or the appearance of the respiratory epithelium. After 12 weeks of exposure to carbon only, tracheas showed many mucous cells, some dying cells with loss of microvilli, and some congestion in the alveoli. The same forms of damage, but slightly more severe, were observed in the animals exposed to carbon plus acid. After 20 weeks, the observed morphological changes were slightly fewer than at 12 weeks, indicating the maximal response for this concentration may have been reached. The parameter showing the greatest difference between control and exposed animals was the bactericidal activity of the lungs when exposed to *Klebsiella:* in comparison to unexposed control mice, mice exposed to carbon or acid–carbon aerosols had significantly enhanced bacterial counts in the lungs, but little significant difference was noted between the carbon and acid–carbon groups. Thus, it appears possible that a factor in lung bactericidal activity, and in the functioning of the macrophages, is the presence of engulfed particles in the macrophages, especially when these ingested particles have adsorbed toxic chemicals. Numerous phagocytized carbon particles could be seen in the alveolar macrophages examined in this study.

E. Immunotoxicity

The study of serum immunoglobulin levels in experimental animals is of interest because several studies of human lung disorders reported concomitant changes in the concentration of immunoglobulins. Chronic obstructive pulmonary emphysema was accompanied by elevated serum IgA levels (Biegel and Krumholz, 1968); patients with farmer's lung disease exhibited high levels of IgA and IgG (Roberts *et al.*, 1973). Exposure to NO$_2$ was accompanied by elevated levels of serum IgA and IgM in humans (Kosmider *et al.*, 1973), whereas in laboratory studies with mice, NO$_2$ exposure resulted in increases in the IgM and IgG levels but a decrease in IgA (Ehrlich *et al.*, 1975). The relation between immunoglobulin levels and pulmonary disease is not well established; indeed, it is as yet unclear whether changes in immunoglobulin levels reflect a cause of or a response to pollutant-mediated infection.

In the acid–carbon study cited above (Fenters *et al.*, 1979) the IgG serum levels of the mice exposed to acid and carbon were depressed, as were the IgG_{2a} and IgG_1 levels at intermediate stages of the testing, whereas IgG_{2b} levels were elevated in both the carbon and acid–carbon exposure groups. These results should be weighed cautiously, because the studies began with young mice and the sera levels in the control animals varied widely during the course of the testing.

Another test of the immunotoxicity of inhaled particles is the change in number of antibody-producing spleen cells after pollutant exposure. In these tests, animals are immunized intraperitoneally with sheep red blood cells immediately after exposure; 4 days after immunization, the spleens are removed and assayed for plaque-forming units by the Jerne hemolytic plaque technique.

Tests of the effects of cadmium and nickel chloride on antibody-producing spleen cells revealed that, at sufficiently high concentrations, inhalation of both metals significantly reduces the number of plaque-forming units, with cadmium being more toxic on a molar basis (Graham *et al.*, 1975). In a similar study, mice were injected with cadmium chloride, chromium chloride, nickel chloride, nickel sulfate, or nickel oxide (Graham *et al.*, 1978). Only for nickel chloride was a significant linear dose–response relationship apparent, whereas nickel oxide did not cause any immunosuppression. The cadmium and chromium chlorides elicited but slight immunosuppression at the highest concentrations tested, a somewhat surprising result in light of the overall toxicity of cadmium in all other types of tests. One would believe that this difference can be attributed to the different route of administration of the metals; because of the large volume of air processed by an animal and despite natural clearance mechanisms, administration of even low concentrations of aerosols may deliver more metal to the spleen than can injection. Nonetheless, comparison of the responses to total dosages of cadmium and nickel administered by the two routes revealed that the inhalation route is more sensitive by at least a factor of 200.

V. *IN VIVO* TESTING BY INTRATRACHEAL INSTILLATION

The *in vivo* inhalation facilities described above have been designed to simulate closely the conditions under which humans encounter polluted ambient air. The main disadvantage of such facilities is that they are

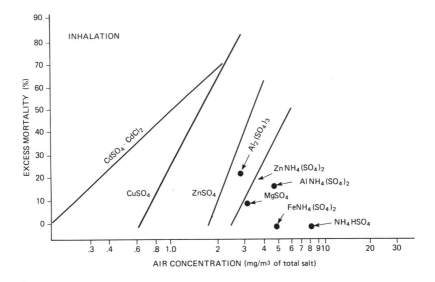

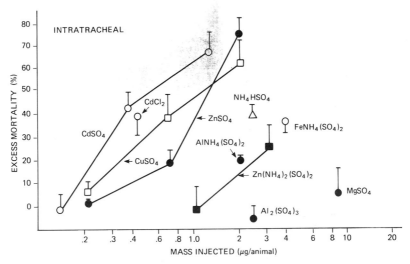

Fig. 6. Comparison of dose–response curves for mice administered various metallic compounds by inhalation (top) or by intratracheal injection (bottom). The similarity of the two graphs indicates the intratracheal route of administration elicits effects similar to those elicited by inhalation. The upper and lower portions of the figure are aligned so that equivalent lung levels of the compounds can also be compared by the two methods.

rather expensive to construct and operate; hence, they are not readily available to most laboratories. Thus, efforts have been made to develop alternate methods of pollutant exposure that maintain the lung as the primary target. One such method is intratracheal instillation, whereby a catheter is inserted in the trachea, and a measured amount of test substance is sprayed directly into the respiratory tract. Besides the economic advantages, this technique offers several other benefits: (1) the actual delivered doses can be measured accurately; (2) relatively small amounts of the test substance are required; (3) large particles can be tested; (4) large quantities of nontoxic substances can be administered; and (5) many engineering and human safety problems can be avoided.

To evaluate the reliability of this administration route, a study was performed in which the effects of 23 compounds administered by intratracheal instillation and in inhalation exposure chambers were compared. The complete experimental details are described by Hatch *et al.* (1981). Briefly, 4- to 5-week-old female CD-1, COBS mice were anesthetized with 5% halothane, suspended by the incisors on a wire, and administered 10 μl of the test substance (or saline solution for the controls) with a syringe. The amount of test compound administered to each animal was matched with inhalation studies by calculating the expected lung dosage from the minute ventilation rate of the mice and the pollutant concentration used for inhalation studies. One hour after instillation, to allow recovery from the anesthetic, the mice were exposed for 20 minutes to an aerosol of *Streptococcus* in the standard infectious exposure chambers.

For the 23 compounds tested, four that showed greater than 40% enhancement of mortality upon aerosol exposure produced similar mortality increases when comparable amounts were injected intratracheally. Five other chemicals produced smaller (20–30%) mortality enhancement upon injection, but not all produced such an effect with inhalation. The remaining 14 compounds elicited no mortality enhancement by either route of administration. Figure 6 illustrates the dose–response relationships for some representative chemicals tested by both methods. In general, it may be concluded that both routes of administration yield closely similar responses for a given compound.

VI. MECHANISMS OF POLLUTANT ACTION

A number of possible mechanisms have been proposed to explain how various gaseous and particulate pollutants act to reduce the ability of the lung to defend itself against the growth of inhaled microbes. These effects include (1) the production of edema, (2) depression of mucociliary activity, (3) alteration of macrophage activity, and (4) suppression of the immune system. It is likely that to some extent all these mechanisms contribute to a greater incidence of respiratory disease in animals exposed to unclean air. Each of these mechanisms has been studied in relationship to *in vivo* infectivity model systems, and their interlocking relationships and relative significances are only just becoming clear.

A. Production of Edema

An influx of edema fluid into the lung could provide an ideal culture medium for microbial growth. Ozone, long known as an edemagenic agent, has been studied intensively for its ability to enhance respiratory infection. Classic studies have shown that repeated exposures to lower dosages of ozone produce a tolerance against subsequent edema production from a higher dose of ozone, thereby protecting the animal from later, acute, toxic concentrations of the gas (Stokinger *et al.*, 1956; Matzen, 1957; Fairchild, 1967; Coffin and Gardner, 1972b). Infectivity studies have shown that, compared to control animals, animals exposed to ozone levels that elicit tolerance still exhibit enhanced mortality; however, this enhancement is only approximately half that observed with animals given ozone only once, prior to bacterial challenge and without the opportunity to develop tolerance (Gardner and Graham, 1977). If edema were a primary causal factor mediating respiratory infection, the responses of the tolerant and control groups would be expected to be similar. A second reason why edemagenesis does not appear solely responsible for pollutant enhancement of disease is that such enhancement has been shown at levels (Coffin and Gardner, 1972a; Miller *et al.*, 1978a) below those that induce edema in experimental animals (Alpert *et al.*, 1971b).

B. Mucociliary Escalator

A second mechanism by which environmental contaminants may impair host defenses is through depression of the mucociliary escalator. As noted in Section IV,B, administration of cadmium or nickel aerosols resulted in a reduction of the beating frequency of hamster tracheal cilia. Such an effect would be likely to reduce the amount of inhaled bacteria that could be removed from the lungs, thereby increasing the bacterial residence time and, hence, the chance of infection developing to the point of overwhelming the cellular defenses. The metal concentrations required to produce a significant reduction in beating frequency were nevertheless higher than those eliciting mortality increases in the *in vivo* infectivity studies with mice. In addition, studies with healthy animals administered radiolabeled microorganisms indicated that, in a 4-hour period, only about 15–20% of the total bacteria taken in were physically removed from the lungs via this mechanism (Green and Kass, 1964). Thus, mechanical clearance cannot be considered the sole or dominant pulmonary defense.

C. Alveolar Macrophages

Considerable accumulated evidence indicates that the primary pulmonary defense against foreign microbes resides with the alveolar macrophages, the large, mobile scavenger cells whose function is to engulf and remove old and dying cells and bacteria. Several lines of investigation point to this conclusion. An *in vitro* model system (Green and Goldstein, 1966; Kim *et al.*, 1976) was developed in which mice were exposed to *Staphylococcus;* the location of the bacteria was determined (histologically) from the right lung; and the bactericidal activity was determined from the left lung. These studies demonstrated that phagocytosis of the *Staphylococci* by the macrophages precedes and roughly parallels the actual clearance of the bacteria for approximately 2 hours, after which phagocytosis reaches a plateau but bactericidal activity continues to increase. Thus, it would appear that the most critical macrophagic function (i.e., that whose impairment would most likely enhance mortality) is bactericidal activity. Studies described in earlier sections indicated that exposure to such pollutants as ozone, nitrogen dioxide, and cadmium chloride reduces the number, viability, phagocytic ability, and enzymatic activity in macrophages obtained by pulmonary lavage.

One means by which pollutants reduce the viability of the macrophages is direct attack on the macrophages' outer membranes. Scanning electron microscopy studies (Aranyi *et al.*, 1977; Waters *et al.*, 1975) of alveolar macrophages exposed to gases or metals show a marked and substantial destruction of the leafy outer surfaces of these cells.

A second toxic effect may be on the pulmonary, acellular lining substance in which the macrophages are immersed. When normal alveolar macrophages are suspended in normal, acellular lavage fluid, they remain generally stable (15% lysis) over a 4.5-hour test period. This stability is only mildly affected when nontoxic gases (nitrogen, carbon dioxide, or clean air) are bubbled through the lavage fluid for 30 minutes, but is severely affected (65% lysis) when ozone is used (Gardner, 1971). Similarly, if the same normal cells are placed in lavage fluid from animals exposed to ozone for 3 hours, cell lysis again increases. By contrast, when the phagocytic ability of lavaged macrophages was measured, no difference was seen when the macrophages were suspended in normal lavage fluid or in fluid treated with ozone, and the phagocytic ability was enhanced most when normal macrophages were maintained in lavage fluid harvested from rabbits exposed to ozone (Gardner, 1971). In these studies, the phagocytic indices were obtained by counting only the intact macrophages; fewer cells remained in the ozone-treated fluid than in the normal fluid. These studies combined provide support to the notion that pollutants may impair the functioning of the macrophages as much by altering their protective fluid as by direct attack on the cells.

Coupled with this attack on the membrane surface is destruction of the membrane receptor sites. For example, 30-minute *in vitro* exposures to nickel chloride or cadmium chloride caused significant dose-related increases in the binding of antibody-sensitized erythrocytes to the Fc receptor (Hadley *et al.*, 1977). Similarly, lectin-treated alveolar macrophages from rabbits exposed to ozone and nitrogen dioxide exhibited increased rosetting of untreated autologous erythrocytes, again indicating loss of receptor function.

D. Suppression of Immune Responses

As noted in Section IV,E, exposure to various pollutants causes changes both in the serum immunoglobulin levels and in the number of antibody-producing spleen cells in test animals. Although the precise mechanisms by which alveolar macrophages stimulate and enhance the immune response of the lymphocyte system are still imperfectly understood, it is possible that chemotactic factors are released and recognized by the macrophages and lymphocytes, respectively. The disruption of this line of chemical communication, at the surface either of the macrophage or the lymphocyte, would lead to a reduced immune response and, ultimately, to an enhanced possibility of infection overwhelming this last line of pulmonary defense and entering the bloodstream. Although there is now abundant evidence that macrophages are not only involved in immunological reactions of various kinds but also play a major role in the recognition of foreign material and the subsequent induction of the immune response, lung immunology is still poorly understood. There are only a few studies specifically designed to examine the effects of air pollutants on the systemic and pulmonary immune systems. The interaction of environmental chemicals with the immune system may result in undesirable effects of three types: (1) those determined by immunodeficiency or immune suppression, (2) those determined by alterations of host defense mechanisms, and (3) those determined by allergy or hypersensitivity. Increasing evidence indicates that exposure to certain chemicals may depress an individual's responsiveness and increase his susceptibility to infectious agents.

VII. IMPLICATIONS OF THE ANIMAL MODEL FOR HUMAN HEALTH EFFECTS

The ultimate goal of the animal research described above is knowledge of the mechanisms and effects of air pollutants in altering human respiratory disease defenses. A critical question about any animal studies related to human health risks is how far the results may be extrapolated between species. Proper extrapolation requires three elements: verification that the phenomenon does indeed occur in both species, identification of the mechanisms of action, and quantitative comparison of the magnitude of the effect in different species. Below are discussed the correlations that may be drawn between small animal and human respiratory defense impairment and the work that remains to complete and quantify extrapolation to humans.

Bacteria, unlike viruses, are not stringently selective regarding the organism infected, and it is clear that many microbes can infect both humans and mice. In addition, the principal lines of defense (mucociliary escalator, alveolar macrophages, and immune responses) are similar in both species. The animal experiments reviewed in previous sections demonstrate conclusively that certain pollutants cause a deterioration in mouse respiratory defenses; although similar experiments are more difficult to perform on human subjects and must be of an acute nature, epidemiological studies can establish an association, if not a cause-and-effect relationship, between human health effects and ambient air pollution.

Whereas community epidemiological studies can be most difficult to perform on account of the complex and changing mixtures of chemicals in the air, a few such studies have been published and report an association between pollutant exposure and increased, acute respiratory disease incidence. For example, young children living in an environment with low levels of NO_2 were observed to be more susceptible to respiratory infections (Melia *et al.*, 1977, 1979; Speizer *et al.*, 1980). These studies, not only suggestive by themselves, are also supported and clarified by the animal studies described in Section III,B, which indicate that NO_2, particularly in short-term peaks, was probably the major causative agent in the epidemiological studies.

A second association that has long been recognized from human health surveys is between cigarette smoking and incidence of respiratory infection (Haynes *et al.*, 1965; Parnell *et al.*, 1966; Peters and Ferris, 1966; Spurgash *et al.*, 1968). Cigarette smoke, like a number of other inhaled particles (e.g., Pb_2O_3, cotton dust, and quartz), can promote an influx of macrophages and polymorphonuclear leukocytes into the lung alveoli (Reynolds and Newball, 1974; Warr and Martin, 1974; Brain *et al.*, 1978). The macrophage cytoplasm becomes filled with pigmented residues from the smoke (Harris *et al.*, 1970) and responds poorly to a chemotactic agent (Demarest *et al.*, 1979). The phagocytic functioning of the macrophages is reduced significantly (Gee *et al.*, 1979; Haroz and Mattenberger-Kneber, 1977), and cigarette smoke also exhibits a depressant effect on the antibactericidal activity of these defense cells (Green and Carolin, 1967).

Again, laboratory studies on animals have provided correlation and explanations for human epidemiological studies. Thus, animal studies such as those reviewed above (which permit a complete evaluation of the disease and allow researchers choices of exposure modes, chemical agents, animal and microbial species, and pollutant concentrations) allow researchers to evaluate qualitatively the effects of pollutants without subjecting humans to infectious diseases.

The final goal of this research, quantification of the deterioration of human disease resistance by air pollutants, can be approached indirectly. Because many physiological mechanisms are common to both animals and man, it can be hypothesized that if a pollutant causes a particular response in several animal species, it will likely cause a similar effect in man. That is, if the equivalent dose is delivered to a specific target region of the lung in a mouse or a man, a similar response is most likely to occur. The problem is then estimating with confidence how much of the inhaled dose actually reaches the critical target site.

To make such estimates requires mathematical models that will predict regional deposition of the pollutant in the lung. These models must incorporate the anatomic and ventilatory parameters of the species as well as the physicochemical properties of the pollutant and should be based on the physical laws that govern transport and removal of the pollutant. The factors influencing the ability to make these estimates are different for gases and particles.

For particles, the major influencing factors include the size, shape, and density of the particles, which in turn determine their aerodynamic properties. The breathing characteristics and anatomic structure of the test species also determine or influence the fractional deposition in the various regions of the respiratory tract. Individuals may be at greater risk while exercising or doing heavy labor due to either increased pollutant dosages or altered patterns of regional deposition associated with increased ventilation. Some of these changes can occur when one switches from nasal to oronasal breathing with increased ventilatory demand.

For gases, the dosimetry is somewhat different and is more specific to the properties of the individual gas. The major factors affecting the uptake of gases in the respiratory tract are the morphology of the respiratory tract, the route of breathing (nasal versus oronasal), the depth and rate of breathing, the physiological properties of the gas, the physical processes that govern gas transport, and the physicochemical properties of the aqueous material on the surface of the airways. All these components must be known in order to develop the necessary data base for extrapolation of animal data to man.

In our laboratory, we are developing extrapolation models that will permit animal data, such as presented here, to be used for predicting human responses to pollutants such as ozone and nitrogen dioxide (Miller *et al.*, 1978b, 1979, 1981). Once such extrapolation models are perfected, environmental toxicologists will be better able to assess with some accuracy the full impact of air pollutants on public health and the level of pollutants to which humans can be exposed safely.

VIII. SUMMARY

The *in vivo* infectivity model system has been used to test the effects of dozens of environmental chemicals on small animals. The ultimate goal of this research is to provide an assessment of the risks to human health caused by chemicals; hence, these tests have been designed to simulate the concentrations and route of exposure likely to be encountered by humans in urban or working settings. Below are discussed some general conclusions that may be offered at the current stage of this ongoing research.

One important conclusion, which appears generally true for most compounds studied extensively, is that high concentrations of pollutants are, for even a brief exposure period, more hazardous with respect to disease resistance than long-term exposure to low levels of contaminant. This reflects the time course of infectivity in the animal model: in mice, a level of approximately 10^5 bacteria per lung is required for the infection to become established and to invade the bloodstream; faced with a challenge of even a few hundred bacteria, only a few hours of impaired macrophage activity are sufficient for the invasive bacteria to reproduce to uncontrollable numbers (Miller *et al.,* 1978a). Thus, even though prolonged exposure to pollutants does tend to gradually increase the occurrence of respiratory infection, even greater risk is associated with short-term exposure to concentrated pollutants that can disable the defenses sufficiently to permit infection.

For particulate pollutants, size as well as composition was seen to be a factor affecting toxicity, with smaller particles showing greater toxicity (Aranyi *et al.,* 1979, 1981). This may be caused in part by the observed increase in the phagocytosis of the smallest particles. It should be noted that on the surface of "real world" particles various noxious chemicals are adsorbed, and with smaller particles there is a corresponding increase in surface area per unit mass. Hence, a greater amount of material could be released or solubilized from the smaller particle than from larger particles having the same total mass. Because the toxicity of many of these substances can be correlated to the soluble fractions, it becomes toxicologically important to understand this relationship. Furthermore, with finely divided particles there is also an increased probability of contact and thus phagocytosis by the alveolar macrophage. If the chemical is cytotoxic, one can expect that a proportionately greater number of lung cells will be impaired.

The high sensitivity of the *in vivo* infectivity testing system allows some measure of the relative potency of various airborne pollutants. Although it is still premature to claim "safe" or "dangerous" concentration levels (for humans) for compounds eliciting positive infection enhancement in mice, it is possible to suggest that compounds that produce no effects in mice are unlikely to entail major risks to the host defense systems of humans. Furthermore, because humans and mice have the same basic kinds of cellular defense mechanisms, it is possible to order compounds into groups of *relative* toxicity. For example, cadmium and ozone represent the relatively most toxic metallic and gaseous compounds, whereas iron oxide and sulfur dioxide may be included in the least dangerous categories for host defense. Again, it should be emphasized that respiratory infectivity represents but one of a variety of effects to be considered in the total scope of a compound's toxicity; this may represent, however, the most subtle, most sensitive, and perhaps most frequent, toxic effect of air pollutants.

ACKNOWLEDGMENT

The author wishes to thank Dr. Charles J. Alden, Northrop Services, Inc.-Environmental Sciences, for assistance in the preparation of this manuscript.

REFERENCES

Adalis, D., Gardner, D. E., Miller, F. J., and Coffin, D. L. (1977). Toxic effects of cadmium on ciliary activity using a tracheal ring model system. *Environ. Res.* **13**, 111–120.

Adalis, D., Gardner, D. E., and Miller, F. J. (1978). Cytotoxic effects of nickel on ciliated epithelium. *Am. Rev. Respir. Dis.* **118**, 347–354.

Adkins, B., Jr., Richards, J. H., and Gardner, D. E. (1979). Enhancement of experimental respiratory infection following nickel inhalation. *Environ. Res.* **20**, 32–42.

Adkins, B., Jr., Luginbuhl, G. H., Miller, F. J., and Gardner, D. E. (1980). Increased pulmonary susceptibility to streptococcal infection following inhalation of manganese oxide. *Environ. Res.* **23**, 110–120.

Alpert, S. M., Gardner, D. E., Huret, D. J., Lewis, T. R., and Coffin, D. L. (1971a). Effects of exposure to ozone on defensive mechanisms of the lung. *J. Appl. Physiol.* **31**, 247–252.

Alpert, S. M., Schwartz, B. B., Lee, S. D., and Lewis, T. R. (1971b). Alveolar protein accumulation: A sensitive indicator of low level oxidant toxicity. *Arch. Intern. Med.* **128**, 69–73.

Aranyi, C., Fenters, J., Ehrlich, R., and Gardner, D. E. (1977). Scanning electron microscopy of alveolar macrophages after exposure to oxygen, nitrogen dioxide and ozone. *Environ. Health. Perspect.* **16**, 180.

Aranyi, C., Miller, F. J., Andres, S., Ehrlich, R., Fenters, J., Gardner, D. E., and Waters, M. (1979). Cytotoxicity to alveolar macrophages of trace metals absorbed on fly ash. *Environ. Res.* **20**, 14–23.

Aranyi, C., Bradof, J., Gardner, D. E., and Huisingh, J. L. (1981). *In vitro* and *in vivo* evaluation of potential toxicity of industrial particles. *In* "Short-term Bioassays in the Analysis of Complex Environmental Mixtures" (M. D. Waters, S. S. Sandhu, J. L. Huisingh, L. Claxton, and S. Nesnow, eds.), pp. 493–505. Plenum, New York.

Biegel, A. A., and Krumholz, R. A. (1968). An immunoglobulin abnormality in pulmonary emphysema. *Amer. Rev. Respir. Dis.* **97**, 217–222.

Brain, J. D., Golde, D. W., Green, G. M., Massara, D. J., Valberg, P. A., Ward, P. A., and Werb, Z. (1978). Biological potential of pulmonary macrophages. *Am. Rev. Respir. Dis.* **118**, 435–443.

Coffin, D. L. (1970). Study of the mechanisms of the alteration of susceptibility to infection conferred by oxidant air pollutants. *AEC Symp. Ser.* **18**, 259–264.

Coffin, D. L., and Blommer, E. J. (1970). Alteration of the pathogenic role of the *Streptococci* group in mice conferred by previous exposure to ozone. *In* "Aerobiology" (I. H. Silver, ed.), pp. 54–61. Academic Press, New York.

Coffin, D. L., and Gardner, D. E. (1972a). Interaction of biological agents and chemical air pollutants, *Ann. Occup. Hyg.* **15**, 219–234.

Coffin, D. L., and Gardner, D. E. (1972b). Role of tolerance in protection of the lung against secondary insults. *In* "Proceedings of the International Symposium of Occupational Physicians of the Chemical Industry," pp. 344–364. Ludwigshafen, Germany.

Coffin, D. L., Blommer, E. J., Gardner, D. E., and Holtzman, R. S. (1968). Effect of air pollution on alteration of susceptibility to pulmonary infections. *In* "Proceedings of the Third Annual Conference on Atmospheric Contaminants in Confined Space," pp. 75–80. Wright-Patterson Air Force Base, Aerospace Medical Research Laboratories, Dayton, Ohio.

Coffin, D. L., Gardner, D. E., and Blommer, E. J. (1976). Time-dose response in infectivity model system. *Environ. Health Perspect.* **13**, 11–15.

Demarest, G. B., Hudson, L. D., and Altman, L. C. (1979). Impaired alveolar macrophage chemotaxis in patients with acute smoke inhalation. *Amer. Rev. Respir. Dis.* **119,** 279–286.

Ehrlich, R. (1963). Effect of air pollutants on respiratory infection. *Arch. Environ. Health* **6,** 638–642.

Ehrlich, R. (1966). Effect of nitrogen dioxide on resistance to respiratory infection. *Bacteriol. Rev.* **30,** 604–614.

Ehrlich, R. (1975). Interaction between nitrogen dioxide and respiratory infection. *In* "Scientific Seminar on Automotive Pollutants," U.S. EPA Publ. 600-9-75-003, Vol. 1, pp. 1–9. Nat. Tech. Inf. Serv., Springfield, Virginia.

Ehrlich, R. (1980). Interaction between environmental pollutants and respiratory infections. *Environ. Health Perspect.* **35,** 89–100.

Ehrlich, R., and Henry, M. C. (1968). Chronic toxicity of nitrogen dioxide. I. Effect on resistance to bacterial pneumonia. *Arch. Environ. Health* **17,** 860–865.

Ehrlich, R., Silverstein, E., Maigetter, R., Fenters, J. D., and Gardner, D. E. (1975). Immunologic response in vaccinated mice during long-term exposure to nitrogen dioxide. *Environ. Res.* **10,** 217–223.

Ehrlich, R., Findlay, J. C., Fenters, J. D., and Gardner, D. E. (1977). Health effects of short-term inhalation of nitrogen dioxide and ozone mixtures. *Environ. Res.* **14,** 223.

Ehrlich, R., Findlay, J. C., and Gardner, D. E. (1978). Susceptibility to bacterial pneumonia in animals exposed to sulfates. *Toxicol. Lett.* **1,** 325–330.

Fairchild, F. J., II (1967). Tolerance mechanisms. Determinants of lung responses to injurious agents. *Arch. Environ. Health* **74,** 111–126.

Fairchild, G. A., Roan, J., and McCarroll, J. (1972). Effect of sulfur dioxide on the pathogenesis of murine influenza infection. *Arch. Environ. Health* **25,** 174–182.

Fairchild, G. A., Kane, P., Adams, B., and Coffin, D. (1975). Sulfuric acid and *Streptococci* clearance from respiratory tract of mice. *Arch. Environ. Health* **30,** 538–545.

Fenters, J. D., Bradof, J. N., Aranyi, C., Ketels, K., Ehrlich, R., and Gardner, D. E. (1979). Health effects of long-term inhalation of sulfuric acid mist-carbon particle mixtures. *Environ. Res.* **19,** 244–257.

Gardner, D. E. (1971). Environmental influence on living alveolar macrophages. Ph.D. Thesis, University of Cincinnati, Ohio.

Gardner, D. E. (1980a). Influence of exposure patterns of nitrogen dioxide on susceptibility to infectious disease. *In* "Nitrogen Oxides and Their Effects on Health" (S. D. Lee, ed.), pp. 267–288. Ann Arbor Sci. Publ., Ann Arbor, Michigan.

Gardner, D. E. (1980b). Dysfunction of host defenses following nickel inhalation. *In* "Nickel Toxicology" (S. S. Brown and F. W. Sunderman, Jr., eds.), pp. 120–124. Academic Press, New York.

Gardner, D. E. (1980c). Impairment of pulmonary defenses following inhalation exposure to cadmium, nickel and manganese. *Aerosols Sci. Med. Technol.—Biomed. Influence Aerosol.—Conf. 7th, 1979* pp. 120–122.

Gardner, D. E., and Graham, J. A. (1977). Increased pulmonary disease mediated through altered bacterial defenses. *ERDA Symp. Ser.* **43,** 1–21.

Gardner, D. E., Miller, F. J., Illing, J. W., and Kirtz, J. M. (1977a). Alterations in bacterial defense mechanisms of the lung induced by inhalation of cadmium. *Bull. Eur. Physiopathol. Respir.* **13,** 157–174.

Gardner, D. E., Miller, F. J., Illing, J. W., and Kirtz, J. M. (1977b). Increased infectivity with exposure to ozone and sulfuric acid. *Toxicol. Lett.* **1,** 59–64.

Gardner, D. E., Miller, F. J., Blommer, E. J., and Coffin, D. L. (1979). Influence of exposure mode on the toxicity of NO_2. *Environ. Health Perspect.* **30,** 23–29.

Gee, J. B. L., Boynton, B. R., Khandivala, A. S., and Smith, G. J. (1979). Pulmonary alveolar macrophage function: Some effects of cigarette smoke. *In* "Assessing Toxic Effects of Environmental Pollutants" (S. D. Lee and J. B. Mudd, eds.), pp. 77–85. Ann Arbor Sci. Publ., Ann Arbor, Michigan.

Goldstein, E., Jordan, G. W., MacKenzie, M. R., and Osebold, J. W. (1976). Methods for evaluating the toxicological effects of gaseous and particulate contaminants on pulmonary microbial defense systems. *Annu. Rev. Pharmacol. Toxicol.* **16,** 447–464.

Graham, J. A., and Gardner, D. E. (1977). Effects of metals on pulmonary defense mechanisms against infectious disease. *In* "Proceedings of the Conference on Environmental Toxicology," pp. 565–575. Wright-Patterson Air Force Base, Dayton, Ohio.

Graham, J. A., Gardner, D. E., Waters, M. D., and Coffin, D. L. (1975). Effect of trace metals on phagocytosis by alveolar macrophages. *Infect. Immun.* **11,** 1278–1283.

Graham, J. A., Miller, F. J., Daniels, M. F., Payne, E. A., and Gardner, D. E. (1978). Influence of cadmium, nickel and chromium on primary immunity in mice. *Environ. Res.* **16,** 77–87.

Green, G. M. (1973). Lung defense mechanisms. *Med. Clin. North Am.* **57,** 547–562.

Green, G. M., and Carolin, D. (1967). The depressant effects of cigarette smoke on the *in vitro* antibacterial activity of alveolar macrophages. *N. Engl. J. Med.* **276,** 421–427.

Green, G. M., and Goldstein, E. (1966). A method for quantitating intrapulmonary bacterial inactivation in individual animals. *J. Lab. Clin. Med.* **68,** 669–677.

Green, G. M., and Kass, E. H. (1964). Factors influencing the clearance of bacteria by the lung. *J. Clin. Invest.* **43,** 769–776.

Hadley, J. G., Gardner, D. E., Coffin, D. L., and Menzel, D. B. (1977). Effect of ozone and nitrogen dioxide exposure of rabbits on binding of autologous red cells to alveolar macrophages. *In* "International Conference on Photochemical Oxidants and Its Control," U.S. EPA Publ. 600-3-77-001, Vol. 1, pp. 505–512. Natl. Tech. Inf. Serv., Springfield, Virginia.

Haroz, R. K., and Mattenberger-Kneber, L. (1977). Effect of cigarette smoke on macrophage phagocytosis. *ERDA Symp. Ser.* **43** (CONF-760927), 36–57.

Harris, J. O., Swenson, E. W., and Johnson, J. E. (1970). Human alveolar macrophages: Comparison of phagocytic ability, glucose utilization and ultrastructure in smokers and non-smokers. *J. Clin. Invest.* **49,** 2086–2096.

Hatch, G. E., Slade, R., Boykin, E., Hu, P. C., Miller, F. J., and Gardner, D. E. (1981). Correlation of effects of inhaled versus intratracheally injected metals on susceptibility to respiratory infection in mice. *Am. Rev. Respir. Dis.* **124,** 167–173.

Haynes, W. F., Jr., Krstulovic, V. J., and Bell, A. L. L., Jr. (1965). Smoking habit and incidence of respiratory tract infections in a group of adolescent males. *Am. Rev. Respir. Dis.* **93,** 730–735.

Henry, M. C., Findlay, J., Spangler, J., and Ehrlich, R. (1970). Chronic toxicity of NO_2 in squirrel monkeys. *Arch. Environ. Health* **20,** 566–570.

Illing, J. W., Miller, F. J., and Gardner, D. E. (1980). Decreased resistance to infection in exercised mice exposed to NO_2 and O_3. *J. Toxicol. Environ. Health* **6,** 843–851.

Kass, E. H., Green, G. M., and Goldstein, E. (1966). Mechanisms of antibacterial action in the respiratory system. *Bacteriol. Rev.* **30,** 488–497.

Ketels, K. V., Bradof, J. N., Fenters, J. D., and Ehrlich, R. (1977). SEM studies of the respiratory tract of mice exposed to sulfuric acid mist-carbon particle mixtures. *In* "Scanning Electron Microscopy," Vol. II, pp. 519–526. IIT Res. Inst., Chicago, Illinois.

Kim, M., Goldstein, E., Lewis, J. P., Lippert, W., and Warshauer, D. (1976). Murine pulmonary alveolar macrophages: Rates of bacterial ingestion, inactivation and destruction. *J. Infect. Dis.* **133,** 310–320.

Kosmider, S., Misiewicz, A., Felius, E., Drozdz, M., and Ludyga, K. (1973). Experimentalle und klinische Untersuchungen über den Einfluss der Stickstoffoxyde auf die Immunität. *Int. Arch. Arbeitsmed.* **31,** 9–23.

Lauweryns, J. M., and Baert, J. H. (1977). Alveolar clearance and the role of the primary lymphatics. *Am. Rev. Respir. Dis.* **115,** 625–683.

Matzen, R. N. (1957). Development of tolerance to ozone in reference to pulmonary edema. *Am. J. Physiol.* **190,** 84–88.

Melia, R. J. W., Glorey, C. DuV., Altman, D. S., and Swan, A. V. (1977). Association between gas cooking and respiratory disease in children. *Br. Med. J.* **2,** 149–152.

Melia, R. J. W., Glorey, C. DuV., and Chinn, S. (1979). The relation between respiratory illness in primary school children and the use of gas for cooking. I. Results from a national survey. *Inst. J. Epidemiol.* **8,** 133–141.

Miller, F. J., Illing, J. W., and Gardner, D. E. (1978a). Effects of urban ozone levels on laboratory induced respiratory infections. *Toxicol. Lett.* **2,** 163–169.

Miller, F. J., Menzel, D. B., and Coffin, D. L. (1978b). Similarity between man and laboratory animals in regional pulmonary deposition of ozone. *Environ. Res.* **17,** 84–101.

Miller, F. J., McNeal, C. A., Kirtz, J. M., Gardner, D. E., Coffin, D. L., and Menzel, D. B. (1979). Nasopharyngeal removal of ozone in rabbits and guinea pigs. *Toxicology* **14,** 273–281.

Miller, F. J., Graham, J. A., and Gardner, D. E. (1981). The changing role of animal toxicology in support of regulatory decisions. *In* "The Health Issues in Air Quality Control" (in press).

Parnell, J. L., Anderson, D. O., and Kinnis, C. (1966). Cigarette smoking and respiratory infections in a class of student nurses. *N. Engl. J. Med.* **274,** 979–984.

Pecora, D. V., and Yegian, D. (1958). Bacteriology of the lower respiratory tract in healthy and chronic diseases. *N. Engl. J. Med.* **258,** 71–74.

Peters, J. M., Ferris, B. G., Jr. (1966). Smoking and morbidity in a college-age group. *Am. Rev. Respir. Dis.* **95,** 783–789.

Purvis, M. R., and Ehrlich, R. (1963). Effect of atmospheric pollutants on susceptibility to respiratory infection. II. Effect of nitrogen dioxide. *J. Infect. Dis.* **113,** 72–76.

Reynolds, H. Y., and Newball, H. M. (1974). Analysis of proteins and respiratory cells obtained from human lungs by bronchial lavage. *J. Lab. Clin. Med.* **84,** 559–573.

Roberts, R. C., Wenzel, F. J., and Emanuel, D. A. (1973). Serum immunoglobulin levels in farmer's lung disease. *J. Allergy Clin. Immunol.* **52,** 297–302.

Speizer, F. E., Ferris, B. G., Jr., Bishop, Y. M. M., and Spengler, J. (1980). Respiratory disease rates and pulmonary function in children associated with NO_2 exposure. *Am. Rev. Respir. Dis.* **121,** 3–10.

Spurgash, A., Ehrlich, R., and Petzold, R. (1968). Effect of cigarette smoke on resistance to respiratory infection. *Arch. Environ. Health* **16,** 385–391.

Stockinger, H. R., and Scheel, L. D. (1962). Ozone toxicity: Immunochemical and tolerance producing aspects. *Arch. Environ. Health* **4,** 327–334.

Stockinger, H. R., Wagner, W. D., and Wright, P. G. (1956). Studies of O_3 toxicity. I. Potentiating effects of exercise and tolerance development. *Arch. Ind. Health* **14,** 158–162.

Thomas, G. B., Fenters, J. D., Ehrlich, R., and Gardner, D. E. (1981). Effects of exposure to PAN on susceptibility to acute and chronic bacterial infection. *Toxicol. Environ. Health* **8,** 559–574.

U.S. Environmental Protection Agency (1977). "Air Quality—1975 Annual Statistics Including Summaries with Reference to Standards," EPA-450/2-77-002. USEPA, Washington, D.C.

Warr, G. A., and Martin, R. R. (1974). Chemotactic responsiveness of human alveolar macrophages: Effects of cigarette smoking. *Infect. Immunol.* **9,** 769–771.

Waters, M. D., Gardner, D. E., Aranyi, C., and Coffin, D. L. (1975). Metal toxicity for rabbit alveolar macrophages *in vitro. Environ. Res.* **9,** 32–47.

3

Toxic Effects of SO_2 on
the Respiratory
System

Marc J. Jaeger

This chapter is written as an introduction to the topic of air pollution by sulfur products. It dwells in some length on certain particularly important aspects, but neglects what is of lesser importance in the view of the author. Some aspects that are difficult to present because of intrinsic mathematical or biological problems are treated in some detail. The chapter does not try to be comprehensive, because several fairly comprehensive reviews have been published recently (Air Quality Criteria for Sulfur Oxides, 1970; Rall, 1974; Ferris, 1978a; American Thoracic Society, 1978; Green, 1980). For simplicity all concentrations are given in ppm (parts per million). When expressed as a density, 1 ppm SO_2 is equal to 2857 $\mu g/m^3$.

AIR POLLUTION

I. SOURCES OF SO$_2$

Sulfur is a component of all living material, because it is part of certain essential amino acids. It is found, therefore, in fossil fuels, where it is concentrated by the very process of fossilization; however, this concentration varies: for example, Virginia coal is known for its low sulfur content (1–2.5%); Middle West coal averages about 6% sulfur. Upon combustion, the sulfur is oxidized to SO$_2$, which escapes with the exhaust fumes.

Gasoline and kerosene are almost free of sulfur, because of the refining process involved in their production. Automobiles are, therefore, not a primary source of SO$_2$ pollution. However, any combustion of unpreprocessed coal or heavy oil in industrial plants is likely to become a major source of SO$_2$, primarily emitted into the air if the exhaust gases are not scrubbed (see Table I).

The amount of SO$_2$ produced by industrial plants is staggering and may be illustrated with an example: an American city of 100,000 inhabitants requires a power plant of about 200 MW to generate power for the needs of residential living and supporting facilities alone. Typically, such a plant burns the equivalent of one train of coal (100 cars loaded with 20 tons each) per week. If prime coal from Virginia, containing 1% sulfur, is used, the plant will emit each week the equivalent of 20 tons of SO$_2$ into the air surrounding the city. If we assume the city is located in a geographical depression with stagnant ambient air, significant levels of

TABLE I

Estimates of Sulfur Oxide Emissions (United States) 1940–1970[a] (10⁶ tons/year)

Source category	1940	1950	1960	1968	1969	1970
Fuel combustion in stationary sources	16.8	18.3	17.5	24.7	25.0	26.5
Transportation	0.7	1.0	0.7	1.1	1.1	1.0
Solid waste disposal	Neg[b]	0.1	0.1	0.1	0.2	0.1
Industrial process losses	3.8	4.2	4.7	5.1	5.9	6.0
Agricultural burning	Neg	Neg	Neg	Neg	Neg	Neg
Miscellaneous	0.2	0.2	0.3	0.3	0.2	0.3
Total	21.5	23.8	23.3	31.3	32.4	33.9
Total controllable[c]	21.3	23.6	23.0	31.0	32.2	33.6

[a] From Urone (1976).
[b] Negligible (less than 0.05 × 10⁶ tons/year).
[c] Miscellaneous sources not included.

SO$_2$ pollution (0.01 ppm or 28 $\mu g/m^3$) may be expected to develop within 2 weeks. Of course, such stagnant conditions are rare, but the example may indicate the acute threat to the welfare of our cities. Obviously, winds may eliminate the threat to the immediate area, only to carry the polluted air to regions miles downwind of the plant. In large urban agglomerations, the high density of the population results in a high density of power plants and even greater pollution. Present trends in producing electrical energy in this country indicate that in the next decade the energy needs will be covered increasingly by the use of coal, much of which has a high sulfur content. Thus, for example, private and public utilities in the state of Florida plan to expand by 1988 electrical generating power by 55% (to 36,600 MW). It is expected that the SO$_2$ emissions will be approximately 40 tons/installed MW generation capacity (Urone, 1981).

It should be added that SO$_2$ emissions are not a novel occurrence. Large amounts of SO$_2$ are emitted by volcanoes, and it has been estimated that even today, worldwide production of SO$_2$ by volcanoes and other natural sources such as the oceans surpasses industrial emissions. The problem of air pollution is, in part, a problem of distribution, the industrial plants' emitting close to populated areas and the disposal of much of the SO$_2$ as acid rain in locations downwind from the sites of emission. Typical wind patterns play a crucial role in depositing millions of tons of acid.

II. ATMOSPHERIC CHEMISTRY

SO$_2$ as emitted by a power plant is a reactive gas. It reacts with water to produce sulfurous and sulfuric acid. In pure air of moderate humidity, this reaction is slow, because of the relatively low probability of an SO$_2$ molecule meeting a gaseous water molecule. The conversion to equilibrium may take approximately 40 days. But dusty city air contains particulates of all kinds, some metallic, which catalyze the reaction so that equilibrium may be reached in 5 hours. Sunlight and humidity are other factors that favor such reactions. The end product of these reactions is a gas mixture that is, in part, aerosolized and in which SO$_2$ gas and sulfates are in equilibrium in a ratio of approximately 4.9 to 1. It is this mixture that has to be studied as to its effect on health. A more complete, referenced description of atmospheric chemistry is found in Urone (1976). This reference also discusses the important topic of the scavenging of SO$_2$, which is not treated here.

III. PRESENT URBAN LEVELS

The description of the concentration of a pollutant such as SO_2 in the ambient air of a city is, for most nonspecialists in the field, a statistical nightmare. Indeed, the sampling at a given site is likely to vary up and down, depending on the concentration of the pollutant in a particular, sampled gas volume. Typically, the level may rise during periods of heavy industrial activity during the day and decline again at night. This pattern of daily peaks and valleys depends on weather conditions, sunlight, and wind as well as on seasons, because certain seasons may combine high power consumption for heating or air conditioning with high humidity and/or poor wind conditions. The record obtained from a sampling site over a year looks much like the skyline of a mountain range. A certain pattern of highs and lows is visible but difficult to describe quantitatively in terms of cyclic events that show repetitive symmetries. Peak values, important to the mountain climber, are if isolated, probably less descriptive of the pollution level than lower values that are maintained for some period of time. The sampling time, i.e., the time over which levels of SO_2 are integrated, influences the results considerably (Fig. 1). Moreover, the highest value for SO_2 recorded during an

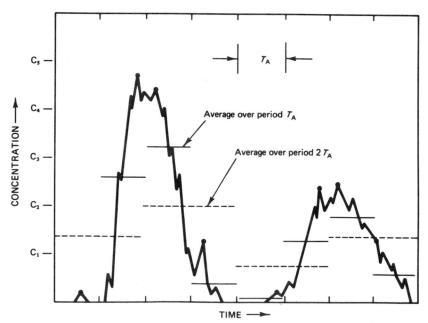

Fig. 1. Graphic representation of effect of instantaneous measure of concentration and time averaging of concentration. (From Urone, 1976.)

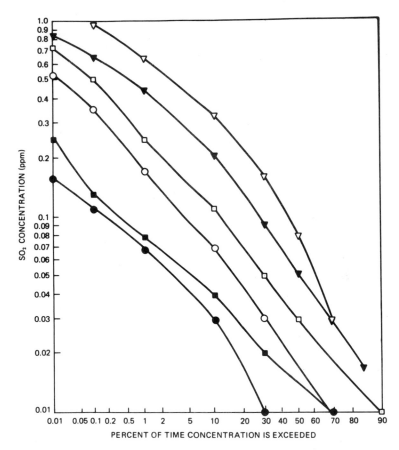

Fig. 2. Frequency distribution of SO₂ levels in selected cities in the United States 1962–1967. Top to bottom: Chicago, Illinois; Philadelphia, Pennsylvania; St. Louis, Missouri; Cincinnati, Ohio; Los Angeles, California; San Francisco, California. (From Urone, 1976.)

observation lasting 1 day may be 0.01 ppm, but the highest reading obtained from measurements taken over 1 year may be much higher, because the likelihood of a very high value is greater when the observation time is extended. Thus, the concentration of a pollutant like SO₂ in a given city has to be expressed statistically by the use of a probability plot (see Fig. 2). Such plots show that, in San Francisco, SO₂ exceeds a level of 0.1 ppm approximately 0.1% of the time, whereas the SO₂ level exceeds 0.1 ppm 70% of the time in Chicago. The probability plots for the two cities are similar in that the concentration–probability relationships are approximately linear and parallel for both cities in the log–log

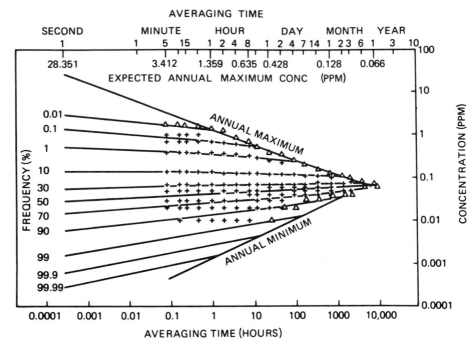

Fig. 3. Concentration, averaging time, and frequency of NO_x in Washington, D.C. (From Larsen, 1969.)

representation. The original statistical analysis upon which this probability plot was based was presented by Larsen (1969) with data for nitrogen oxides in Washington, D.C. Because that analysis presents a complete diagram of the distribution of a typical pollutant, it is presented as Fig. 3.

IV. STANDARDS

The primary standards set by the American Environmental Protection Agency as safeguards against exposure to SO_2-polluted air are listed in Table II. The table includes for comparison data on a number of other pollutants. Some of the statistical principles that govern the time distribution of pollutants are part of the description. Thus, the annual standard, 0.03 ppm, is much lower than the maximal allowed daily standard (0.14 ppm). This is in qualitative agreement with the general pattern shown in Figs. 2 and 3.

A horizontal line drawn on Fig. 2 at a level of 0.14 ppm allows us to compare the present national standard for 24 hours with actual pollu-

TABLE II

Primary and Secondary Ambient Air Quality Standards[a]

Pollutant	National primary	National secondary[b]
Sulfur oxides	80 μg/m^3 (0.03 ppm)[c] ann. arith. mean 365 μg/m^3 (0.14 ppm) max. 24 hour avg.[d]	60 μg/m^3 (0.02 ppm) ann. arith. mean (repealed, 1977) 260 μg/m^3 (0.1 ppm) max. 24 hour avg. (repealed, 1977) 1300 μg/m^3 (0.5 ppm) max. 3 hour avg.
Particulate matter	75 μg/m^3 annual 260 μg/m^3 maximum 24 hour avg.	60 μg/m^3 annual 150 μg/m^3 maximum 24 hour avg.
Carbon monoxide	10 μg/m^3 (9 ppm) max 8 hour avg. 40 μg/m^3 (35 ppm) max 1 hour avg.	Same as primary Same as primary
Photochemical	235 μg/m^3 (0.120 ppm) max (revised 1979) 1 hour avg.	Same as primary
Hydrocarbons	160 μg/m^3 (0.24 ppm) max. 3 hour avg. (6–9 a.m.)	Same as primary
Nitrogen	100 μg/m^3 (0.05 ppm) ann. arith. mean	Same as primary

[a] From Code of Federal Regulations Title 40 (1978).
[b] Adapted as Florida's Ambient Air Standards.
[c] ppm given on volume basis.
[d] All maximum averages are not to be exceeded more than once.

tion levels in different cities. Several cities exceed that standard some of the time, some only once or twice a year ($\leq$ 7.3% of the time, e.g., Cincinnati), some many times a year (e.g., Chicago, Philadelphia). The question that obviously arises is what effects such an exposure has on health.

V. HEALTH EFFECTS OF SO$_2$

This section, the heading of which corresponds to the title of the chapter itself, is the most important. It is divided into three subsections describing three scientific approaches to investigating the effects of SO$_2$

on health. These approaches are (1) epidemiology, (2) controlled human experiments, and (3) animal work.

It may be useful first to describe the three approaches and to weigh their contributions in general terms. Animal work is possibly the best-known method of evaluating toxic effects of substances, including air pollutants. Much work has been done on laboratory rats, because the rat is an animal that is easy to breed and maintain. The rat's behavior, anatomy, and variations are almost as well known as man's, making it an ideal experimental animal. Experiments have been done exposing rats for short and long periods to low, intermediate, and high doses of SO_2 and sulfates. Many authors expose the animal for relatively short periods of time (days or weeks) to relatively high doses of the pollutant under the assumption that the risk of this exposure is similar to the risk incurred by human populations exposed for years to levels possibly 10 or 100 times lower than the ones used in the animal work. This assumption is probably only partially correct. Another factor that may limit the value of conclusions drawn from animal work are the differences in reactions to various injuries and stresses among animals of different species. Thus, certain species are susceptible to certain human diseases, whereas others are not. The guinea pig, for example, is exquisitely susceptible to tuberculosis, whereas the rat is not at all. In its airways, the rat lacks the mucus-producing glands that may contribute significantly to the defense against irritation by SO_2 in humans. The question may then be asked how findings in animals should be interpreted and applied to man. Present knowledge does not give a full and satisfactory answer to this question. David P. Rall (1979) has given probably the best review of the problem to date. His analysis may be paraphrased by saying that animal work does not adequately evaluate the risk that exposure to polluted air imposes on such a heterogeneous population as smoking and nonsmoking, male and female humans of all ages. But animal work can offer important information on the mechanisms by which SO_2 may be harmful to man. This will be elaborated below in Section V,C.

Experiments performed on human volunteers come much closer to providing answers to the toxicity of SO_2 in man, but for obvious reasons, even these experiments have many limitations. In general, only very few subjects are studied, which makes it difficult to conclude that the population at large would react in the same way as the selected volunteers. The duration of exposure is limited to a few hours or, exceptionally, a few days. Finally, experiments that might mimic harmful situations are excluded, because of ethical reasons. Thus, no information is obtained on many urgent questions.

The burden of providing answers to the question of exposure and risk is put, therefore, on epidemiologists. They study health patterns in various populations and try to link their observations with simultaneous ⬛dings of pollution levels. These studies are extremely time consum⬛d arduous, but have yielded more recently some magnificent re⬛ Epidemiology will be discussed first, because it provides the most ⬛rtant information.

A. Epidemiology of SO$_2$ Effects

Epidemiologists were first to analyze and explain effects of high levels of SO$_2$ on man and to initiate the demand for more study and some regulating, legal controls. In 1930, an acute episode of high air pollution was generated in the Meuse Valley, a highly industrial region of Belgium, and resulted in a number of deaths. In London, postwar industrial development and certain residential heating customs (such as burning coal in small fireplaces) resulted in 1 weeklong episode of heavy pollution during the winter of 1954–1955 and another during the winter of 1962–1963. Subsequent studies and comparisons of death statistics indicated an excess of 4000 deaths for the two episodes (Ministry of Health, 1954; Scott, 1963). The hospital records indicate that the greatest excess mortality was among the aged and those suffering from chronic lung or heart ailments. Similar, but less cataclysmic episodes of heavy pollution occurred in Donora, Pennsylvania, New York City, and Osaka, Japan. Scientific analysis of those stunning catastrophies was much hampered by the lack or inadequacy of technical equipment for determining the level of pollutants in the ambient air during those episodes. Later extrapolations suggested that weeklong levels of SO$_2$ exceeding 0.4 ppm (about three times the primary 24-hour standard) and of particulate matter exceeding 2000 μg/m^3 (seven times the primary standard for a 24-hour average) were responsible for those catastrophies. It is disquieting that such levels are reached, even for periods of only a few hours, in some American cities (Fig. 2). Public outrage and efforts by the scientific community led to a change of residential heating technology in Britain: coal fireplaces were replaced by oil furnaces with better combustion controls. The level of pollution decreased progressively to the point where it is now. More recently, the use of natural gas from the North Sea Fields has further alleviated the pollution in British cities. The efforts of the scientific community led to the installation of monitoring stations for prevention, alarm, and investigation. It is nonetheless astounding that information on pollution levels was made available to the public only recently, 25 years after the recognition of the dangers of pollution.

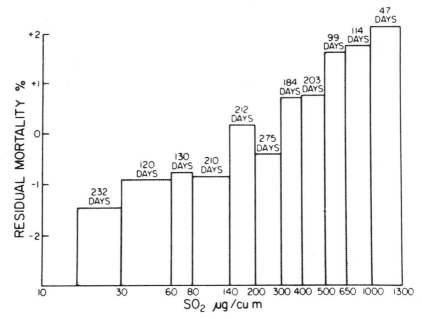

Fig. 4. Means of residual mortality plotted against SO_2 concentration in New York–New Jersey. (From Buechley *et al.*, 1973.)

Among the first statistical analyses to correlate mortality data obtained from hospitals with simultaneous air quality figures are those of Buechley *et al.* (1973), who studied deaths from all causes in the New York metropolitan area from 1962 to 1966. The investigators adjusted their data mathematically for the effects of other important factors such as extremes of temperature, seasonal cycles, and small epidemics of influenza. After these corrections, their figures show impressive increments in the residual mortality with increases in the level of SO_2 studied over a 1826-day period. When the SO_2 level exceeded 0.19 ppm, a value only slightly above the present primary standard, the residual mortality was 2% above the expected mean (Fig. 4). This study is among the best and most quoted of the mortality studies published in the early 1970s. Yet the significance of the study was reduced, because data from only one station monitoring pollution in New York City were available so that the authors had to assume pollution was distributed evenly over the entire metropolitan area, from which the mortality data were gathered.

Buechley's work does not address itself to the important question of whether, in addition to being a deadly threat to some presumably already chronically ill persons, SO_2 is also a health hazard to the general population, causing diseases, loss of work time, and perhaps chronic damage to certain organs. This question is studied in more recent work, of which three examples will be presented.

Bouhuys *et al.* (1979) studied three populations comprising a total of 8675 individuals living in three different environments. The first group lived in Lebanon, Connecticut, a rural community away from urban centers. The second group lived in Ansonia, Connecticut, a highly industrialized area, that had experienced, in the decade preceding the study, record high levels of air pollution. The third group resided in Winnsboro, South Carolina, a small country town. Some of these residents formed a special group, because they were exposed daily to heavy occupational pollution at a textile mill. Some selected data from this study are summarized in Table III. They show that during the time of

TABLE III

SO₂ Level, Prevalence of Symptoms, and Lung Function in Four Cities

	Lebanon	Ansonia	Winnsboro residents	Winnsboro cotton-mill workers
Annual mean SO₂ level (ppm)	.0035	.0047	.0012	.0012
Total suspended particulate (μg/m³)	41.0	62.8	48.2	48.2
Results in middle-aged males				
Prevalence of usual phlegm (more than 3 months/ year)				
Nonsmokers	4.5%	18.0%	16.0%	17.0%
Smokers	32.0%	20.0%	29.0%	56.0%
Lung function (FEV₁ residual in liters)				
Nonsmokers	+.115	−.162	−.145	−.199
Smokers	−.429	−.326	−.366	−.568
Results in middle-aged females				
Prevalence of usual phlegm (more than 3 months/ year)				
Nonsmokers	0%	11%	8%	22%
Smokers	16%	8%	16%	33%
Lung function (FEV₁ residual in liters)				
Nonsmokers	−.049	−.022	−.087	−.281
Smokers	−.175	−.197	−.265	−.495

the study, pollution levels were not much lower in rural Lebanon than in industrial Ansonia; by contrast, the textile workers in South Carolina were living in much "cleaner" air. The authors studied symptoms indicative of chronic lung involvement (phlegm) and lung-function tests expressed as residuals, i.e., differences between the measured values and values predicted on the basis of standard tables for healthy individuals of the same sex, age, and body size. If negative, the residual expresses a loss of lung function. The data are not easy to evaluate, but the following summary may be attempted:

1. There seems to be a higher prevalence of respiratory symptoms in urban residents (Ansonia) than in rural dwellers (Lebanon, Winnsboro). This trend is exemplified most clearly in nonsmoking women, but is also seen in nonsmoking males. There is, however, no appreciable decrease in lung function in those nonsmoking individuals who show symptoms to support the suspicion that city dwelling has a detrimental influence. These results were partly supported, partly contradicted, by findings in male and female children living in the same environments.

2. Smoking has a strong effect that is most prominent in males and females living in rural areas, where air pollution is a lesser problem Smoking results in both an increase of symptoms and a decrease in lung function in all groups studied, i.e., in both urban and rural residents as well as cotton-mill workers.

3. The effects on health of occupational pollution are most apparent in nonsmoking women, who exhibit an appreciable increase of symptoms and decrease of lung function. Smoking cotton-mill workers have the highest decrease in lung function.

The reader is referred to the original publication for a more complete presentation of results and a more finely tuned discussion.

Recently, Ferris *et al.* (1979) published a study of 18,079 residents in six communities in the United States. The sample was reduced to 11,040 individuals, 25–74 years old, by eliminating from the study individuals who were not eligible for such reasons as insufficient length of residence. This study will be used as a second example of modern epidemiological work. Selected results from this study are shown for comparison in Table IV. They may be summarized also as follows:

1. Comparisons between nonsmokers in cities with apparently fairly clean air (Portage, Wisconsin; Topeka, Kansas; Watertown, Massachusetts) and those in cities where pollution clearly exceeds legal standards (St. Louis, Missouri; Steubenville, Ohio) show a moderate decline of lung function in residents of polluted areas. The data are not unequivocal, however, and female nonsmokers living in the city with the heaviest pollution have no noticeable reduction in lung function.

TABLE IV

SO$_2$ Levels, Particulates, and Lung Function in 11,000 Residents of Six Cities[a]

	Port	TPKA	Kingst	Watrn	STL	STBVL
SO$_2$ annual mean (ppm)	.0033	.0014	.013	.0045	*.040*[b]	*.038*
Total suspended particulates (μg/m^3)	23	60	73	42	*140*	*182*
FEV$_1$ residual (liters) for males						
Nonsmokers	+.02	−.1	−.04	+.02	−.06	−.12
Smokers[c]	−.50	−.52	−.62	−.56	−.63	−.66
FEV$_1$ residual (liters) for females						
Nonsmokers	+.04	−.02	−.08	−.06	−.11	−.01
Smokers	−.20	−.29	−.39	−.18	−.38	−.32

[a] Port, Portage, Wisconsin; TPKA, Topeka, Kansas; Kingst, Kingston-Harriman, Tennessee; Watrn, Watertown, Massachusetts; STL, St. Louis, Missouri, STBVL, Steubenville, Ohio.

[b] The italicized values exceed the National Pollution Air Quality Standard.

[c] The smokers smoked at least 21 cigarettes/day.

2. Strong evidence shows that compared to nonsmokers, people who smoke at least 21 cigarettes per day have a marked decrease in lung function. Among men and women with similar smoking habits, the decrease is more marked in males.

It is difficult to compare data from individuals living under different environmental conditions with those from people indulging in different smoking habits or exposed to various degrees of air pollution. As a common yardstick, the following hypothesis is offered: studies of the same authors (Ferris *et al.*, 1979) show that, beginning with age 20, lung function (FEV$_1$) decreases by 0.0285 liter/year in asymptomatic, male nonsmokers and by 0.0268 liter/year in asymptomatic, female nonsmokers. This is a physiological process, which may be subsumed under the general term "aging." Using this "normal" decline of lung function with time, one may attempt to compare the loss of lung function in clean-air cities with that for polluted cities for smokers and nonsmokers. The results of this comparison are found in Table V: for example, the numbers indicate that a hypothetical, standardized male of 50 would have a lung function corresponding to that of a man 50.7 years of age, if he lives in one of the less polluted cities and is a nonsmoker. However, if he lives in a polluted area, his lung function would correspond to that of a man 53.2 years old. For a heavy smoker (21 or more cigarettes per

TABLE V

Lung Function Losses Expressed in Equivalent Years[a]

	Derived from Ferris's work		Derived from Bouhuy's work	
	Nonsmokers	Smokers	Nonsmokers	Smokers
Males				
Clean areas	50.7	68.5	54.0	65.1
Polluted areas	53.2	72.6	56.8	62.8
Females				
Clean areas	50.5	58.2	51.8	56.5
Polluted areas	52.4	63.1	50.8	57.4

[a] The computation is based on the following relationship (males):

$$\frac{(30 \times .0285) + \text{residual}}{.0285} + 20$$

For females the annual decline is .0268. This equation assumes that the decline of lung function begins at the age of 20. If an imaginary, 50-year-old male has no residual, his lung function expressed in equivalent years is $20 + (30 \times .0285)/.0285 = 50$. To compute equivalent years for clean air cities the average residuals of Portage, Wisconsin; Topeka, Kansas; and Watertown, Massachusetts, were used. For polluted cities the average residuals of Steubenville, Ohio and St. Louis, Missouri, were used.

day), the figures are 68.5 and 72.8 years, respectively. Similar figures are found for females. The right-hand side of Table V shows a similar computation based on Bouhuys' data presented earlier. The similarity of the figures is convincing evidence that measurements indeed reflect the effect of current air quality on health, an effect slighter than that of smoking but nonetheless noticeable.

The third study (Aubry *et al.*, 1979) compared statistics on health in three areas (Beauharnais, Pointe-aux-Trembles, and St. Eustache) of the greater Montreal region with measurements of air pollution. Table VI is taken from this work; it shows only some of the available data describing pollution levels and the health of the 300 participants, and the reader is referred to the original paper for complete information. This study is remarkable for several reasons: it provides data from one single climatic area; some of the studies quoted earlier compare cities with differences in climate as well as air pollution. It offers results from several, independent lung-function tests rather than one alone. It uses advanced statistics to combine all parameters into one single, discriminatory test procedure.

The average level of pollution in Montreal is comparatively low; a level of 0.05 ppm SO_2 is exceeded only 6.6% of the days in the region with the highest level of SO_2 (compare with Fig. 2). The number of males suffering from chronic productive cough is, however, very high.

The percentage of cigarette smokers varied from 60 to 73%, being lowest in the region with least pollution. Among women, the percentage of cigarette smokers varied from 32 to 38%, being highest in the region with lowest pollution. The statistical analysis indicated that symptoms and lung function did not differ among the inhabitants of the three regions after corrections were made for age and differences in smoking habits. By contrast, effects of cigarette smoking on health could easily be detected by statistical analysis.

The three examples of epidemiological studies presented above document the immense progress accomplished in this field. But it may be useful, at this point, to outline briefly some of the steps that made this progress possible.

First, the techniques for measuring lung function have been refined. Even though the test for measuring the forced expired volume of the first second has been known since 1948 (Tiffeneau and Pinelli), new and improved techniques had to be developed to make the test applicable to field studies. Only recently have standardized testing procedures been developed and found any wide acceptance (Ferris, 1978b). Even so, results cannot always be compared, because researchers use different equipment.

TABLE VI

Air Pollution and Health in Three Urban Communities in the Greater Montreal Region

	Beauharnais	Pte.-aux-Trembles	St. Eustache
Particulate matter ($\mu g/m^3$)	131	94	84
SO_2 (ppm) (24 hour avg.)	.021	.043	.005
Symptoms: males			
Cough 3 months/year (%)	43	30	18
Phlegm 3 months/year (%)	22	23	11
Symptoms: females			
Cough 3 months/year (%)	12	16	10
Phlegm 3 months/year (%)	11	8	4
Lung function: males			
FEV_1[a]/FVC (%)	72.8	74.8	74.5
CC[b]/TLC (%)	47.9	46.9	46.3
Lung function: females			
FEV_1/FVC (%)	79.5	77.5	77.9
CC/TLC (%)	47.9	46.7	46.2

[a] FEV_1, volume expired in 1 second in forced expiratory vital capacity (VC) maneuver.
[b] CC, closing capacity in single-breath nitrogen washout expressed in % of total lung capacity.

Next, improved prediction equations that determine the "standard" value of a lung function for a given "normal" individual have been developed. Until relatively recently, standard tables were based on measurements made indiscriminately on smokers and nonsmokers, because the effect of smoking on lung function had not yet been recognized. Newer predictions are based on data obtained from nonsmokers alone; however, some other external factors that may affect lung function have not always been kept constant or eliminated from groups of subjects studied as controls. It is indeed difficult to find people who have always lived in an atmosphere of pristine cleanliness, and air pollution may occur in areas that do not appear suspect. It was noted recently (Melia *et al.*, 1977) that gas stoves produce such significant amount of nitrogen oxides that the atmosphere in a typical American or continental kitchen contains levels of this oxidizing pollutant that are as high as those found in heavily polluted cities. Individuals exposed daily in their homes to appreciable levels of nitrogen oxides cannot be used as controls in a study to assess the possible effects of air pollution. Even though some of the modern prediction equations may not be perfect, they are much improved. An important aspect of this improvement has been the inclusion of elderly people in the selection of control subjects. In earlier work, normal control values were derived on the basis of data obtained solely from middle-aged adults.

Even greater progress has been accomplished in standardizing symptoms and the recording of medical history. The long evolution leading to the definition of the symptoms of chronic bronchitis cannot be recalled here (Fletscher *et al.*, 1976); suffice it to note that standardized questionnaires have been established to allow (and compel) authors to use the same methods when questioning individuals on smoking habits, previous exposures to pollution, and symptoms, thus allowing comparisons among various sets of data (Ferris, 1978b). In summary, immense technical and methodological background work had to be completed before the newer, large-scale epidemiological studies could be successfully completed.

Polluted air contains a host of various chemicals, some diluted, some suspended as aerosols. Thus, arises the problem of correlating symptoms or lung-function tests with the level of one particular pollutant or with a particular combination of pollutants. The difficulty was alluded to in Tables II, III, and VI, in which levels of suspended particulates have been listed together with SO_2 levels. The particulates confound the issue, because their elevation is sometimes parallel to the elevation of SO_2. A few studies in which attempts were made to correlate epidemiological observations with several pollutants may be cited to further exemplify this point; these studies were concerned with effects of air pollution on patients.

Patients with chronic bronchitis were found to have increases of respiratory symptoms when the SO$_2$ level exceeded 0.1 ppm. This level, it may be recalled, is very close to the 24-hour primary standard (Table I) and is exceeded frequently in American cities (Fig. 2). But the level of suspended particulates was elevated in this study to 350 μg/m^3, a value exceeding the standard given in Table I. There was a better correlation of the symptoms to particulates than to SO$_2$ (Lawther *et al.*, 1970), but the results were difficult to interpret, because some of the elevations of particulates occurred at the same time as the changes of SO$_2$.

Similar conclusions were reached in a study that evaluated symptoms in workers over a 10-year period in London. During that time, the level of particulates gradually decreased to less than 50% of the concentration at the beginning of the study, whereas the SO$_2$ concentration did not change (Fletscher *et al.*, 1976, p. 272). The study found a reduction of symptoms, which was caused, presumably, by the decline of particulates.

In yet another study, the frequency of asthma attacks in 20 asthmatics was found to be related weakly to the levels of SO$_2$ and particulates. The relationship was less strong than the relationship of attacks to ambient temperature variations (Cohen *et al.*, 1972).

It is obviously very difficult to draw conclusions from the epidemiological studies reviewed. There seem to be two main lessons, both well documented and supported by strong scientific evidence. The first is that air pollution with SO$_2$, SO$_4$, and particulates that exceed 24-hour standards by a factor of three or more for a period of 1–2 weeks may have disastrous effects culminating in the death of thousands of people, mostly elderly and those afflicted with chronic heart or lung diseases. The second lesson is that present levels of pollution are apparently relatively innocuous. Large surveys (Tables II and III) show a small decline in lung function and a small increase of symptoms in populations living in polluted areas. But the effects are so slight that the authors are cautious in attributing them to pollution. In the words of Ferris *et al.* (1979), "Any differences found between cities in a cross-sectional examination of adults may be attributed to past events and such variables as socioeconomic status. Nevertheless, this preliminary analysis leads us to believe that our measurements may indeed reflect effects of current air quality on health."

I perceive a certain contradiction between the two lessons. The levels of SO$_2$ that caused catastrophies are indeed only moderately higher than the levels that have apparently only a barely detectable effect on health. It is as if moderate levels of pollution are well tolerated, whereas an elevation above a threshold becomes acutely dangerous. The discrepancy may perhaps be related to the differences in response of elderly, sick people as compared to young, healthy individuals who make up the majority in a general population. Moreover, some effects may be attributed to the presence of other pollutants such as chemically inert particles which may act as catalysts in air chemistry or may act as vehicles in the transport of pollutants deep within the lungs. Some of these mechanisms are mentioned in the next section.

B. Controlled Human Studies

Controlled human experiments allow us to determine the effects of clearly defined levels of chemically pure pollutants on man. They permit us to determine the lowest concentration at which an effect is detected and to investigate effects of mixtures of different pollutants. Moreover, one can evaluate effects of a given pollutant on specially selected groups of individuals, such as asthmatics, who may be particularly sensitive to respiratory irritants.

In a landmark study, Frank *et al.* (1962) exposed 11 healthy volunteers to 1, 5, and 13 ppm SO_2 in air for 10–30 minutes. No effect was found at 1 ppm, but airway resistance was increased at 5 and 13 ppm. In another study, the appropriateness of the 3-hour standard was ascertained by exposing 40 asthmatics and 40 normal subjects to 0.5 ppm gaseous SO_2 in air (Jaeger *et al.*, 1979). Ten lung-function tests were used to assess possible alterations in lung function, but none showed any significant change in normal subjects, and only one test, the midmaximal, forced expiratory flow rate, was marginally lowered in asthmatic subjects. This marginal decrease, was, however, statistically significant. The asthmatics had been selected to represent a groupx of patients with relatively mild disease. Two asthmatics and one normal subject with apparent sensitivity to irritants experienced transient bronchoconstriction after the exposure.

Some of the precautions observed in this study are typical for this kind of work and may be briefly mentioned. All subjects served as their own controls, i.e., they were exposed for one 3-hour period to clean air and for another to 0.5 ppm SO_2. The sequence of control versus exposure was randomized, and neither the subjects nor the attending physician knew the type of experiment being conducted. Experiments on the same individual were conducted at the same time of day, i.e., if the control exposure to clean air took place in the morning from 9:00–12:00, the

SO$_2$ exposure was scheduled for the same hours. This precaution eliminated any possible effect of daily variations of lung function attributable
to circadian rhythms.

The daily variation of lung function was found to exceed any change
that could be attributed to the SO$_2$. Thus, the midmaximal flow rate in
asthmatics varied during air exposures from 2.56 to 2.80 liters/second, a
change of 0.24 units, whereas the average difference between SO$_2$ and
clean-air exposures was 0.07 units. The circadian rhythms are of considerable practical interest to epidemiology, because they determine procedures and methods. An example of a circadian variation of lung function is given below.

Kerr and co-workers (Kerr, 1973) studied the effect of Baltimore
smog on airway resistance in six healthy volunteers. The volunteers were
exposed in a chamber for 3 consecutive days to filtered air and then, for
the same length of time, to polluted air. In a second series, the exposures
to filtered and polluted air were reversed, i.e., the subjects were first
exposed to polluted air, then to clean air. All subjects show a circadian
rhythm of airway resistance with minimal values in the early afternoon
and peak values in early morning. The periods (time from one peak to
the next) seem fairly uniform for all subjects, but the amplitude of the
variation changes from subject to subject; some subjects consistently had
a large amplitude, others consistently a low amplitude. The average
obtained over the period of polluted-air breathing was actually slightly
lower than the average obtained when nonpolluted air was breathed,
implying a virtual beneficial effect of pollution. Kerr's data serve to
stress the importance of circadian rhythms for epidemiological studies:
Lung function changes consistently during the day and in the early
afternoon reaches values that clinicians or epidemiologists would term
"improved" or "better." In almost everybody, but especially in asthmatics, airway resistance decreases during the first part of the day and
increases at night; most lung-function tests are affected by this bron-
chodilatation. It is therefore essential that scientists studying normal
controls or individuals in polluted cities take into account the daily variation. To take a hypothetical example, if subjects serving as controls were
studied predominantly in the afternoon, whereas the subjects exposed to
pollution were tested predominantly in the morning, the decrease found
in lung function could not be attributed entirely to pollution. It seems
that in the past, the pitfalls of ignoring circadian variables have not
always been avoided; many authors do not refer to the time of day
measurements were made in different groups of subjects.

A study by Nadel and co-workers (1965) has elucidated the mechanism by which SO_2 produces a change in airway resistance in man. They showed that the effect is blocked by atropine. As a consequence, one has to assume that most or all effects noted in controlled human subjects result from a reflex bronchoconstriction. The afferent limb of the reflex arc originates presumably from sensory receptors (sometimes referred to as "cough" receptors) in the upper airways; the efferent limb is mediated by vagal and sympathetic pathways.

Studies by Frank *et al.* (1969) and Anderson *et al.* (1974) complement the findings of Nadel beautifully. They demonstrated that because of its high solubility SO_2 is absorbed almost totally in the upper airways. If SO_2 containing air is blown into a nostril, while the subject closes the pharynx, air collected from the mouth opening contains almost no detectable SO_2. This finding has been confirmed in animal experiments, and one has to assume that most of the inhaled SO_2 is deposited in the nose and in the upper airways, where it stimulates sensitive nerve endings and receptors, but that it does not reach deeper parts of the lungs. Upon deposition in the mucus layer covering the airways, SO_2 is presumed to combine chemically with water to produce H_2SO_3 and H_2SO_4. These acids may be the actual stimuli of the receptors.

Lawther *et al.* (1975) analyzed the effect of increased ventilation. They found that 1 ppm of SO_2, which usually has no detectable effect on lung function, results in an increase of airway resistance when the subject hyperventilates. It is assumed that hyperventilation results in a higher acid load on pharyngeal receptors and that SO_2 may reach more deeply seated bronchial receptors, such as those in the carina and large bronchi, to mediate the enhanced effects.

Ambient air polluted with SO_2 contains, in general, sulfuric acid in the form of aerosolized particulates (e.g., SO_4) as discussed in Section II. The fate of these particulates in the respiratory system is quite different from that of the SO_2. Particulates are deposited according to their size: large particulates are retained in the nose, which functions as a scrubber; small particulates reach into small bronchi; and very small particulates are in part deposited in the alveoli. Much of the SO_4 contained in the atmosphere of polluted cities assumes the form of small particulates, which are deposited presumably deep inside the lungs. Sackner *et al.* (1978) have presented a very complete study of SO_4 aerosols in man: normal and asthmatic subjects breathing an H_2SO_4 mist of 1000 $\mu g/m^3$ for 10 minutes showed no alterations in lung function when they were tested with a host of sophisticated procedures. There were no delayed reactions and the results in asthmatics did not differ from those in normal subjects. These results were confirmed by similar experiments performed on unanesthetized sheep and anesthetized dogs, in which the exposure levels were up to 14 times as high as those in the experiments with humans and extended for up to 4 hours. Thus, single, relatively brief exposures to H$_2$SO$_4$ aerosols do not produce any detectable effects in man or animals.

Several authors have studied the effect of a combination of two pollutants on man. The most studied combination, which is prevalent in many American cities, is SO$_2$ plus ozone (O$_3$). The prevalent sunshine in the southern United States favors the natural occurrence of ozone for several reasons: among these, some minute amounts of suspended metallic particulates catalyze the formation of ozone from oxygen in the presence of strong sunlight. Initial reports indicated that ozone and SO$_2$ combined had a potentiated effect, i.e., the effect of the combination was greater than the effect either of the two components had separately and higher than the sum of the components' separate effects (Hazucha and Bates, 1975). More recently, authors have disclaimed this potentiated effect (Bedi *et al.*, 1979).

Even though the observation of a synergistic effect of ozone and SO$_2$ has not been confirmed, synergistic action of known pollutants is not yet ruled out. Indeed, it seems that present information derived from controlled human studies does not explain completely the questions left open by epidemiological studies. Controlled experiments seem to indicate that ambient levels of approximately 0.4 ppm have been responsible for air pollution's greatest catastrophies in the early 1950s. The prolonged exposure during those episodes may have contributed to the fatal effects of SO$_2$. But combinations of pollutants such as SO$_2$, SO$_4$, and inert particulates, with or without oxidizing substances, may also have played a role.

The finding that SO$_2$ is deposited in the upper airways does not diminish its danger as a chronic, bronchoconstrictive irritant, but eliminates it as a single cause of lung-tissue destruction. The very complete studies of Sackner *et al.* seem to exclude SO$_4$ as an irritant causing acute changes of lung function, but do not, of course, exclude possible destructive effects of SO$_4$ in the lungs when inhaled for long periods of time. This question could be answered, perhaps, by animal work.

Controlled human experiments confirm that subjects with particularly sensitive airways, such as asthmatics known to respond strongly to all kinds of irritants, show greater effects from SO$_2$ than normal, healthy individuals. The increased response appears in two ways: Averaged results show that asthmatics as a group react differently than normal subjects. Moreover, some asthmatics respond with a delayed, transient bronchoconstriction. In the observations listed above, the asthmatics who developed mild attacks after SO$_2$ exposure did not respond differently than other asthmatics during the exposure and were not responsible for the change in average values of lung-function tests. Thus, the controlled studies illustrate how ambient pollution with SO_2 may affect a population by affecting different individuals in different ways.

C. Animal Experiments

The bronchial narrowing observed in man has been confirmed in animals. The response occurs within minutes of the inhalation of SO_2 and produces an increase in airway resistance that can be sensed by an intrathoracic pressure device and by flow measurement at the mouth. The increase of resistance vanishes within minutes of the removal of SO_2. The quickness of onset and decline are suggestive of a reflex.

Most interesting are experiments performed with long-term exposure, experiments that cannot be performed on human subjects. Guinea pigs were exposed continuously for 12 months to 5.7 ppm SO_2 with no identifiable effects in the lungs at autopsy (Alarie et al., 1970). Similar experiments conducted on monkeys over a period of 18 months at levels up to 1.3 ppm produced no pathological changes either (Alarie, 1972). Rats showed, however, a depression of the ciliary clearing mechanism when exposed for 1 week to 1 ppm SO_2 (Ferin and Leach, 1973). In another series of experiments, hamsters were treated initially with papain, which caused severe destruction of lung tissue. Microscopic analysis showed the destruction to be similar to human emphysema. The animals were then exposed intermittently for more than 10 weeks to very high levels of SO_2 (650 ppm) in an effort to mimic the exposure of patients with emphysema to the varying levels of air pollution. Only mild changes were observed (Goldring, 1970). Hamsters are, however, apparently more tolerant of SO_2 than other animals, even though they are rather susceptible to the toxic effects of papain.

In another series of experiments, the effect of SO_2 mixed with particulates that promote the formation of sulfuric acid was studied (Amdur, 1969). The effects of SO_2 were enhanced and signs of synergism elicited. Air humidity was shown to be a very important factor in the formation of aerosols containing sulfuric acid. Below 70% humidity, synergism was reduced considerably.

Thus, animal experiments provide useful background support for the concepts developed on the basis of human work. Prolonged exposure to levels of SO_2 more than 100 times higher than those found in polluted air fails to produce pronounced, deleterious, pathological changes in animals. This observation is certainly valuable and may allay the fears of people particularly anxious about the effects of air pollution on health.

Yet the results must be regarded with caution because animals of different species and man react quite differently to toxic stimuli.

VI. CONCLUSION

Summarizing remarks have been made at the end of each of the last three subsections in an effort to draw conclusions from the available data. This conclusion will therefore be kept brief. The authors who have studied SO_2 and its derivatives most carefully have concluded that the levels of SO_2 found presently in American cities (even those with the severest pollution) are so low that their effect on health is questionable or marginal. The standards presently in force seem to provide adequate protection for the general population, including individuals with heart or lung disease. Subjects with hypersensitive airways, such as asthmatics, may suffer exacerbations during peaks of pollution by SO_2.

One cannot forget, however, that levels of SO_2 only moderately higher (three- to fourfold) than those presently existing have resulted, when coinciding with other detrimental factors such as high particulate levels and high humidity, in catastrophic episodes of greatly increased mortality. The present vigilance toward pollution with SO_2 cannot therefore be decreased, but there seems to be no urgent necessity to reduce the present levels of SO_2 in ambient air.

ACKNOWLEDGMENTS

I would like to thank Dr. E. Schlenker for critically reading the manuscript.
Supported by ICAAS, University of Florida and Fogarty International Center, NIH.

REFERENCES

Air Quality Criteria for Sulfur Oxides (1970). Publ. AP-50. NAPCA.
Alarie, Y. (1972). Long-term continuous exposure to SO_2 in monkeys. *Health* **24,** 115.
Alarie, Y., Ulrich, C. E., Busey, W. B., Swann, H. E., and MacFarland, H. N. (1970). Long-term continuous exposure of guinea pigs to SO_2. *Arch. Environ. Health* **21,** 769–777.
Amdur, M. O. (1969). Toxologic appraisal of particulate matter, oxides of sulfur, and sulfuric acid. *J. Air Pollut. Control Assoc.* **19,** 638.
American Thoracic Society (1978). "Health Effects of Air Pollution." American Lung Association, New York.
Anderson, I., Lundquist, G. R., Jensen, P. L., and Proctor, D. F. (1974). Human response to controlled levels of SO_2. *Arch. Environ. Health* **28,** 31–39.
Aubry, F., Gibbs, G. W., and Becklake, M. R. (1979). Air pollution and health in three urban communities. *Arch. Environ. Health* **34,** 360–368.

Bedi, J. F., Folingsbee, L. J., and Horvath, S. M. (1979). Effect of ozone, sulfur dioxide, heat and humidity on the pulmonary function of young male nonsmokers. *Am. Rev. Respir. Dis.* **119,** 200.

Bouhuys, A., Beck, G. J., and Schoenberg, J. B. (1979). Priorities in prevention of chronic lung disease. *Lung* **156,** 129–148.

Buechley, R. W., Riggan, W. B., Hasselblad, V., and Van Bruggen, J. B. (1973). SO$_2$ levels and perturbations in mortality. A study in New York—New Jersey metropolis. *Arch. Environ. Helath* **27,** 134.

Code of Federal Regulations Title 40 (1978). "Protection of the Environment," Parts 50–59. U.S. Govt. Printing Office, Washington, D.C.

Cohen, A. A., Bromberg, S., Buechley, R. W., Heiderscheit, L. I., and Shy, C. M. (1972). Asthma and air pollution and a coal-fueled power plant. *Am. J. Public Health* **62,** 1181.

Ferin, J., and Leach, L. J. (1973). Effect of SO$_2$ on lung clearance of T$_i$O$_2$ particles in rats. *Am. Ind. Hyg. Assoc. J.* **34,** 260.

Ferris, B. G., Jr. (1978a). Health effects of exposure to low levels of regulated air pollutants. A critical review. *J. Air Pollut. Control Assoc.* **28,** 482–497.

Ferris, B. G., Jr. (1978b). Epidemiology standardization project. *Am. Rev. Respir. Dis.* **118,** Suppl., 1–120.

Ferris, B. G., Jr., Speizer, F. E., Spengler, J. D., Dockery, D., Bishop, M., Wolfson, M., and Humble, C. (1979). Effects of sulfur oxides and respirable particles on human health. *Am. Rev. Respir. Dis.* **120,** 767–779.

Fletscher, C. M., Peto, R., Tinker, C., and Speizer, F. E. (1976). "The Natural History of Chronic Bronchitis and Emphysema." Oxford Univ. Press, London and New York.

Frank, N. R., Anders, M. O., Worcester, J., and Whittenberger, J. L. (1962). Effects of acute controlled exposure to SO$_2$ on respiratory mechanics in healthy male adults. *J. Appl. Physiol.* **17,** 252–258.

Frank, N. R., Yoder, R. E., Brain, J. D., and Yokoyama, E. (1969). SO$_2$ absorption by nose and by mouth under conditions of varying concentrations and flow. *Arch. Environ. Health* **18,** 315–322.

Goldring, I. P. (1970). Pulmonary effects of SO$_2$ exposure in Syrian hamster. *Arch. Environ. Health* **21,** 32.

Green, A. E. S. (1980). "Coal Burning Issues." Univ. of Florida Presses, Gainesville.

Hazucha, M., and Bates, D. V. (1975). Combined effects of ozone and sulfur dioxide on human pulmonary function. *Nature (London)* **257,** 50–51.

Jaeger, M. J., Tribble, D., and Wittig, H. J. (1979). Effect of .5 ppm SO$_r$ on the respiratory function of normal and asthmatic subjects. *Lung* **156,** 119–127.

Kerr, H. D. (1973). Diurnal variation of respiratory function independent of air quality. *Arch. Environ. Health* **26,** 144–155.

Larsen, R. T. (1969). A new mathematical model of air pollution concentration. *J. Air. Pollut. Control Assoc.* **19,** 24.

Lawther, P. J., Waller, R. E., and Henderson, M. (1970). Air pollution and exacerbations of bronchitis. *Thorax* **25,** 525.

Lawther, P. J., MacFarlane, A. J., Waller, R. E., and Brooks, A. G. F. (1975). Pulmonary function and SO$_2$. *Environ. Res.* **10,** 355–367.

Melia, R. J. W., Florey, C., Altmann, D. S., and Swan, A. V. (1977). Association between gas cooking and respiratory disease in children. *Br. Med. J.* **2,** 149–152.

Ministry of Health (1954). "Mortality and Morbidity during the London Fog of 1952," Rep. No. 95. H. M. Stationery Office, London.

Nadel, J. A., Salem, H., Tamplin, B., and Tokiwa, Y. (1965). Mechanism of bronchoconstriction during inhalation of sulfur dioxide. *J. App. Physiol.* **20,** 164–167.

Rall, D. P. (1974). Review of health effects of sulfur oxides. *Environ. Health Perspect.* **8,** 97–121.

Rall, D. P. (1979). Relevance of animal experiments to humans. *Environ. Health Perspect.* **32,** 297–300.

Sackner, M. A., Ford, D., Fernandez, R., Cipley, J., Perez, D., Feinhart, M. Michelson, E., Schreck, R., and Wanner, A. (1978). Effect of sulfuric acid aerosol on cardiopulmonary function of dogs, sheep, and humans. *Am. Rev. Respir. Dis.* **118,** 497–510.

Scott, J. A. (1963). The London Fog of 1962. *Med. Off.* **109,** 250.

Tiffeneau, R., and Pinelli, A. (1948). Régulation bronchique de la ventilation pulmonaire. *J. Fr. Med. Chir. Thorac.* **2,** 221.

Urone, P. (1976). Primary air pollutants, gaseous, their occurrence, sources, and effects. *In* "Air Pollution" (A. C. Stern, ed.), Vol. 1, pp. 23–75. Academic Press, New York.

Urone, P. (1981). *In* "Impact of Increased Coal Used in Florida," pp. 2–1 to 2–29. ICAAS, University of Florida, Gainesville.

4

The Effect of Gaseous Pollutants on Breathing Mechanics and Airway Reactivity

William M. Abraham

I. INTRODUCTION

A variety of noxious substances enter the body through the airways, frequently producing slight changes in normal respiratory function. To

107

AIR POLLUTION
Copyright © 1982 by Academic Press, Inc.
All rights of reproduction in any form reserved.
ISBN 0-12-483880-4

assess the effects exposures to various pollutants have on airways, a variety of measurements of pulmonary function have been employed. Thus far, exposures to pollutants have been shown to cause changes in breathing pattern (Lee *et al.*, 1980), bronchoconstriction (Nadel *et al.*, 1965; Lawther *et al.*, 1975; Sheppard *et al.*, 1980), increases in airway reactivity to nonspecific bronchoconstrictors (Lee *et al.*, 1977; Golden *et al.*, 1978; Holtzman *et al.*, 1979; Abraham *et al.*, 1980a), decreases in tracheal mucous velocity (Abraham *et al.*, 1980b) and pulmonary mucociliary clearance (Phalen *et al.*, 1980; Ferin and Leach, 1973), mucus hypersecretion (Phipps, 1981), and increased susceptibility to respiratory infections (Illing *et al.*, 1980; Fairchild *et al.*, 1972; Henry *et al.*, 1970). Furthermore, it has been suggested that some of these effects may be accentuated in both atopic and/or asthmatic individuals (von Nieding and Wagner, 1979; Orehek *et al.*, 1976), and in animal models of allergic airway disease (Abraham *et al.*, 1980c, 1981a). The discussion in this chapter has been limited to alterations in airway mechanics that occur with exposures to ozone (O_3), nitrogen dioxide(NO_2), and sulfur dioxide (SO_2) and to the mechanisms that might be responsible for these changes in breathing mechanics. Although the literature is reviewed to support the discussion, the review is by no means inclusive.

II. MEASUREMENTS OF BREATHING MECHANICS

Before reviewing the experimental findings and discussing the possible mechanisms involved in the pollutant-induced changes in airway mechanics, it might be useful to discuss briefly the tests and measurements that are used most frequently to study the effects of pollutants on the airways. The tests described are used to estimate changes in the force necessary for the movement of air into and out of the lungs (i.e., resistance). As such, these tests require a knowledge of volume, airflow, and pressure or the change in volume as a function of time. Resistance to gas flow is a function of airway caliber (from Poiseuille's law, resistance α 1/radius^4). Thus, resistance measurements are indirect measurements of airway size and can be used to approximate airway smooth muscle function.

A. Airway Resistance (Dubois *et al.*, 1956b; Comroe *et al.*, 1977)

Airway resistance (R_{aw}) is one of the most commonly measured parameters in humans. The determination of R_{aw} requires the simultaneous measurements of airflow ($\dot{V}$) and the pressure difference

)etween the alveoli (P_A) and the mouth (P_m). From these measurements, R_{aw} is calculated:

$$R_{aw} = \frac{P_m - P_A}{\dot{V}} \tag{1}$$

Airflow and mouth pressure are easily measured by means of a pneumotachograph and by a catheter in the breathing tube, respectively. The pneumotachograph and the subject are contained within a constant-volume body plethysmograph. The body plethysmograph makes use of Boyle's law (relating pressure and volume of gases) to measure alveolar pressure directly during the breathing cycle. The constant volume body plethysmograph is also used to measure thoracic gas volume (V_{tg}). Thoracic gas volume (the amount of gas in the lungs) is an important variable in the determination of R_{aw}, because changes in lung volume affect R_{aw}, so an increase in V_{tg} decreases R_{aw} (Briscoe and DuBois, 1958). The relationship between R_{aw} and V_{tg} is curvilinear, whereas the relationship between the reciprocal of R_{aw}, i.e., the airway conductance (G_{aw}), and V_{tg} approaches linearity. Therefore, to standardize measurements, R_{aw} and G_{aw} are reported in terms of specific resistance, i.e.,

$$SR_{aw} = R_{aw} \times V_{tg} \tag{2}$$

or specific conductance, i.e.,

$$SG_{aw} = 1/R_{aw} \times 1/V_{tg} \tag{3}$$

respectively.

B. Forced Expiratory Volumes (Comroe *et al.*, 1977)

A second method that can be used to estimate changes in resistance to airflow in humans is the measurement of forced expiratory volumes. Of the volumes measured, the forced expiratory volume in 1 second (FEV_1) is probably the one most frequently used to assess airway constriction (bronchoconstriction). For this measurement, the subject is asked to make a maximal inspiration and then exhale as rapidly as possible. The volume change measured during the first second of expiration is used to estimate the resistance to airflow (i.e., the degree of airway narrowing). The FEV_1 measurement is simple, but the deep inspiration prior to the maneuver may interfere with the results of the test, especially in asthmatic individuals (Orehek *et al.*, 1981).

C. Pulmonary Resistance (Comroe *et al.*, 1977)

Pulmonary resistance (R_L) is determined by measuring the pressure differential between the intrapleural space (P_{pl}) and the mouth (P_m), i.e., transpulmonary pressure, simultaneously with airflow and tidal volume.

Pleural pressure can be measured either directly, by introducing catheters, needles, and/or balloons into the pleural space, or indirectly, by means of a balloon catheter placed in the lower third of the esophagus, because pressure changes in the esophagus have been shown to reflect pressure changes in the intrapleural space. Thus, pulmonary resistance (R_L) is defined by

$$R_L = \frac{P_m - P_{pl}}{\dot{V}} \tag{4}$$

Pulmonary resistance (R_L) is generally 20% greater than airway resistance (R_{aw}), because the transpulmonary pressure (i.e., $P_m - P_{pl}$) recorded during airflow overcomes not only resistance to gas flow (R_{aw}) but also the pulmonary tissue resistance and the elastic recoil of the lung.

In practice, the pressure needed to overcome the elastic recoil of the lung is corrected for, so the R_L value given is the resistance to overcome gas flow plus the pulmonary tissue resistance. This correction can be made in one of two ways: (1) the pressure needed to overcome the elastic recoil of the lung (i.e., lung compliance) can be subtracted, electrically, from R_L (Mead and Whittenberger, 1953), or (2) by knowing the pressure–volume curve of the lungs is linear over the tidal-volume range, one can measure the changes in transpulmonary pressure and flow at an isovolume point (Cook *et al.*, 1957), thus canceling the effect of the elastic component.

In animals, the measurement of R_L remains a standard procedure, but in man it has largely been replaced by the measurements of R_{aw} and FEV_1.

III. AIRWAY (BRONCHIAL) REACTIVITY (SEE BOUSHEY *ET AL.*, 1980)

Physical stimuli (cold air and/or exercise), pharmacological agents (methacholine, carbachol), mediators of anaphylaxis (histamine, prostaglandin $F_{2\alpha}$), noxious gases (O_3, NO_2, SO_2), and particulates (sulfates, nitrates) produce bronchoconstriction in some individuals. The severity of the response (i.e., the degree of bronchoconstriction) to these different stimuli indicates the sensitivity of the airways of a particular subject. This sensitivity or irritability of the airways is termed airway reactivity. Airway hyperreactivity is a reduction in the tolerance, or an increase in the degree of bronchospasm, to some of these stimuli. Airway hyperreactivity appears to be a sensitive indicator of existing airway abnormalities. For example, patients with allergic airway disease (Fish *et al.*, 1976) or with chronic bronchitis (Klein and Salvaggio, 1966; Laitinen, 1974) or individuals recovering from acute viral infections (Empey *et al.*, 1976) have been shown to have bronchial hyperreactivity. Because bronchial hyperreactivity is an above-normal sensitivity of the airways to a variety of stimuli caused by some pathological or acquired condition, standardized tests (bronchoprovocation tests) have been designed and used to investigate the onset, potentiation, and pathogenesis of a variety of abnormal airway conditions.

A. Bronchoprovocation Tests (see Wanner, 1980)

A bronchial provocation test is a standardized test used to determine the sensitivity of the airways of a person or an experimental animal to a particular physical, chemical, or pharmacological agent. The most widely used agents for testing airway reactivity are either parasympathomimetic agents, e.g., methacholine and carbachol, or mediators of immediate hypersensitivity reactions, e.g., histamine and prostaglandin $F_{2\alpha}$. These agents are given as aerosols under controlled conditions (see Section III,B,1). Airway reactivity to these agents is determined by plotting changes in airflow resistance against increasing doses of the inhaled bronchoconstricting agent. This results in a dose–response curve (Fig. 1). As indicated previously, the most commonly used measurements are SG_{aw} and FEV_1, in humans, and R_L or SG_L, in animals.

Dose–response curves yield two types of information (Orehek *et al.*, 1977): the reactivity, which is defined as the slope of the curve, and the sensitivity, which is the dose of the bronchoconstrictor required to alter the measured parameter of airway function by a predetermined amount (Fig. 1). Although patients with allergic airway disease differ from normal subjects in terms of bronchial reactivity (i.e., slope of the dose–response curve), most investigators measure airway responsiveness in terms of airway sensitivity, i.e., the dose of provocative agent that results in a predetermined change in airway function. A 20% decrease in FEV_1 or a 35% decrease in SG_{aw} is commonly used as an end point for human studies.

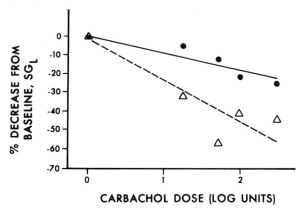

Fig. 1. Dose–response curves with carbachol provocation before (●) and 24 hours after (△) a 2-hour exposure to 0.5 ppm ozone in conscious sheep. Data are expressed as percent decrease from baseline in specific conductance of the lung (SG_L). Linear regression analysis of the data indicated that the preexposure dose–response curve is best described by $y = -10.5x - 2.8$, whereas $y = -18.9x - 6.2$ described the 24-hour test. The differences in the slopes of the two curves suggest that airway reactivity of the sheep has been altered by the ozone exposure. The enhanced response to the largest carbachol dose indicates that airway sensitivity is also increased after ozone exposure. (Data redrawn from Abraham *et al.*, 1980a.)

B. Problems Associated with Bronchoprovocation Tests (Orehek, 1980)

Because the results of bronchoprovocation tests are assessed quantitatively, techniques must be standardized. Particular consideration should be given to (1) the technique of drug administration; (2) the bronchoconstricting drug used, and (3) the method of measuring the response and expressing the results.

1. It is important that the technique used to deliver the bronchoconstricting aerosol be made as uniform as possible. For example, airflow to the nebulizer should be maintained constant to minimize variations in particle size and/or amount of aerosol delivered. Differences in particle size, as determined by characteristics of the nebulizer, are important in determining whether large or small airways are affected. Because most of the resistance to airflow is in the larger airways, the degree of response to any given bronchoconstrictor depends on particle size. Changes in breathing pattern during aerosol inhalation may also affect the retention and distribution of the bronchoconstricting agent, thereby altering results. Controlling ventilation during the delivery of

the bronchoconstricting agent is easily done in experimental animals by means of a ventilator (Abraham *et al.*, 1980a), but is still a problem in human subjects.

2. There are two basic types of bronchoprovocation tests: noncumulative and cumulative. The noncumulative technique requires that after a dose of bronchoconstrictor aerosol, the increase in resistance return to baseline before the next higher dose of drug is delivered. For the cumulative test, the drug effect should remain stable between consecutive doses of bronchoconstrictor.

The decision of which type of test to use is linked to the choice of bronchoconstrictor. For the noncumulative test, the bronchoconstricting agent should be relatively fast acting, e.g., histamine or acetylcholine; the cumulative dose–response curve requires a compound that produces a more stable response, e.g., methacholine or carbachol.

3. As we have previously discussed, dose–response curves generated by bronchoprovocation tests yield two types of information: reactivity (slope of the dose–response curve) and sensitivity (predetermined end point of the test). There is no significant correlation between the airway reactivity and airway sensitivity in either normal subjects or asthmatic patients, thus reinforcing the theory that accurate assessment of bronchoprovocation tests requires both determinations (Orehek *et al.*, 1977).

C. Importance of Bronchoprovocation Tests in Assessing Pollutant-Induced Airway Abnormalities

We have already mentioned that bronchoprovocation tests can be used clinically to detect irritable airways. It is important to discuss experimental evidence suggesting that bronchoprovocation tests are more sensitive than standard pulmonary-function measurements in detecting pollutant-induced airway effects.

Experiments on normal sheep compared the sensitivity of three tests in detecting airway disturbances resulting from a 2-hour exposure to 0.5 ppm O_3 (Abraham *et al.*, 1980a). The tests were (1) airway reactivity to aerosolized carbachol, (2) tracheal mucous velocity, and (3) pulmonary resistance. The order of sensitivity of these tests was airway reactivity > tracheal mucous velocity > pulmonary resistance. Similar results were obtained in normal dogs (Lee *et al.*, 1977) and normal humans (Golden *et al.*, 1978) after exposure to 0.5–0.6 ppm O_3. Thus, it is possible that even in the absence of changes in airflow resistance, airway reactivity can increase.

D. Mechanisms Contributing to Airway Hyperreactivity (Boushey *et al.*, 1980)

Several mechanisms may be responsible for changes in airway reactivity (as defined by bronchoprovocation tests). These mechanisms include (1) decreased airway caliber, (2) increased sensitivity of airway irritant receptors to nonspecific stimuli, (3) imbalance of the normal motor control in the airways, i.e., alterations in the relative contributions of the adrenergic, cholinergic, and purinergic (nonadrenergic inhibitory system) pathways, (4) increased sensitivity of the airway smooth muscle to neurotransmitters or chemical mediators, (5) alteration of the pharmacological receptors in the airway smooth muscle, and (6) increased permeability of the airway mucosa to bronchoconstricting agents, allowing increased penetration of the stimuli to the airway smooth muscle.

It is unlikely that a single mechanism is generally responsible for airway hyperreactivity. Rather, it would appear that different combinations of these mechanisms are involved in the different forms of airway hyperreactivity. Thus, the airway hyperreactivity associated with a pathological condition such as asthma may have contributing mechanisms different from those that are responsible for pollutant-induced airway hyperreactivity. The current literature provides indications of some underlying factors that can influence changes in breathing mechanics and bronchial reactivity as a result of pollutant exposures.

IV. POLLUTANT-INDUCED CHANGES IN BREATHING MECHANICS AND BRONCHIAL REACTIVITY

A. Ozone

Results of studies in anesthetized dogs (Lee *et al.*, 1977), conscious sheep (Abraham *et al.*, 1980a), and man (Golden *et al.*, 1978; Holtzman *et al.*, 1979) suggest that a 2-hour exposure to 0.5–0.6 ppm O_3 results in increased airway reactivity to aerosols of either histamine or carbachol. The airway hyperreactivity was greatest immediately after exposure in the human studies, whereas the greatest increase in both animal studies was observed 24 hours after the O_3 exposure. In all studies, airway hyperreactivity was observed without significant increases in the post-O_3 baseline-resistance values, indicating that an alteration in resting airway caliber was not a contributing factor.

The O_3-induced increases in airway reactivity to aerosolized histamine in both the dog and human studies were prevented by pretreatment with atropine sulfate. Because atropine modified the response, it was suggested that the O_3-induced increase in airway reactivity was caused by stimulation of cholinergic post ganglionic pathways.

Holtzman et al. (1979) compared the airway response to aerosolized histamine after O_3 in both normal and in atopic subjects. In addition, the normal subjects were also provoked with methacholine. It is important to note that histamine and methacholine produce bronchoconstriction via different means. Histamine is thought to cause bronchoconstriction both directly, by stimulating histamine H_1 receptors on the tracheal smooth muscle, and indirectly, by exciting vagal irritant receptors, which results in reflex bronchoconstriction. Although there is disagreement as to the relative contributions of the two mechanisms, the existence of both pathways is accepted. Methacholine (as well as other cholinergic agonists) is thought to act directly on the smooth muscle and to have little, if any, reflex effects.

Holtzman and co-workers (1979) found that O_3 increased airway reactivity to both aerosolized histamine and methacholine in normal subjects and to histamine in atopic subjects. Because O_3 induced airway hyperreactivity in both normal subjects and atopic individuals, this response (at least at the O_3 concentration used, 0.6 ppm) could not have been related to the subjects' atopic histories (Holtzman et al., 1979). It is interesting, however, to note that O_3 exposure increased the airway response to histamine in the nonatopic subjects to 110% and in the atopic individuals by 244%. Although this difference was not significant, it suggests that the atopic individuals were more sensitive to the O_3 stress than those who were nonatopic. Furthermore, Holtzman et al. (1979) reconfirmed that atropine blocked the post-O_3 response to histamine, thus reinforcing the suggestion that O_3-induced changes in postganglionic cholinergic pathways were responsible for the observed increase in airway reactivity. Therefore, it is possible that O_3 influences the muscarinic receptors on the bronchial smooth muscle, possibly by increasing their numbers or their sensitivity to cholinergic agonists.

Although these results support the hypothesis that alterations of cholinergic postganglionic pathways play a role in O_3-induced airway hyperreactivity, they are not inconsistent with O_3-induced changes in airway epithelial permeability or alterations of pharmacologically defined receptors on the airway smooth muscle. Increased airway permeability would allow increased access of inhaled bronchoconstrictors to reach the cholinergic receptors on the airway smooth muscle, thus enhancing the post-O_3 bronchoconstrictor response.

Gordon and Amdur (1980) studied the airway responses of guinea pigs to O_3. These investigators gave subcutaneous injections of histamine (0.125 mg/kg) after the ozone exposure to assess the airway reactivity. When compared to an air (control) group, the guinea pigs exposed to 0.1, 0.2, 0.4, and 0.8 ppm ozone all had significantly greater decreases in dynamic compliance in response to the postexposure histamine challenge. Also, at 0.8 ppm O_3, resistance changes in response to histamine injection were significantly increased. These results are interesting in that they add support to the hypothesis that O_3, as may other pollutants, alters the pharmacological receptors on the airway smooth muscle.

Prolonged Exposures to Ozone

Although many investigators have used airway reactivity to assess the short-term effects of exposure to O_3, changes in forced expiratory parameters (FEV_1, FVC) have been utilized to follow changes in airway mechanics that occur during prolonged exposures to O_3.

Farrell *et al.* (1979) exposed normal human subjects to 0.4 ppm O_3 for 3 hours per day for 5 days. FVC, FEV_1, and SG_{aw} were monitored over the 5-day period. In comparison to exposure to air alone, exposure to O_3 resulted in a significantly decreased FEV_1 on days 1–4, FVC on days 1–3, and SG_{aw} on days 1 and 2. These authors concluded that although exposure to 0.4 ppm O_3 produces changes in airway function, adaptation can take place.

Folinsbee *et al.* (1980) used a slightly different protocol to study prolonged exposures. A 5-day protocol was used with air exposures occurring on days 1 and 5 and exposure to O_3 on days 2 and 4. Three groups were studied: group 1, exposed to 0.20 ppm; group 2, exposed to 0.35 ppm; and group 3, exposed to 0.50 ppm O_3. Group 1 showed no significant effects. In groups 2 and 3, FEV_1 was significantly depressed on days 2 and 3, but the response was abolished on day 5. These results also support the adaptive response.

In an extension of Folinsbee's study, Horvath *et al.* (1981) hypothesized that persons who are unusually sensitive to O_3 would require a longer time to adapt than they would to lose their adaptation. The results indicated that the number of O_3 exposures (12.5 minutes, 0.5 ppm) required to produce adaptation ranged from 2 to 7 and that adaptation in the most sensitive individuals (defined as those who showed a >10% decrease in FEV_1) was variable, but, on the average, adaptation ranged from 7 to 20 days.

Kulle *et al.* (1981) investigated whether bronchial reactivity to methacholine shows adaptation with repeated O_3 exposure. As expected, these workers found that standard pulmonary function measurements showed an adaptive response with repeated daily exposures (3 hours, 0.4

ppm) to O_3. However, these investigators also reported that the same adaptive phenomenon occurred with bronchial reactivity to methacholine.

Adaptation may be one of the reasons some investigators have not always found significant changes in pulmonary mechanics after acute O_3 exposures. Indeed, this idea was proposed by Hackney and co-workers (1977) when they found that in comparison to residents of relatively clean areas of Canada, residents of Southern California (where high ambient concentrations of O_3 are prevalent) had diminished responses to O_3.

Finally, it has been suggested that because vitamin E acts as an antioxidant, it might protect the airways against the effects of O_3: however, Hackney et al. (1981) could not demonstrate significant protection from ozone (2 hours, 0.5 ppm) in a group of young test subjects given supplemental vitamin E.

As yet, the mechanism responsible for the O_3-induced increase in airway resistance is unknown. Dixon and Mountain (1965) reported that exposure to O_3 resulted in endogenous histamine release. Quite possibly, as is apparent for NO_2 (see below), O_3-stimulated mediator releases could be a factor in O_3-induced bronchoconstriction. Similarly, mediator depletion (Abraham et al., 1981b) or other biochemical adaptations (Mustafa and Tierney, 1978) that might occur with exposure might play a role in the adaptive response observed with long-term exposure.

B. Nitrogen Dioxide

As with O_3, pulmonary function after NO_2 exposure has been evaluated in terms of pulmonary mechanics and airway reactivity.

Folinsbee et al. (1978) exposed young, healthy, exercising males to 0.62 ppm NO_2, but found no consistent changes in pulmonary function. Hackney and co-workers (1978) exposed humans to 1 ppm NO_2 for 2 hours on 2 successive days. These investigators reported that the only demonstrable change was a minor decrease in forced vital capacity after the second exposure. Von Nieding and associates (1973) found significant increases in airway resistance in healthy subjects after exposures to NO_2. The increased resistance could be blocked by pretreatment with antihistamine but not by a sympathomimetic bronchodilator (von Nieding and Krekeler, 1971). Similar results were obtained in patients suffering from chronic nonspecific lung disease (von Nieding and Wagner, 1979). In this study, inhalation of as little as 1.5 ppm NO_2 for 5 minutes resulted in a significant increase in airway resistance. Lower concentrations had no significant effect. Again, antihistamine was effective in preventing the NO_2-induced increase in airway resistance, whereas atropine and oreiprenaline were ineffective. These results suggest that NO_2 increases airway resistance by releasing histamine.

NO_2 also increases bronchial reactivity. Biel and Ulmer (1976) exposed humans to NO_2 concentrations as high as 7.5 ppm. Mild bronchoconstriction occurred in subjects breathing 2.5 ppm NO_2 for 14 hours, whereas bronchial hyperreactivity to aerosolized acetylcholine was seen after the subjects breathed 5 ppm NO_2 for 14 hours and 7.5 ppm for 3 hours.

In sheep, exposure to 7.5 ppm NO_2 for 4 hours produced bronchial hyperreactivity in 5 of 10 animals (Abraham et al., 1980b). Exposure to 15 ppm NO_2 for 4 hours increased pulmonary resistance in 7 of 10 animals immediately after exposure, whereas two of the three that did not bronchoconstrict had increased airway responsiveness to aerosolized carbachol.

Perhaps the most provocative study was performed by Orehek and co-workers (1976). These investigators studied the effects of a 1-hour exposure to 0.1 ppm NO_2 on airway function and bronchial reactivity in "mild asthmatic" subjects. In 13 of 20 subjects, the exposure to NO_2 resulted in slight but significant increases in airway resistance and bronchial reactivity to aerosolized carbachol, whereas 7 subjects showed no change.

This study is of great importance because it suggested that individuals with abnormal airway function (in this instance, the subjects were asthmatic) may be responsive to levels of pollutants that do not readily affect those with normal airway function. The background for such a theory arose from epidemiological data that suggest that patients with airway diseases are more severely affected by pollutant episodes. Because this question is of considerable importance, the most recent work in terms of pollutant effects on the airways has been designed to determine whether individuals with allergic airway disease (Holtzman et al., 1979; Sheppard et al., 1980) or other airway disturbances (von Nieding and Wagner, 1979) are more severely influenced by pollutant insults than are normal persons.

C. Sulfur Dioxide

Although a whole chapter has been devoted to SO_2, a few brief comments are in order. SO_2 can cause bronchoconstriction. Nadel and associates (1965) suggested that the initial bronchoconstriction is reflex in nature because they were able to prevent the bronchoconstrictor effect with atropine pretreatment. Lawther et al. (1975) pointed out that this bronchoconstriction may also be abolished with time. These two results are not inconsistent, because it is possible that irritant receptors, which are thought to be stimulated initially by SO_2, adapt with time.

Sheppard and co-workers (1980) compared the airway response to 10 minutes of SO_2 breathing in seven asthmatic, seven atopic, and seven normal subjects. Furthermore, they assessed the significance of the parasympathetic pathway in this response by repeating the exposure after the subjects were pretreated with atropine sulfate. The results showed that SO_2 at 1, 3, and 5 ppm significantly increased SR_{aw} in the asthmatic individuals whereas the atopic and nonatopic groups responded only to 5 ppm SO_2. Atropine pretreatment abolished the response, suggesting involvement of the cholinergic pathways.

In agreement with human studies with NO_2 (Orehek et al., 1976) and SO_2 (Sheppard et al., 1980), we find that sheep previously sensitized to *Ascaris suum* antigen (allergic sheep) are more sensitive to SO_2 than their normal counterparts (Abraham et al., 1980c, 1981a). Allergic sheep have antigen-induced airway responses similar to those occurring in patients with allergic airway disease (Wanner et al., 1979). Thus, airway challenge with *Ascaris suum* antigen results in bronchoconstriction, hyperinflation,

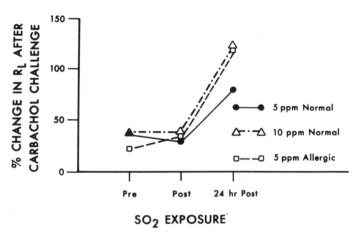

Fig. 2. Percentage change in pulmonary resistance (R_L) to a standard 0.25% carbachol challenge in normal and allergic sheep before and after a 4-hour exposure to SO_2. Immediately after exposure, the airway sensitivity to carbachol inhalation (when compared to the preexposure response) was unchanged in any group. However, 24 hours after exposure, allergic sheep exposed to 5 ppm SO_2 showed a significant enhancement in airway response to carbachol. The change was also significantly different from the change observed in the normal sheep exposed to 5 ppm SO_2. Airway hyperreactivity to carbachol was observed when normal sheep were exposed to 10 ppm SO_2. (Data redrawn from Abraham et al., 1980c, 1981a.)

and arterial hypoxemia (Wanner *et al.*, 1979). Exposure of both normal and allergic sheep to 5 ppm SO_2 resulted in enhancement of airway reactivity in only the allergic group (Fig. 2). Significant enhancement of airway reactivity in normal sheep occurred only after exposure to 10 ppm SO_2 (Abraham *et al.*, 1980c).

V. THE ROLE OF AIRWAY DAMAGE IN POLLUTANT-INDUCED AIRWAY RESPONSES

Two trends traverse the literature on gaseous pollutants: (a) that exposures to pollutants can produce bronchoconstriction and airway hyperreactivity in normal humans and animals, and (2) that persons with airway disease (Orehek *et al.*, 1976; von Nieding and Wagner, 1979; Sheppard *et al.*, 1980) as well as animals with allergic airway disease (Abraham *et al.*, 1980c, 1981a; Osebold *et al.*, 1980; Matsumura, 1970) are more severely affected by exposures to pollutants than are their normal counterparts. As indicated previously (Section III,D), there may be many factors responsible for these trends; however, a case can be made for one factor being common to all conditions, i.e., pollutant-induced airway damage.

That airway damage plays a role in the pollutant-induced effects on airways observed in both animals and man is supported by the following evidence.

Pollutant insults have been shown to increase airway mucosal permeability to a variety of molecular weight compounds (Matsumura, 1971; Davis *et al.*, 1980) as well as to enhance intracellular transport of substances (Vai *et al.*, 1980) in normal animals. The changes in permeability have been reported to be transient (Davis *et al.*, 1980) whereas the alterations in intracellular transport may remain for extended periods (Vai *et al.*, 1980). Thus, it is possible that these pollutant-induced morphological changes are, in part, responsible for both increased reflex bronchoconstriction, by exposing and/or sensitizing vagal irritant receptors, and airway hyperreactivity to inhaled bronchoconstrictors, by allowing greater amounts of the bronchoconstrictor agent to reach the target organ (i.e., smooth muscle). In addition to the aforementioned studies, experimental evidence (Fig. 3) from our laboratory supports this theory of increased permeability.

The enhanced response of patients and animals with airway abnormalities to pollutants is also consistent with this idea of epithelial damage. Cutz and associates (1978) have described changes in the airway mucosa of asthmatic patients that are similar to the morphological

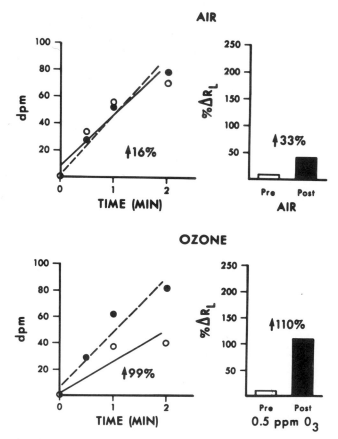

Fig. 3.   Timed appearance of [³H]histamine in plasma (from arterial blood) during a controlled inhalation challenge with 1% histamine containing 200 μCi/ml of [³H]histamine. The aerosol (mass median aerodynamic diameter 4.3 μm, geometric standard deviation 2.4) was delivered for 2 minutes at a rate of 20 breaths/minute with a tidal volume of 500 ml. [³H]histamine levels expressed as disintegrations per minute (dpm). Values are corrected for background and standardized per ml of aerosol delivered. The rate of appearance of the [³H]histamine in the plasma was used to estimate airway permeability. Note that airway permeability to [³H]histamine increases by 16% 1 day after a 2-hour exposure to air (●) when compared to the preexposure (○) rate. This increase in [³H]histamine appearance was associated with a 33% increase in the change in pulmonary resistance (%ΔR_L) produced by histamine inhalation. These results are compared to those observed in the same animal before (○) and 1 day after (●) a 2-hour exposure to 0.5 ppm O_3. After O_3 exposure there is a 99% increase in the rate of appearance of [³H]histamine. At the same time, the increase in airway response to histamine inhalation is 110%. These results suggest a possible association between the rate of appearance of [³H]histamine, i.e., permeability of the airway to [³H]histamine, and the airway bronchoconstrictor response. (From Abraham *et al.*, 1981c.)

changes that have occurred with pollutant insults, which suggests that the airways of asthmatic patients are, in terms of morphology, already compromised. In addition, antigen-induced bronchoconstriction in allergic monkeys has been reported to be linked with increased airway mucosal permeability to inhaled [^{3}H]histamine (Boucher *et al.*, 1979). If this were also the case in patients with allergic airway disease, then allergic episodes could render the airways of asthmatic individuals more susceptible to subsequent pollutant-induced airway injury. Thus, the more severe the injury, the more severe the response. In this regard, it is important to note that the pollutant-induced effects on the airways of atopic individuals who have no history of asthma do not appear to differ significantly from those of normal subjects (Holtzman *et al.*, 1979; Sheppard *et al.*, 1980).

VI. CONCLUSIONS

It is well documented that exposures to pollutant atmospheres can produce abnormalities in airway mechanics; however, neither the mechanisms responsible for these abnormalities nor the long-range implications of such insults are known. These are obviously important considerations for future studies. Furthermore, it is important to note that in this chapter we have considered only the effects of gaseous pollutants on airway mechanics. As indicated in Section I, there are many other airway functions altered by pollutant insults. Neither the interactions nor the long-range effects of these airway disturbances have been considered. Hopefully, future research will provide answers to such questions.

ACKNOWLEDGMENT

This work was supported in part by NIEHS Grant 02668.

REFERENCES

Abraham, W. M., Januszkiewicz, A. J., Mingle, M., Welker, M., Wanner, A., and Sackner, M. A. (1980a). The sensitivity of bronchoprovocation and tracheal mucous velocity in detecting airway responses to O$_3$. *J. Appl. Physiol.* **48,** 789–793.
Abraham, W. M., Welker, M., Oliver, W., Jr., Mingle, M., Januszkiewicz, A. J., Wanner, A., and Sackner, M. A. (1980b). Cardiopulmonary effects of short-term nitrogen dioxide in conscious sheep. *Environ. Res.* **22,** 61–72.

Abraham, W. M., Oliver, W., Jr., Welker, M. J., King, M. M., Chapman, G. A., Yerger, L., Maurer, D. R., Sielczak, M., Wanner, A., and Sackner, M. A. (1980c). Sulfur dioxide induced airway hyperreactivity in allergic sheep. *Am. J. Ind. Med.* **1**, 383–390.

Abraham, W. M., Oliver, W., Jr., Welker, M. J., King, M. M., Wanner, A., and Sackner, M. A. (1981a). Differences in airway reactivity in normal and allergic sheep after exposure to sulfur dioxide. *J. App. Physiol.* **51**, 1651–1656.

Abraham, W. M., Yerger, L., Oliver, W., Jr., King, M. M., Chapman, G. A., and Wanner, A. (1981b). The effect of ozone on antigen induced bronchospasm in allergic sheep. *Am. Rev. Respir. Dis.* **123**(II), 152.

Abraham, W. M., Yerger, L., Delehunt, J. C., and Oliver, W., Jr. (1981c). Association between airway reactivity and airway permeability to inhaled ^{3}H-histamine in conscious sheep. *Physiologist* **24**, 29 (abstr.).

Biel, M., and Ulmer, W. T. (1976). Wirkung von NO_2 in Mak-Bereich auf Atemmechanic und bronchiole Acetylcholinempfindlichkeit bei normalen Personnen. *Int. Arch. Occup. Environ. Health* **38**, 31–44.

Boucher, R. C., Pare, P. D., and Hogg, J. C. (1979). Relationship between airway hyperreactivity and hyperpermeability in *Ascaris* sensitive monkeys. *J. Allergy Clin. Immunol.* **64**, 197–201.

Boushey, H. A., Holtzman, M. J., Sheller, J. R., and Nadel, J. A. (1980). Bronchial hyperreactivity. *Am. Rev. Respir. Dis.* **121**, 380–413.

Briscoe, W. A., and DuBois, A. B. (1958). Relationship between airway resistance, airway conductance and lung volume in subjects of different age and body size. *J. Clin. Invest.* **37**, 1279–1285.

Comroe, J. H., Jr., Forster, R. E., DuBois, A. B., Briscoe, W. A., and Carlsen, E. (1977). Mechanics of breathing. *In* "The Lung—Clinical Physiology and Pulmonary Function Tests," pp. 162–204. Year Book Med. Publ., Chicago, Illinois.

Cook, C. D., Sega, S., Sutherland, J. M., Cherry, R. B., Mead, J., McIlrod, M. B., and Smith, C. A. (1957). Studies of respiratory physiology in the newborn infant. III. Measurement of mechanics of respiration. *J. Clin. Invest.* **36**, 440–448.

Cutz, E., Levison, H., and Cooper, D. M. (1978). Ultrastructure of airways in children with asthma. *Histopathology* **2**, 407–421.

Davis, J. D., Gallo, J., Hu, E. P. C., Boucher, R. C., and Bromberg, P. A. (1980). The effects of ozone on respiratory epithelial permeability. *Am. Rev. Respir. Dis.* **121**(II), 231 (abstr.).

Dixon, J. R., and Mountain, J. T. Role of histamine and related substances in development of tolerance to edemagenic gases. *Toxicol. Appl. Physiol. Pharmacol.* **7**, 756–766.

DuBois, A. B., Botelho, S. Y., Bedell, G. N., Marshall, R., and Comroe, J. H., Jr. (1956a). A rapid plethysmographic method for measuring thoracic gas volume: A comparison with a nitrogen washout method for measuring functional residual capacity in normal subjects. *J. Clin. Invest.* **35**, 322–326.

DuBois, A. B., Botelho, S. Y., and Comroe, J. H., Jr. (1956b). A new method for measuring airway resistance in man using a body plethysmograph: Values in normal subjects and in patients with respiratory disease. *J. Clin. Invest.* **35**, 327–335.

Empey, D. W., Laitinen, L. A., Jacobs, L., Gold, W. M., and Nadel, J. A. (1976). Mechanisms of bronchial hyperreactivity in normal subjects after respiratory tract infections. *Am. Rev. Respir. Dis.* **113**, 131–139.

Fairchild, G. A., Roan, J., and McCarrol, J. (1972). Atmospheric pollutants and the pathogenesis of viral respiratory infection: Sulfur dioxide and influenza infection in mice. *Arch. Environ. Health* **25**, 174–182.

Farrell, B. P., Kerr, H. D., Kulle, T. J., Sauder, L. R., and Young, J. L. (1979). Adaptation in human subjects to the effects of inhaled ozone after repeated exposure. *Am. Rev. Respir. Dis.* **119,** 725–730.

Ferin, J., and Leach, L. J. (1973). The effect of SO_2 on lung clearance of TiO_2 particles in rats. *Am. Ind. Hyg. Assoc. J.* **34,** 260–263.

Fish, J. E., Rosenthal, R. R., Batra, G., Menkes, H., Summer, W., Permutt, S., and Norman, P. (1976). Airway responses to methacholine in allergic and non-allergic subjects. *Am. Rev. Respir. Dis.* **113,** 579–586.

Folinsbee, L. J., Horvath, S. M., Bedi, J. F., and Delehunt, J. D. (1978). Effect of 0.62 ppm NO_2 on cardiopulmonary function in young nonsmokers. *Environ. Res.* **15,** 199–205.

Folinsbee, L. J., Bedi, J. F., and Horvath, S. M. (1980). Respiratory responses in humans repeatedly exposed to low concentrations of ozone. *Am. Rev. Respir. Dis.* **121,** 431–439.

Golden, J. A., Nadel, J. A., and Boushey, H. A. (1978). Bronchial hyperirritability in healthy subjects after exposure to ozone. *Am. Rev. Respir. Dis.* **118,** 287–294.

Gordon, T., and Amdur, M. O. (1980). Effect of ozone on respiratory response of guinea pigs to histamine. *J. Toxicol. Environ. Health* **6,** 185–195.

Hackney, J. D., Linn, W. S., Karuza, S. K., Buckley, R. D., Law, D. C., Bates, D. R., Hazucha, M., Pengelly, L. D., and Silverman, F. (1977). Effects of ozone exposure in Canadians and Southern Californians: Evidence for adaptation? *Arch. Environ. Health* **32,** 110–115.

Hackney, J. D., Thiede, F. C., Linn, W. S., Pedersen, E. E., Spier, C. E., Law, D. C., and Fischer, D. A. (1978). Experimental studies on human health effects of air pollutants. IV. Short-term physiological and clinical effects of nitrogen dioxide exposure. *Arch. Environ. Health* **33,** 176–181.

Hackney, J. D., Linn, W. S., Buckley, R. D., Jones, M. P., Wightman, L. H., Karuza, S. K., Blessey, R. L., and Hislop, H. J. (1981). Vitamine E supplementation and respiratory effects of ozone in humans. *J. Toxicol. Environ. Health* **7,** 383–390.

Henry, M. C., Findlay, J., Spangler, J., and Ehrlich, R. (1970). Chronic toxicity of NO_2 in squirrel monkeys. III. Effects on resistance to viral and bacteria infection. *Arch. Environ. Health* **20,** 566–570.

Holtzman, M. J., Cunningham, J. H., Sheller, J. R., Irsigler, G. B., Nadel, J. A., and Boushey, H. A. (1979). Effect of ozone on bronchial reactivity in atopic and non-atopic subjects. *Am. Rev. Respir. Dis.* **129,** 1059–1067.

Horvath, S. M., Gliner, J. A., and Folinsbee, L. J. (1981). Adaptation to ozone: Duration and effect. *Am. Rev. Respir. Dis.* **123,** 496–499.

Illing, J. W., Miller, F. J., and Gardner, D. E. (1980). Decreased resistance to infection in exercised mice exposed to NO_2 and O_3. *J. Toxicol. Environ. Health* **6,** 843–851.

Klein, R. C., and Salvaggio, J. E. (1966). Nonspecificity of the bronchoconstricting effect of histamine and acetyl-beta-methylcholine in patients with obstructive airway disease. *J. Allergy* **37,** 158–168.

Kulle, T. J., Kerr, H. D., Farrell, B. P., Sander, L. R., Bremel, M. S., and Smith, D. M. (1981). Adaptation to ozone on pulmonary function and bronchial reactivity in human subjects. *Am. Rev. Respir. Dis.* **123**(II), 127 (abstr.).

Laitinen, L. A. (1974). Histamine and methacholine challenge in the testing of bronchial reactivity. *Scand. J. Respir. Dis., Suppl.* **86,** 1–47.

Lawther, P. H., MacFarlane, A. J., Waller, R. E., and Brooks, A. G. F. (1975). Pulmonary function and sulfur dioxide: Some preliminary findings. *Environ. Res.* **10,** 355–367.

Lee, L.-Y., Bleecker, E. R., and Nadel, J. A. (1977). Effect of ozone on bronchomotor response to inhaled histamine aerosol in dogs. *J. Appl. Physiol.* **43,** 626–631.

Lee, L.-Y, Djokic, T. D., Dumont, C., Graf, P. D., and Nadel, J. A. (1980). Mechanism of ozone-induced tachypenic response to hypoxia and hypercapnia in conscious dogs. *J. Appl. Physiol.* **48,** 163–168.

Matsumura, Y. (1970). The effect of ozone, nitrogen dioxide and sulfur dioxide on the experimentally induced allergic respiratory disorder in guinea pigs. III. The effect on the occurrence of dyspneic attack. *Am. Rev. Respir. Dis.* **102,** 444–447.

Matsumura, Y. (1971). The effect of ozone, nitrogen dioxide and sulfur dioxide on the experimentally induced allergic disorder in guinea pigs. II. The effect of ozone on the absorption and retention of antigen in the lung. *Am. Rev. Respir. Dis.* **102,** 438–443.

Mead, J., and Whittenberger, J. L. (1953). Physical properties of human lungs measured during spontaneous respiration. *J. Appl. Physiol.* **5,** 779–796.

Mustafa, M., and Tierney, D. F. (1978). Biochemical and metabolic changes in the lung with oxygen, ozone and nitrogen dioxide toxicity. *Am. Rev. Respir. Dis.* **118,** 1061–1090.

Nadel, J. A., Salem, H., Tamplin, B., and Tokiwa, Y. (1965). Mechanism of bronchoconstriction during inhalation of sulfur dioxide. *J. Appl. Physiol.* **20,** 164–167.

Orehek, J. (1980). Airway response to inhaled bronchoconstrictor drugs: Clinical measurement and standardization. *In* "Airway Reactivity: Mechanism and Clinical Relevance" (F. E. Hargreave, ed.), pp. 201–215. Astra Pharmaceutical Canada, Ltd., Ontario.

Orehek, J., Massari, J. P., Gayrard, P., Grimaud, C., and Charpin, J. (1976). Effect of short-term, low level nitrogen dioxide exposure on the bronchial sensitivity of asthmatic patients. *J. Clin. Invest.* **57,** 301–307.

Orehek, J., Gayrard, P., Grimaud, C., and Charpin, J. (1977). Airway response to carbachol in normal and asthmatic subjects. Distinction between bronchial sensitivity and reactivity. *Am. Rev. Respir. Dis.* **115,** 937–943.

Orehek, J., Nicoli, M. M., Delpierre, S., and Beaupré, A. (1981). Influence of the previous deep inspiration on the spirometric measurement of provoked bronchoconstriction in asthma. *Am. Rev. Respir. Dis.* **123,** 269–272.

Osebold, J. W., Gershwin, L. J., and Zee, Y. C. (1980). Studies on the enhancement of allergic lung sensitization by inhalation of ozone and sulfuric acid aerosol. *J. Environ. Pathol. Toxicol.* **3,** 221–234.

Phalen, R. F., Kenoyer, J. L., Crocker, T. T., and McClure, T. R. (1980). Effects of sulfate aerosols in combination with ozone on elimination of tracer particles inhaled by rats. *J. Toxicol. Environ. Health* **6,** 797–810.

Phipps, R. J. (1981). The airway mucociliary system. "Respiration Physiology" (J. G. Widdicombe, ed.), Vol. III, No. 23, pp. 240–243. Univ. Park Press, Baltimore, Maryland.

Sheppard, D., Wong, W. S., Uehara, C. F., Nadel, J. A., and Boushey, H. A. (1980). Lower threshold and greater bronchomotor responsiveness of asthmatic subjects to sulfur dioxide. *Am. Rev. Respir. Dis.* **122,** 873–878.

Vai, F., Fournier, M. F., Lafuma, J. C., Touaty, E., and Pariente, R. (1980). SO$_2$-induced bronchopathy in the rat: Abnormal permeability of the bronchial epithelium *in vivo* and *in vitro* after anatomic recovery. *Am. Rev. Respir. Dis.* **121,** 851–858.

von Nieding, G., and Krekeler, H. (1971). Pharmakologische Beeinflussung der akuten NO$_2$-wirkung auf die Lugenfunktion von gesunden und kranken mit einer chronischen Bronchitis. *Int. Arch. Arbeitsmed.* **29,** 55–63.

von Nieding, G., and Wagner, H. M. (1979). Effects of NO_2 on chronic bronchitics. *Environ. Health Perspect.* **29,** 137–142.

von Nieding, G., Krekeler, H., Fuchs, R., Wagner, M., and Koppenhagen, K., (1973). Studies of the effects of NO_2 on lung function influence on diffusion, perfusion and ventilation in the lungs. *Int. Arch. Arbietsmed.* **31,** 61.

Wanner, A. (1980). Interpretation of pulmonary function tests. *In* "Diagnostic Techniques in Pulmonary Disease" (M. A. Sackner, ed.), Vol. 1, pp. 353–360. Dekker, New York.

Wanner, A., Mezey, R. J., Reinhart, M. E., and Eyre, P. E. (1979). Antigen-induced bronchospasm in conscious sheep. *J. Appl. Physiol.* **47,** 917–922.

5

Carbon Monoxide Toxicity

Guillermo Gutierrez

I. INTRODUCTION

For many years, the accepted theory of carbon monoxide (CO) toxicity stated that hemoglobin binds preferentially to CO, reducing the amount of oxygen (O_2) that this molecule can deliver to the tissues. Recently, other mechanisms of toxicity have been proposed, including the direct effect of the dissolved CO in blood on the cellular respiratory apparatus.

II. O_2 TRANSPORT IN BLOOD

Scientific interest in the physiological effect of CO began in the late 1880s with the discovery that the transport of O_2 from the lungs to the tissues took place with the aid of hemoglobin, a very complex molecule

127

AIR POLLUTION
Copyright © 1982 by Academic Press, Inc.
All rights of reproduction in any form reserved.
ISBN 0-12-483880-4

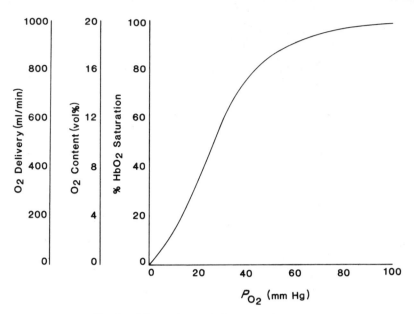

Fig. 1. The oxyhemoglobin saturation curve.

found inside the red blood cells. This molecule has the remarkable property of binding to O_2 in the lungs, where the concentration of O_2 is high, and releasing it in the tissues, where cellular consumption keeps the concentration of O_2 low. This transport process takes place with the loss in transit of only a small percentage of the bound O_2. How hemoglobin accomplishes this is explained by Fig. 1, which shows the O_2–hemoglobin dissociation curve for humans (Comroe *et al.*, 1962; Comroe, 1965). In this figure, the ordinate represents the percentage of hemoglobin bound to O_2, and the abscissa represents the partial pressure (P) of O_2 dissolved in blood. In the lung capillaries, P_{O_2} in blood is approximately 100 mm Hg, which corresponds to a hemoglobin saturation of 98%. This saturation does not change appreciably as the O_2 pressure is lowered, until the steep portion of the curve is reached. This lies in the 30–60 mm Hg P_{O_2} region and happens to coincide with the range of P_{O_2} found in the tissues. The saturation of the hemoglobin molecule may drop from 90 to 60% during the release of O_2 to the tissues; it then returns to the lungs through the venous system to continue the cycle.

The amount of O_2 carried by hemoglobin in blood is a function of its concentration, expressed in grams per 100 ml blood, the percent saturation, and the maximum capacity for carrying O_2. The latter can be obtained by noting that four molecules of O_2 bind in a sequential fashion to one molecule of hemoglobin. This sequential binding process is responsible for the characteristic sigmoid shape of the dissociation curve (Adair, 1925; Arturson, 1974). Because the molecular weight of hemoglobin is 64,800 (Antonini *et al.*, 1971), the amount of O_2 carried by fully saturated hemoglobin can be calculated as:

$$4 \times 22{,}400/64{,}800 = 1.38 \text{ ml (STPD)/g}$$

The O_2-carrying capacity of blood is:

$$Q_{O_2} = 1.38 \times (\% \text{ saturation}) \times (\text{hemoglobin concentration} \atop \text{in g/100 ml blood}) \qquad (1)$$

For an average hemoglobin concentration of 15 g/100 ml with a saturation of 97%, as seen in arterial blood, the amount of O_2 transported by hemoglobin in 100 ml of blood is:

$$Q_{O_2} \text{ arterial} = 1.38 \times 0.97 \times 15 = 20 \text{ ml/100 ml (or vol \%)}$$

The total amount of O_2 delivered by hemoglobin to the tissues is the product of the cardiac output (C.O.) in ml/minute and the O_2-carrying capacity:

$$O_2 \text{ delivered to the tissues} = \frac{Q_{O_2} \times \text{C.O.}}{100} \text{ (ml } O_2\text{/minute)} \qquad (2)$$

For an average C.O. of 5000 ml/min this quantity becomes:

$$O_2 \text{ delivery} = \frac{20 \times 5000}{100} = 1000 \text{ ml/minute}$$

Assuming a basal metabolic state with a CO_2 production of 200 ml/minute and a respiratory quotient (R.Q.) of 0.8, the amount of O_2 extracted by the tissues is:

$$O_2 \text{ extraction} = CO_2 \text{ production/R.Q.} \qquad (3)$$

$$O_2 \text{ extraction} = 200/0.8 = 250 \text{ ml/minute}$$

Subtracting the O_2 extracted by the tissues from the amount delivered by arterial blood and dividing by the cardiac output yields the O_2 content of venous blood.

$$Q_{O_2} \text{ venous} = \frac{1000 - 250}{5000} \ 100 = 15 \text{ vol \%}$$

This corresponds to a O_2 of 40 mm Hg and a hemoglobin saturation of approximately 72%. Venous blood is then carried to the lungs and is again oxygenated in the pulmonary capillaries, where the cycle begins anew.

In addition to that bound to hemoglobin, a small quantity of O_2 can be found dissolved in blood:

$$\text{Dissolved } O_2 = 0.003 \times P_{O_2} \text{ (mm Hg) in ml/100 ml} \qquad (4)$$

Under physiological conditions, only 0.3 vol % O_2 are transported to the tissues in this fashion; however, this transport mode can become significant when hyperbaric oxygenation is used.

III. CLASSICAL THEORY OF CO TOXICITY

It was J. S. Haldane who, in 1895, first attempted to study the effect of CO in humans by exposing himself to a known concentration of CO and recording his reaction (Haldane, 1895a,b). He found the effects to be similar to those produced by the lack of O_2. Then in cooperation with Lorrain-Smith (Haldane *et al.*, 1897) he postulated what has become the keystone of our present understanding of the interaction among O_2, hemoglobin, and CO in blood, i.e., Haldane's principle:

$$\frac{[HbCO]}{[HbO_2]} = \frac{MP_{CO}}{P_{O_2}} \qquad (5)$$

where [HbCO] = blood concentration of carboxyhemoglobin; [HbO$_2$] = blood concentration of oxyhemoglobin; P_{CO} = partial pressure of CO in blood; P_{O_2} = partial pressure of O_2 in blood; M = Haldane constant. This principle was derived by noting that the dissociation curves of oxy- and carboxyhemoglobin are almost identical when the P_{CO} scale is multiplied by a factor M, which is generally accepted to be in the range of 190–250. Thus, the affinity of hemoglobin for CO is M times greater than for O_2. As a consequence of this increased affinity, small amounts of CO can saturate a large number of hemoglobin molecules, effectively reducing the O_2-carrying capacity of the blood (Campbell, 1929; Roughton, 1945; Roughton *et al.*, 1957, 1958). This was the accepted theory of CO toxicity until it was noted that people whose total amount of available

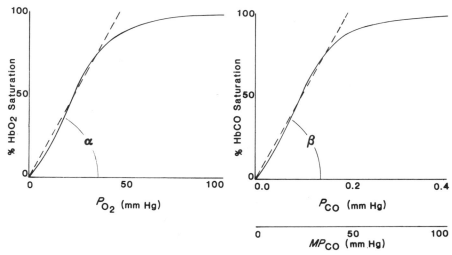

Fig. 2. The oxy- and carboxyhemoglobin saturation curves and their relationship to Haldane's principle.

hemoglobin has been similarly reduced by anemia rather than CO intoxication did not manifest any of the hypoxic symptoms of CO poisoning (Haldane, 1927).

This finding led Douglas and Haldane in 1912 to postulate what has been called the "Haldane effect" (Douglas *et al.*, 1912). They claimed that CO not only reduced the total O_2-carrying capacity of hemoglobin, but also shifted the oxyhemoglobin curve toward the left, thereby placing the steep or unloading part of the curve in a region where P_{O_2} was below that usually found in the tissues. This results in the available hemoglobin binding to the O_2 in the lung capillaries but not releasing it to the tissues (Forster, 1970; Joels *et al.*, 1958; Ledwith, 1978; Sharma, 1976).

Figure 2 shows the dissociation curves for O_2 and CO. Linearizing the middle portions and multiplying the P_{CO} scale by Haldane constant M produces similar slopes for both curves: slope of Hb–P_{O_2} curve = slope of Hb–MP_{CO}

$$\tan \alpha = \tan \beta$$

This relationship yields Haldane's equation:

$$\frac{HbCO}{MP_{CO}} = \frac{HbO_2}{P_{O_2}}$$

This principle was postulated to be valid only when the amount of reduced or unbound hemoglobin present is negligible (Haldane and Priestley, 1935; Haldane, 1912). This statement provides a second equation that must accompany Haldane's principle:

$$HbCO + HbO_2 = Hb_{total} \tag{6}$$

In other words, the amount of hemoglobin bound to O_2 plus that bound to CO must equal the total amount of hemoglobin available in blood. Dividing equation (6) by Hb_{total} results in:

$$\% \ HbCO + \% \ HbO_2 = 100\% \tag{7}$$

The above condition is quite restrictive and, in fact, was ignored by Roughton (Roughton and Darling, 1944; Roughton, 1964, 1970) who assumed that Haldane's principle described the partition of hemoglobin between O_2 and CO in the presence of reduced hemoglobin. He assumed that the amount of reduced hemoglobin at a given P_{O_2} and P_{CO} is the same as if the P_{O_2} were set equal to $P_{O_2} + MP_{CO}$ in the absence of CO. Then, the amount of reduced hemoglobin can be directly read from the normal oxyhemoglobin dissociation curve. As an example, let us find the P_{O_2} in equilibrium with blood containing 20% HbCO and 30% HbO_2. The total amount of bound hemoglobin is:

$$HbCO + HbO_2 = 20 + 30 = 50\%$$

From the standard oxyhemoglobin dissociation curve, this corresponds to a P_{O_2} of 26. Then

$$P_{O_2} + MP_{CO} = 26 = P_{O_2} \left(1 + \frac{MP_{CO}}{P_{O_2}} \right) \tag{8}$$

From Haldane's principle

$$\frac{MP_{CO}}{P_{O_2}} = \frac{HbCO}{HbO_2} = \frac{20}{30} \tag{9}$$

Substituting into Eq. (8)

$$P_{O_2}(1 + 20/30) = 26 \tag{10}$$

or

$$P_{O_2} = 15.6 \text{ mm Ha}$$

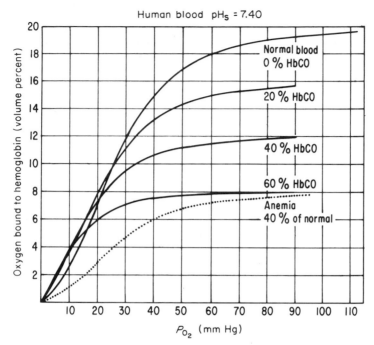

Human blood pH$_s$ = 7.40

Oxygen bound to hemoglobin (volume percent)

Normal blood
0 % HbCO

20 % HbCO

40 % HbCO

60 % HbCO

Anemia
40 % of normal

P_{O_2} (mm Hg)

Fig. 3. Roughton's experimental results based on Haldane's principle.

By comparing the above result with the P_{O_2} of 18 mm Hg corresponding to an HbO$_2$ of 30 in the absence of carbon monoxide, it can be seen that the curve does shift to the left. These equations can be used to predict the change in the oxyhemoglobin curve under different levels of carboxyhemoglobin. Roughton's results, which he verified experimentally (Roughton, 1945), are shown in Fig. 3. From Fig. 4, it can be seen that a person with a blood level of 60% carboxyhemoglobin and requiring 3 vol % O$_2$ to maintain cellular respiration will experience a drop in blood oxygenation from an arterial content of 8 vol % to a venous content of 5 vol %. This corresponds to a reduction in P_{O_2} from 95 to 15 mm Hg. At this low P_{O_2}, there may not be enough of a concentration gradient to push the O$_2$ molecules into the deep tissue layers, resulting in anoxia with concomitant cell damage or necrosis. On the other hand, in a person lacking 60% of the red cell mass from anemia, the venous O$_2$ content will also be 5 vol %; however, because the dissociation curve has not shifted to the left, the venous P_{O_2} will be approximately 35 mm Hg. This is more than adequate to provide an effective O$_2$ concentration gradient between the capillaries and the tissues. This reasoning has been used to explain the difference in symptomatology between CO poisoning and anemia.

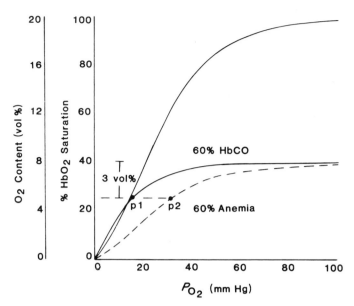

Fig. 4. The effect of carbon monoxide and anemia on the oxyhemoglobin dissociation curve.

IV. SYMPTOMS OF CO TOXICITY

The symptoms of CO poisoning have been correlated to the percentage carboxyhemoglobin in human volunteers undergoing low-level exposure experiments and from blood gas determinations performed in acutely ill patients brought to the emergency room. The HbCO fraction values with their corresponding most commonly encountered symptoms are shown in Table I (Aronow, 1977, 1979; Ayres *et al.*, 1970; Chiodi *et al.*, 1941; Kurt, 1979; Stewart, 1975).

In addition to the acute effects of CO poisoning, there is clinical evidence of CNS damage similar to that produced by anoxic events. During the acute phase, the patient may exhibit coma, myoclonic jerks, and decerebrate rigidity. Tissue examination reveals hyperemia with petechial hemorrhages throughout the brain and areas of necrosis and hemorrhages in the myocardium (Garland and Pearce, 1967). As the patient's condition improves, he may experience cerebellar ataxia, dystonic movements, and changes in personality or cognitive abilities. A 13% incidence of significant neuropsychiatric deficits has been reported in patients receiving a severe CNS insult as a result of CO intoxication. This syndrome usually develops 1–3 weeks after the exposure. Micro-

TABLE I

Symptoms of CO Toxicity and Corresponding Carboxyhemoglobin Values

HbCO in blood (%)	Symptoms
0–1	Asymptomatic
5–10	Chest pains in persons with coronary heart disease
10–20	Visual disturbances, decreased manual dexterity, headache
20–40	Nausea and vomiting, drowsiness, ST segment depression in the ECG
40–60	Stupor, coma, convulsions
>60	Fatal in most instances

scopic examination of the brain may show changes ranging from slight cellular degeneration to tissue necrosis. Most of the changes are found in the cerebral cortex, globus pallidus, and the red zone of the substantia nigra (Wilks and Clark, 1959).

V. DETERMINANTS OF CO UPTAKE

The standards for the maximum allowable concentration of CO in the environment have been determined by estimating the rate of formation of carboxyhemoglobin in blood when a person is exposed for a specified length of time to a particular atmospheric concentration of CO.

The rate of formation of HbCO depends on (1) the level of exposure, (2) the length of time of the exposure, and (3) the transport process of CO from the atmosphere to the hemoglobin molecule inside the erythrocyte. The latter is a very complicated process and is mainly a function of the respiratory rate, lung volumes, barometric pressure, alveolar diffusing capacity, P_{CO} in capillary blood, cardiac output, red cell membrane permeability, and the molecular binding properties of CO, O_2, and hemoglobin.

Several investigators have studied the rate of CO uptake in healthy, young volunteers exposed to small quantities of this gas (Collier, 1976; Peterson and Stewart, 1970; Stadie et al., 1925). From these experiments, empirical formulas have been developed to describe this process (Peterson and Stewart, 1975). Forbes et al. (1945) devised the following equations that take in account the percentage of CO in inspired air, ventilation rate ($\dot{V}_E$), heart rate (HR), and the exposure time in minutes (t):

$$\% \ HbCO = R \times \% \ CO \times t \tag{11}$$

Where the factor R depends upon the activity of the subject as follows:

$R = 3$ Rest ($\dot{V}_E = 6$ liters/minute, HR = 70)
$R = 5$ Light activity ($\dot{V}_E = 9$ liters/minute, HR = 80)
$R = 8$ Light work ($\dot{V}_E = 18$ liters/minute, HR = 110)
$R = 11$ Heavy work ($\dot{V}_E = 30$ liters/minute, HR = 135)

For a particular individual, the predicted level may differ by as much as $\pm 20\%$ from the measured HbCO. In addition, the formula assumes an initial HbCO less than 5%. For a % CO in inspired air of 0.01% or less, the formula is valid up to a concentration of 7% HbCO. This can be extended to 30% when the % CO is below 0.1%. As the inspired % CO is increased, the range of applicability can be increased to the 50–60% HbCO range, although at this point there are scant data to validate the formula's predictive ability.

Pace *et al.* (1946), have proposed the following formula based on a study conducted with 12 healthy male volunteers.

$$\% \text{ HbCO} = \frac{\substack{\text{inspired CO (ppm)} \times \dot{V}_E \\ \text{(liters/minute)} \times t \text{ (minutes)}}}{\substack{4650 \times \text{blood volume} \\ \text{(liters/m}^2 \text{ body surface} \times 3)}} \tag{12}$$

The authors state that this formula is valid for environmental CO concentrations of 100–2000 ppm and up to one-third of the % HbCO equilibrium value.

An attempt was made by Coburn *et al.* (1965) to develop a predictive equation for calculating the rate of change of carbon monoxide in the body, taking into account Haldane's principle and several ventilatory parameters:

$$\frac{d(\text{CO})}{dt} = \dot{V}_{\text{CO}} - \frac{[\text{HbCO}](P_{O_2})\text{cap}}{[\text{HbO}_2]M} \frac{1}{(1/D_L) + (P_B - 47)/\dot{V}_A}$$

$$+ \frac{(P_{\text{CO}})\text{insp}}{(1/D_L) + (P_B - 47)/\dot{V}_A} \tag{13}$$

where $d(\text{CO})$ = rate of change of CO in the body; (P_{O_2})cap = mean pulmonary capillary O_2 pressure; D_L = diffusion capacity of the lungs; P_B = barometric pressure; $\dot{V}_A$ = alveolar ventilation rate; and (P_{CO})insp = inspired CO partial pressure.

It is assumed that the mean capillary O_2 pressure and the oxy-hemoglobin concentration are constant and independent of the concentration of carboxyhemoglobin. This is clearly not the case. In arterial blood where the amount of reduced hemoglobin is very small, for any increase in HbCO there must be a similar drop in HbO_2 concentration. These restrictions drastically reduce the usefulness of the Coburn–Forster–Kane (CFK) except in those cases where the percentage HbCO is small compared to the % HbO_2.

VI. TREATMENT OF CO POISONING (JACKSON AND MENGES, 1980; MYERS, 1979)

As with most environmental toxins, the best treatment lies in preventing the exposure; however, when a comatose patient is brought into the emergency room with acute CO intoxication, the following therapeutic interventions should be considered.

A. Initial Assessment

Adequate ventilation must be provided, the pulse checked, and CPR instituted, if necessary.

B. History, Physical, and Laboratory Work

A quick history regarding possible exposure to CO must be obtained. Was the patient found near or in an automobile. Was he involved in a fire or working with an engine or heater in an enclosed space. In the majority of the cases, this can be done while vital signs are obtained. The next step is to obtain arterial blood gases, glucose levels, and electrolytes while a rapid physical exam is performed. Because some blood gas apparatuses will not measure this quantity routinely, it is sometimes necessary to specify that the carboxyhemoglobin level be determined from the arterial blood gases. Instead, the P_{O_2} is measured, and the hemoglobin saturation is calculated from the standard dissociation curves. This may lead to a false sense of security, because as seen before in Fig. 4, the P_{O_2} does not drop with a decrease in oxyhemoglobin caused by carboxyhemoglobin. It is of interest to note that the classical "cherry red" appearance of the skin is usually not seen in cases of acute CO intoxication.

C. Correction of Acidosis

Frequently, these patient's exhibit a profound metabolic acidosis as a result of the anoxic tissue cells switching to anaerobic metabolism and producing lactic acid in great quantities. This acidosis should be corrected promptly with appropriate amounts of intravenous sodium bicarbonate.

D. Adequate Oxygenation

Oxygen is the most important treatment modality for CO intoxication, and, in a comatose patient, endotracheal intubation should be performed to assure that 100% O_2 is delivered to the lungs. A nonrebreathing mask with 100% O_2 can be used when the HbCO is less than 40% and there are no detectable neurological deficits. The half-life for the elimination of CO from the blood is a function of the inspired O_2 concentration, the alveolar ventilation, the diffusion capacity of the lungs, and the atmospheric pressure. Table II shows approximate half-life values under different inspired O_2 concentrations and atmospheric pressures (Pace *et al.*, 1950).

E. Hyperbaric Oxygenation

The accelerated elimination of HbCO with increasing atmospheric pressure provides the rationale for the use of hyperbaric O_2 therapy in cases of CO intoxication. In addition, breathing O_2 at 3 atm results in an arterial P_{O_2} of approximately 2100 mm Hg. From Eq. (4), the amount of dissolved O_2 in arterial blood is $0.003 \times 2100 = 6.3$ vol %, which should be sufficient to sustain aerobic cellular metabolism by itself. It should be remembered, however, that O_2 delivery depends also on cardiac output and blood flow regulatory mechanisms at the tissue level, and very high levels of P_{O_2} may interfere with these parameters, reducing the effective amount of O_2 carried to the tissues by this route. Another possible advantage of the use of hyperbaric oxygenation is the decrease in anoxic cerebral edema seen experimentally by Sukoff *et al.* (1967).

Some authors recommend the use of hyperbaric O_2 for patients with HbCO levels of 25% or greater if a hyperbaric chamber is readily avail-

TABLE II

Rate of HbCO Elimination from Blood under Different Pressures and Inspired O_2 Concentrations

Half-life of HbCO (minutes)	Inspired O_2 concentration (%)	Ambient pressure (atm.)
249	21	1
47	100	1
25	100	2.5

able (Kindwall, 1975). In hospitals without this facility, the patient should be transferred to a center with a hyperbaric chamber if the HbCO level is greater than 40%. If there are neurological signs or mentation changes, hyperbaric oxygenation should be used regardless of the carboxyhemoglobin level. Current recommendations suggest using 100% O_2 at 3 atm for two half-lives and then 2 atm until the HbCO concentration falls below 10%.

The complications of hyperbaric oxygen include barotrauma and O_2 toxicity in the form of patchy atelectasis. Prolonged exposures can result in microhemorrhages, and fibrosis of lung tissue.

The single most important factor determining the use of hyperbaric oxygenation is the availability of a hyperbaric chamber and personnel trained in its use. In many instances, the time involved from the initial emergency room contact with the patient to the initiation of hyperbaric treatment is approximately one or more HbCO half-lives when treating with 100% O_2 at atmospheric pressure.

There is considerable controversy regarding the use of hyperbaric oxygenation for the treatment of acute CO toxicity. To date, there are no prospective data showing a difference in morbidity or mortality among patients treated with 100% O_2 at atmospheric pressure or with hyperbaric oxygenation. Until these data become available, it is recommended that if a hyperbaric chamber is available, patients with HbCO greater than 40% or manifesting CNS changes should be treated with 100% O_2 at 2–3 atm until the HbCO drops below 10%.

The use of 95% O_2 and 5% CO_2 has been advocated to increase the ventilatory rate by producing a small degree of hypercapnea. It should be remembered that these patients usually have a metabolic acidosis and, unless there is profound CNS depression, are breathing at or near their maximum capacity. Under these circumstances, the addition of 5% is at best ineffective and it could aggravate the acidosis by reducing the CO_2 elimination via the lungs. This therapy is not recommended in the treatment of CO poisoning.

The rate of fall in carboxyhemoglobin should be monitored with frequent blood gases, and, in elderly patients or those with known coronary artery disease, myocardial enzymes with MB bands and the electrocardiogram should be followed serially to rule out a myocardial infarction.

VII. NEW DEVELOPMENTS

The recent work of Goldbaum and co-workers has raised serious questions regarding previous theories of CO toxicity. Their work as well as other studies suggesting a possible direct effect of CO is reviewed in

Chapter 6. In brief, they demonstrated in the series of experiments that dogs injected intraperitoneally with 100% CO or transfused with red blood cells containing 80% carboxyhemoglobin survived indefinitely even though blood levels of HbCO were in the 60–80% range. In another group, dogs given 13% CO to breathe all died within 15 minutes with HbCO levels of 54–90%. These data suggest that mechanisms of toxicity other than hypoxemia may play a fundamental role in CO poisoning (Goldbaum *et al.*, 1975, 1977; Goldsmith and Landaw, 1968; Orellano *et al.*, 1976). A possible mechanism for toxicity is that CO dissolved in blood and in equilibrium with HbCO acts similar to cyanide by interfering directly with the cellular cytochrome system to prevent the exchange of O_2 (Belovolova *et al.*, 1975; Chance *et al.*, 1970; Greenwood *et al.*, 1974; Raybourn, 1978; Wald and Allen, 1957; Wharton, 1976). Other investigators feel that significant binding of CO by the myocardial and skeletal muscle myoglobin may also be responsible for

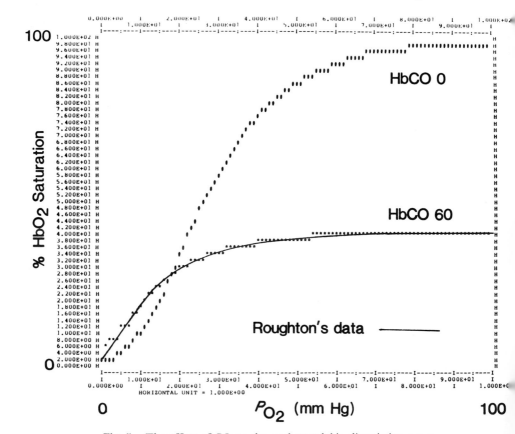

Fig. 5.   The effect of CO on the oxyhemoglobin dissociation curve.

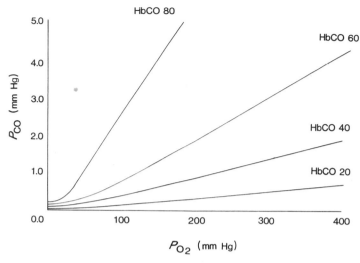

Fig. 6. Predicted P_{CO} at different levels of % HbCO and P_{O_2}.

the apparent lack of correlation between the toxic effects and the HbCO level (Coburn, 1979). Because the toxicity of CO may be related more closely to the P_{CO} rather than the amount bound to hemoglobin, it is critically important to determine this quantity in any experiment attempting to elucidate the mechanisms of CO toxicity. This is particularly important if therapeutic interventions that may alter the relationship between HbCO and P_{CO} (such as hyperbaric oxygenation) are planned. At the present time, it is not technically feasible to measure very small amounts of dissolved CO in blood; therefore, other investigational tools, such as mathematical models, are required to provide a reliable estimate of this quantity (Hill *et al.*, 1977). Only by knowing the P_{CO} in blood under physiological conditions may it become possible to test the above mentioned theories of CO toxicity.

A recently proposed mathematical model has been used to predict the changes in the oxyhemoglobin curve produced by CO (Gutierrez *et al.*, 1981). The equations used by the model are independent of the hemoglobin saturation level and Haldane's principle. The model has been validated *in vitro* with human blood exposed to varying concentrations of O_2 and CO. The predicted values lie within $\pm 7\%$ of the experimental results published by Roughton (1964). A representative curve is shown in Fig. 5. This model can also be used to predict P_{CO} in blood. Figure 6 shows the predicted relationship between O_2 and CO when the level of carboxyhemoglobin is kept constant. This graph shows a parabolic increase in P_{CO} as the rising P_{O_2} displaces the CO molecules from hemoglobin. This relationship has not been experimentally validated because of the major difficulties encountered in measuring very small amounts of CO dissolved in blood.

VIII. CONCLUSIONS

Carbon monoxide intoxication remains a serious therapeutic and public health problem. An extensive amount of research has been conducted in an effort to elucidate its pathophysiology. For many years, it was postulated that the tight bond formed by CO and hemoglobin interfered with the O_2 delivery by the blood, resulting in a hypoxic condition at the tissue level. This theory is now under serious scrutiny as a result of several animal experiments showing prolonged survival of dogs in the face of high carboxyhemoglobin levels caused by CO administered into the peritoneal space. The possibility that CO has a direct toxic effect on the cellular respiratory chain has been proposed, based on *in vitro* experiments demonstrating a bond between CO and the cytochrome system. It is not possible to prove the latter theory, because the amount of dissolved CO in blood is so small that it is technically very difficult to obtain an accurate determination of its concentration. The most likely explanation of the toxic effect of CO probably lies in a combination of tissue hypoxia and direct interference with the cellular respiratory chain. Further animal experiments and mathematical modeling are needed to clarify this issue.

The cornerstone of treating acute CO poisoning continues to be the delivery of 100% O_2 to the lungs. This treatment modality should be instituted as soon as possible after the CO exposure. The use of hyperbaric oxygenation under these circumstances remains a subject of controversy, and prospective, randomized studies are needed before it can be fully endorsed.

REFERENCES

Adair, G. S. (1925). The hemoglobin system. VI. The oxygen dissociation curve of hemoglobin. *J. Biol. Chem.* **63,** 529.
Antonini, E., and Brunori, M. (1971). Hemoglobin and myoglobin in their reactions with ligands. *Front. Biol.* **21,** 260.
Aronow, W. S. (1977). Effect of carbon monoxide on exercise performance in chronic obstructive pulmonary disease. *Am. J. Med.* **63,** 904.
Aronow, W. S. (1979). Effect of carbon monoxide on cardiovascular disease. *Prev. Med.* **8,** 271.
Arturson, G. (1974). The oxygen dissociation curve of normal human blood with special

reference to the influence of physiological effector ligands. *Scand. J. Clin. Lab. Invest.* **34**,(I), 9.

Ayres, S. M., Gianelli, S., and Hueller, H. (1970). Myocardial and systemic responses to carboxyhemoglobin. *Ann. N.Y. Acad. Sci.* **174**, 268.

Belovolova, L. V., Blyumenfel'd, D., Burbaev, Sh., and Vanin, A. F. (1975). Formation of a complex between carbon monoxide and cytochrome c. *Mol. Biol. (Moscow)* **9**, 939.

Campbell, J. A. (1929). Tissue oxygen tension and carbon monoxide poisoning. *J. Physiol. (London)* **68**, 81.

Chance, B., Erecinska, M., and Wagner, M. (1970). Mitochondrial responses to carbon monoxide toxicity. *Ann. N.Y. Acad. Sci.* **174**, 193.

Chiodi, H., Dill, D. B., Consolazio, F., and Horvath, J. M. (1941). Respiratory and circulatory responses to acute carbon monoxide poisoning. *Am. J. Physiol.* **134**, 683.

Coburn, R. F. (1979). Mechanisms of carbon monoxide toxicity. *Prev. Med.* **8**, 310.

Coburn, R. F., Forster, R. E., and Kane, B. P. (1965). Considerations of the physiological variables that determine the blood carboxyhemoglobin concentration in man. *J. Clin. Invest.* **44**, 1899.

Collier, C. R. (1976). Oxygen affinity of human blood in presence of carbon monoxide. *J. Appl. Physiol.* **40**(3), 487.

Comroe, J. H. (1965). "Physiology of Respiration." Year Book Med. Publ., Chicago, Illinois.

Comroe, J. H., Forster, R. E., DuBois, A. B., Briscoe, W. A., and Carlsen, E. (1962). "The Lung," 2nd ed. Year Book Med. Publ., Chicago, Illinois.

Douglas, C. G., Haldane, J. S., and Haldane, J. B. S. (1912). The laws of combination of hemoglobin with carbon monoxide and oxygen. *J. Physiol. (London)* **44**, 275.

Forbes, W. H., Sargent, F., and Roughton, F. J. W. (1945). The rate of carbon monoxide uptake by normal men. *Am. J. Physiol.* **143**, 594.

Forster, R. E. (1970). Carbon monoxide and the partial pressure of oxygen in tissue. *Ann. N.Y. Acad. Sci.* **174**, 233.

Garland, H., and Pearce, J. (1967). Neurological complications of carbon monoxide poisoning. *Q. J. Med.* **144**, 445.

Goldbaum, L. R., Ramirez, R. G., and Absalon, K. G. (1975). What is the mechanism of carbon monoxide toxicity? *Aviat., Space Environ. Med.* **46**(10), 1289.

Goldbaum, L. R., Orellano, T., and Dergal, E. (1977). Studies on the relation between carboxyhemoglobin concentration and toxicity. Joint committee on aviation pathology. XVI. *Aviat., Space Environ. Med.* **48**, 969.

Goldsmith, J. R., and Landaw, S. A. (1968). Carbon monoxide and human health. *Science* **162**, 1352.

Greenwood, C., Wilson, M. T., and Brunori, M. (1974). Studies on partially reduced mammalian cytochrome oxidase. Reactions with carbon monoxide and oxygen. *Biochem. J.* **137**, 205.

Gutierrez, G., Dantzker, D. R., and Rotman, H. H. (1981). Carbon monoxide toxicity. *Physiologist* **24**, 90.

Haldane, J. B. S. (1912). The dissociation of oxyhaemoglobin in human blood during partial CO poisoning. *J. Physiol. (London)* **45**, XXII.

Haldane, J. B. S. (1927). Carbon monoxide as a tissue poison. *Biochem. J.* **21**, 1068.

Haldane, J. S. (1895a). The relation of the action of carbonic oxide to oxygen tension. *J. Physiol. (London)* **18**, 205.

Haldane, J. S. (1895b). The action of carbonic oxide on man. *J. Physiol. (London)* **18**, 430.

Haldane, J. S., and Smith, J. L. (1897). The absorption of oxygen by the lungs. *J. Physiol. (London)* **22**, 231.

Haldane, J. S., and Priestley, J. G. (1935). "Respiration" Oxford Univ. Press, London and New York.

Hill, A. V. (1921). LXII. The combinations of haemoglobin with oxygen and carbon monoxide and the effects of acid and carbon dioxide. *Biochem. J.* **XV,** 577.

Hill, E. P., Hill, J. R., Power, G. G., and Longo, L. D. (1977). Carbon monoxide exchanges between the human fetus and mother: A mathematical model. *Am. J. Physiol.* **232,** H311.

Jackson, D. L., and Menges, H. (1980). Accidental carbon monoxide poisoning. *JAMA, J. Am. Med. Assoc.* **243**(8), 772.

Joels, N., and Pugh, L. (1958). The carbon monoxide dissociation curve of human blood. *J. Physiol. (London)* **142,** 63.

Kindwall, E. P. (1975). Carbon monoxide and cyanide poisoning. *In* "Hyperbaric Oxygen Therapy" (J. C. Davis and T. K. Hunt, eds.), pp. 177–190. University of California, San Francisco.

Kurt, T. L. (1979). Ambient carbon monoxide levels and acute cardiorespiratory complaints: An exploratory study. *Am. J. Public Health* **69,** 360.

Ledwith, J. W. (1978). Determining P_{50} in the presence of carboxyhemoglobin. *J. Appl. Physiol.: Respir., Environ. Exercise Physiol.* **44**(2), 317.

Myers, R. A. (1979). Carbon monoxide poisoning: The injury and its treatment. *JACEP* **8**(11), 479.

Orellano, T., Dergal, E., Alijani, M., Briggs, C., Vasquez, J., Goldbaum, L. R., and Absolon, K. B. (1976). Studies on the mechanism of carbon monoxide toxicity. *J. Surg. Res.* **20,** 485.

Pace, N., Consolazio, W. V., White, W. A., and Behnke, A. R. (1946). Formulation of the principal factors affecting the rate of uptake of carbon monoxide by man. *Am. J. Physiol.* **147,** 352.

Pace, N., Strasman, E., and Walker, E. (1950). Acceleration of carbon monoxide elimination in man by high pressure oxygen. *Science* **111,** 652.

Peterson, J. E., and Stewart, R. D. (1970). Absorption and elimination of carbon monoxide by inactive young men. *Arch. Environ. Health* **21,** 165.

Peterson, J. E., and Stewart, R. D. (1975). Predicting the carboxyhemoglobin levels resulting from carbon monoxide exposures. *J. Appl. Physiol.* **39,** 633.

Raybourn, M. S. (1979). An *in vitro* electrophysiological assessment of the direct cellular toxicity of carbon monoxide. *Toxicol Appl. Pharmacol.* **46,** 769.

Roughton, F. J. W. (1945). The kinetics of the reaction $CO + O_2Hb \rightleftharpoons O_2 + COHb$ in human blood at body temperature. *Am. J. Physiol.* **143,** 609.

Roughton, F. J. W. (1964). Transport of oxygen and carbon monoxide. *In* "Handbook of Physiology" (W. O. Fenn and H. Rahn, eds.), Sect. 3, Vol. I, Chapter 31. Am. Physiol. Soc., Washington, D.C.

Roughton, F. J. W. (1970). The equilibrium of carbon monoxide with human hemoglobin in whole blood in biological effects of carbon monoxide. *Ann. N.Y. Acad. Sci.* **174,** 177.

Roughton, F. J. W., and Darling, R. C. (1944). The effect of carbon monoxide on the oxyhemoglobin dissociation curve. *Am. J. Physiol.* **141,** 17.

Roughton, F. J. W., and Rupp, J. C. (1958). Problems concerning the kinetics of the reactions of oxygen, carbon monoxide and carbon dioxide in the intact red cell. *Ann. N.Y. Acad. Sci.* **75,** 156.

Roughton, F. J. W., Forster, R. E., and Cander, L. (1957). Rate at which carbon monoxide replaces oxygen from combination with human hemoglobin in solution and in the red cell. *J. Appl. Physiol.* **11,** 269.

Sharma, V. S. (1976). Dissociation of CO from carboxyhemoglobin. *J. Biol. Chem.* **251**(14), 4267.

Stadie, W. C., and Martin, K. A. (1925). The elimination of carbon monoxide from the blood. *J. Clin. Invest.* **2,** 177.

Stewart, R. D. (1975). The effect of carbon monoxide on humans. *Annu. Rev. Pharmacol.* **15,** 409.

Sukoff, M. H., Hollin, S. A., and Jacobson, J. H. (1967). The protective effect of hyperbaric oxygenation in experimentally produced cerebral edema and compression. *Surgery* **62,** 40.

Wald, G., and Allen, D. W. (1957). The equilibrium between cytochrome oxidase and carbon monoxide. *J. Gen. Physiol.* **40L,** 593.

Wharton, D. C. (1976). Stoichiometry of carbon monoxide binding by cytochrome c oxidase. *J. Biol. Chem.* **251**(9), 2861.

Wilks, S. S., and Clark, R. T. (1959). Carbon monoxide determinations in postmortem tissues as an aid in determining physiologic status prior to death. *J. Appl. Physiol.* **14,** 313.

6

Physiological Effects of Carbon Monoxide

James J. McGrath

AIR POLLUTION
Copyright © 1982 by Academic Press, Inc.
All rights of reproduction in any form reserved.
ISBN 0-12-483880-4

I. INTRODUCTION

A. History

Although carbon monoxide (CO) was present in large quantities in the earth's primitive atmosphere (Hart, 1979) and produced and consumed by microorganisms long before man discovered fire, it was man's use of fire and the concomitant production of CO in confined areas that made CO an environmental hazard. Although they could not, of course, identify CO, the ancient Greeks and Romans were aware of the toxic phenomenon associated with fire and lack of ventilation and used CO to both execute criminals and commit suicide.

CO poisoning is unique in its close association with the evolution of technology. Poisoning from CO increased drastically in the fifteenth century with the use of coal for domestic heating. The drastic increase resulted directly from inhalation of CO formed by incomplete combustion in home heating appliances and explosions and fires in coal mines.

Illuminating gas (a mixture of hydrogen, carbon monoxide, and methane), introduced as a fuel for domestic heating purposes, increased further the hazard of CO poisoning. In more recent years, the introduction of the internal combustion engine and the development of numerous technological processes that produce carbon monoxide have contributed further to the hazard associated with this gas. The internal combustion engine is now recognized as the principal contemporary anthropogenic source of CO [National Research Council (NRC), 1977].

Several recent reviews discuss the effects of CO on health [NRC, 1977; Wright and Shephard, 1979; Environmental Protection Agency (EPA), 1979; Ginsburg, 1980; Turino, 1981]. This chapter will consider (1) the source of CO, (2) the mechanism of toxicity of CO with emphasis on the direct effects (the indirect effects have been considered in Chapter 5), (3) the cardiovascular effects of CO, (4) the effects of chronic exposure to CO, and (5) the effects of CO at high altitude.

The physiological effects of CO were described first in 1857 by Claude Bernard. He determined that CO produces hypoxia through its reversible combination with hemoglobin to form carboxyhemoglobin. The reaction is rapid, and the affinity of CO for hemoglobin is many times greater than the affinity of hemoglobin for oxygen (Root, 1965). When CO combines with hemoglobin, its oxygen carrying capacity is reduced, and oxygen that is transported is bound more tightly to the hemoglobin molecule. Thus, even though the blood oxygen content may be the same for both hypoxias, the symptoms produced by CO poisoning are more severe than those produced by anemia (Haldane, 1895).

Man is exposed to CO produced and/or emitted by energy conversion processes, tobacco smoking, and his own normal physiological processes. Exposure to CO may occur during a variety of occupational or other human activities.

B. CO Emissions

The most common and ubiquitous pollutant in the lower atmosphere is CO (Jaffee, 1970). Previously, the observed global concentrations of CO were thought to arise from anthropogenic sources. It is now believed that man's activities contribute approximately 10% to total worldwide CO concentrations; however, CO emitted into urban air by man's activities far surpasses natural contributions (NRC, 1977).

In 1979, the Environmental Protection Agency (EPA) estimated that the total emission of CO in the United States in 1977 was 102.7 million metric tons and divided the sources of CO emissions into categories (Table I). Fuel combustion by mobile sources was the largest single emission category, contributing 85.7 million metric tons or 83.4% of the total. Within this category, highway vehicles, including automobiles and trucks, emitted 77.2 million metric tons or 75% of the total, whereas other mobile sources, including aircraft, railroads, and vessels, contributed 8.5 million metric tons or 8.3% of the total.

The second major CO emission category was fuel combustion by stationary sources. This grouping includes all stationary combustion equip-

TABLE I

Nationwide Carbon Monoxide Emission Estimated for 1977 (Environmental Protection Agency, 1979)

Source category	(10^6 metric tons/year)	% of total
Fuel combustion—mobile sources	85.7	83.4
Highway vehicles	77.2	75.2
Aircraft, railroad, vessels, etc.	8.5	8.3
Fuel Combustion—stationary sources	1.2	1.1
Electric utilities	0.3	0.2
Industrial	0.6	0.6
Residential, commercial, and institutional	0.3	0.3
Industrial processes	8.3	8.1
Solid waste combustion	2.6	2.5
Miscellaneous	4.9	4.9
Total	102.7	100

ment such as boilers and stationary internal combustion engines that are used for the production of power and heat. Stationary sources emitted 1.2 million metric tons or 1.1% of the total.

The third major category, industrial processes, comprises all processing industries, including those which manufacture chemicals, refine petroleum, and produce metals and metal products. This category contributed 8.3 million metric tons or 8.1% of total CO emissions.

The fourth major category, solid waste combustion, includes the combustion of wastes in municipal and other incinerators and from open burning of refuse. This grouping contributed 2.6 million metric tons or 2.5 % of total CO emissions.

The miscellaneous category includes emissions from combustion of forest, agricultural, and coal refuse materials and from structural fires. This grouping contributed 4.9 million metric tons or 4.9% of total CO emissions.

C. CO Exposures

Goldsmith (1970) estimates that 40–60% of the adults in any community are affected by cigarette smoking. Cigarette smoke contains up to 4% carbon monoxide by volume (40,000 ppm). The CO inhaled during smoking is diluted with air so that the alveolar gas of smokers may contain 400–500 ppm CO.

Smoke produced by a burning cigarette is divided into several smoke streams (Hoegg, 1972). The smoke drawn directly into the lungs has received the most attention and is termed mainstream smoke. All the other smoke streams are emitted into the environment and are responsible for the health concerns associated with "passive smoking," i.e., the inhalation of smoke by nonsmokers in the vicinity of a smoker. Sidestream smoke generated by the burning end of the cigarette during the *puff interval* contributes approximately 95% of the total smoke, whereas the smolderstream, also generated during the puff interval, contributes 4%. The remaining smoke is emitted through the glowstream (that coming from glowing end of the cigarette during the puff), the effusion stream (that flowing from the sides of the cigarette during the puff), and the diffusion stream (that radiating from the sides of the cigarette between puffs).

The amount of CO produced by cigarette smoking constitutes a minor percentage of the total atmospheric burden. In closed spaces, however, smoking can significantly elevate CO levels [U.S. Department of Health, Education and Welfare (USDHEW), 1979].

The major determinants of CO levels caused by smoking are the size of the space in which smoking occurs, the number and type of tobacco products smoked, and the amount and effectiveness of ventilation.

Hoegg (1972) reported sidestream and mainstream values for CO produced by a machine smoking cigarettes in a 25-m^3 nonventilated chamber. There was a linear relationship between the number of cigarettes smoked and the chamber CO concentration. Smoking 24 cigarettes produced a chamber concentration of 69.8 ppm. Sidestream smoke and mainstream smoke of a single cigarette contained 75.5 and 16.0 ml of CO, respectively. The sidestream–mainstream ratio (*S/M*) for carbon monoxide was 4.7. This study is significant because it separated mainstream smoke from sidestream smoke and demonstrated that sidestream smoke is an important source of CO in a closed environment.

Coburn *et al.* (1964) reported on the effects of smoking on the buildup of CO in confined areas. In hospital wards where smoking was prohibited, CO concentrations averaged 2.2 ppm (but reached to 5 ppm). Air taken from smoke-filled conference rooms contained from 4.3 to 9.0 ppm CO. Ten cigarettes burned in a small room produced an atmospheric concentration of 20 ppm CO.

Lawther and Commins (1970) demonstrated the potential for exposure to CO by "passive smoking." In a chamber, a nonsmoker wearing a CO sensor was seated next to a smoker. The chamber CO concentration rose to 20 ppm in 1 hour after the smoking of seven cigarettes. Transient peaks of up to 90 ppm CO were measured in the breathing zone of the nonsmoker.

Russell *et al.* (1973) studied the absorption of CO from smoke-filled rooms by nonsmokers. These workers reported an increase in mean carboxyhemoglobin (COHb) levels of 1.6–2.6% in 12 nonsmokers sitting in a smoke-filled room with six smokers. Furthermore, they suggested that the amount of CO absorbed by a nonsmoker in these experiments was approximately equal to that absorbed by a person who had smoked and inhaled one cigarette.

A variety of personnel are exposed to elevated CO levels during the course of a workday. People who work in casting, welding, drying or preheating, boiler cleaning, or blast furnace operations or who transport hot coke or slag are exposed to high levels of CO. Occupational groups involved with vehicles, metal processing, chemical processing, stone and glass processing, printing, welding, electrical assembly and repair work, and certain types of graphic artwork have been found to have average carboxyhemoglobin levels in excess of 2% (Wright and Shephard, 1979). Lawther and Commins (1970) reported elevated carboxyhemoglobin levels in policemen, garage workers, toll-booth operators, greenhouse workers, and firemen. They concluded that in all groups, with the possible exception of greenhouse workers, carboxyhemoglobin levels were elevated more by cigarette smoking than by occupational exposure.

CO concentrations in a submarine were reported by Davies (1975). In a nuclear submarine, 75–90% of the total CO load is from tobacco smoking. In a Polaris submarine with a 150-man crew, 100 smokers will produce approximately 200 liters CO in a breathable air volume of 4000 m³. In current nuclear submarine operations, the mean CO level for more than 90% of diving time varies between 5 and 15 ppm. Bondi *et al.* (1978) reported ambient concentrations aboard a submarine of 7 ppm and COHb levels of 2.1, 1.7, and 1.7% in nonsmokers at the start, middle, and end of a 40-day patrol.

Significant amounts of CO can be generated in enclosed areas by ice-resurfacing machines. Johnson *et al.* (1965) reported that operation of a propane-powered ice-resurfacing machine produced average values of 157–304 ppm CO in an ice skating arena. Anderson (1971) measured CO concentration up to 250 ppm in an ice skating rink after an episode of illness among skaters marked by headache and nausea.

The possibility of risking elevated COHb levels without being exposed to CO also exists. Stewart *et al.* (1972) reported on a co-worker who had breathed varnish remover vapors while "stripping" furniture. Further study revealed that the methylene chloride in the varnish metabolized in the body to CO, producing elevated COHb levels. This observation was confirmed by Ratney *et al.* (1974) who reported COHb levels of 9% in factory workers exposed to methylene chloride levels of 180–200 ppm. On the basis of their study, Ratney *et al.* recommended that the allowable limits of exposure to methylene chloride be decreased from 500 ppm to 75–100 ppm to avoid body burdens of CO greater than those allowed persons exposed to exogenous CO.

D. CO Exposure Limits

Exposure limits for CO have been set by a number of agencies for a variety of situations. These limits are summarized briefly in Table II. The continuous exposure limit for CO during 90- and 1000-day flights in U.S. spacecraft is 15 ppm, because higher CO levels might compromise the high levels of judgment and performance required of pilots and other occupants of space vehicles. For U.S. submarines, the contaminant concentration limits for CO are 200 ppm for 1 hour, 200 ppm for 24 hours, and 25 ppm for 90 days. The emergency exposure limits (EEL) for CO recommended to military and space agencies by the Committee on Toxicology—National Research Council are 400 ppm for 60

TABLE II

Carbon Monoxide Exposure Limits

Situation	Concentration (ppm)
Spacecraft	
90-day flights	15
1000-day flights	15
Submarines	
60 minutes	200
24 hours	200
90 days	25
EEL[a]	
10 minutes	1500
30 minutes	800
60 minutes	400
TLV-STEL,[b] 15 minutes	400
TLV-TWA,[c] 8 hours	50
NAAQS[d]	
1 hour	35
8 hours	9

[a] Emergency exposure limits.
[b] Threshold limit value for short-term exposure.
[c] Threshold limit value, time-weighted average.
[d] National ambient air standard set by EPA.

minutes, 800 ppm for 30 minutes, and 1500 ppm for 10 minutes. By definition, an EEL is a limit for an accidental exposure that would not normally be repeated in a lifetime (MacEwen, 1973).

The threshold limit value (short-term exposure limit TLV-STEL) for CO set by the American Conference of Governmental and Industrial Hygienists (ACGIH) (1978) is 400 ppm for 15 minutes. This value is defined as the maximal concentration to which workers can be exposed for a period of up to 15 minutes without suffering ill effects or narcosis of a degree sufficient to increase accident proneness, impair self-rescue, or materially reduce work efficiency, provided no more than four excursions a day are permitted, with at least 60 min between exposure periods, and provided the daily threshold limit value—time weighted average (TLV-TWA) is not exceeded. The TLV-TWA is the time-weighted average concentration to which nearly all workers may be exposed during the course of a working lifetime without suffering adverse effects. The national ambient air standard for CO set by EPA is 9 ppm for 1 hour and 35 ppm averaged over 8 hours (NRC, 1977).

E.   Endogenous Production

In addition to exogenous sources, man is exposed also to small amounts of CO that are produced endogenously. Most of this is accounted for by the normal degradation of the α-methane moieties of red cell hemoglobin. One molecule of CO is formed for every heme decomposed to bile pigment (Sjöstrand, 1970). A small fraction results from the metabolism of nonhemoglobin heme-containing compounds within the liver. Endogenous production of CO accounts for 0.42 ml/hour in healthy males and leads normally to blood COHb levels of less than 1%; however, considerably higher levels are observed in patients with hemolytic anemia (Coburn *et al.*, 1963).

The principal processes that influence the body stores of CO are depicted in Fig. 1 and have been described by Coburn (1970). The model assumes that approximately one-half of the body stores of CO arise from endogenous production and the other half from inhaling ambient air containing 2 ppm CO. Body stores tend to be decreased by excretion via the lungs and metabolism of CO to CO_2. In normal man, the conversion of CO to CO_2 is so slow as to have little effect on body stores. Although metabolism may be an insignificant mechanism of CO elimination compared to excretion by the lungs, Luomanmaki and Coburn (1969) have shown that absolute rates of CO metabolism may be directly proportional to blood COHb and have suggested that metabolism may become more important at higher blood COHb levels.

Inhaling ambient CO and endogenous production are processes that tend to increase body stores. Endogenous CO production may be increased in patients with hemolytic anemias. Luomanmaki and Coburn (1969) reported on hemolytic anemia patients with carboxyhemoglobin values up to 2.74% produced by an endogenous production rate elevated to 3.4 ml CO per hour. These studies demonstrate that normal carboxyhemoglobin levels of 0.5–0.9% can be increased to 2.74% in various disease states. Coburn (1970) suggests that very high carboxyhemoglobin levels (12%) are sufficient to inhibit oxidase systems involved in the degradation of hemoglobin to CO.

Endogenous CO production varies considerably in healthy subjects (Longo, 1977). Thus, healthy men and menstruating women (during the estrogen phase of the menstrual cycle) produce approximately 6.1 and 5.3 μl/hour/kg body weight of CO, respectively. During the progestational or secretory phase of the menstrual cycle, however, endogenous CO production doubles (10.2 μl/hour/kg in healthy women). During pregnancy, CO production further increases to 13.7 μl/hour/kg, and, immediately after pregnancy, CO production rates may exceed 25 μl/

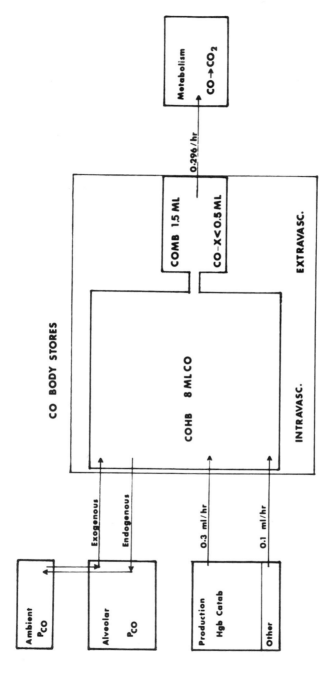

Fig. 1. Processes that influence the body carbon monoxide stores in normal man. (From Coburn, 1970.)

hour/kg. Of the increased rate of CO production in the pregnant woman, 30–40% is caused by the increased red cell mass, whereas the balance is caused by CO production by the fetus. Thus, the normal carboxyhemoglobin level of pregnant women may vary from 0.4 to 1.21%.

Maternal smoking (i.e., women smoking during pregnancy) is probably the most common source of fetal exposure to greater than normal concentrations of CO. Several studies have reported the carboxyhemoglobin levels in the blood of smoking mothers and their newborn infants. These values range from 2 to 10% for the fetuses and 2 to 14% for the mothers (Longo, 1977). It is possible, however, that imprecise measurement techniques may grossly underestimate carboxyhemoglobin levels and that the average values for both smoking mothers and their fetuses could be well above the concentrations reported.

II. DIRECT EFFECTS

A. Introduction

The mechanisms by which CO elicits its toxic effects are well known and are detailed in Chapter 5. In short, CO combines with blood hemoglobin to form carboxyhemoglobin (COHb). This causes a functional anemia, by decreasing the amount of hemoglobin available for oxygen transport and by causing the oxyhemoglobin dissociation curve to shift to the left.

There are, however, reports in the literature suggesting that CO might have, in addition to its hemoglobin-mediated effects, a direct toxic effect on functioning tissue. Peirce et al. (1972) and Goldbaum et al. (1975) have pointed out that the effect of CO intoxication and equivalent anemia are not the same. People whose oxygen transport is reducted 30–50% by COHb formation may die, whereas people whose hemoglobin levels have decreased to less than 50% of normal because of anemia may show no toxic symptoms. Additional evidence supporting the possibility of a direct effect of CO is derived from (1) dosage studies, in which saturation rates and routes of administration are varied, (2) tissue P_{CO} measurements, (3) studies on isolated tissue, and (4) studies of altered lethality.

B. Dosage Studies

The problem of whether CO has a direct toxic effect has been studied by changing hemoglobin saturation rates and by administering CO by different routes.

In an early study, Drabkin *et al.* (1943) produced COHb levels of 75% by exposing dogs to CO by inhalation. The animals collapsed with evidence of cardiac and respiratory failure. Surviving animals had extensive necrotic changes in the brain and heart. When similar levels of COHb were produced by replacing blood with CO-saturated erythrocytes, however, there were no signs characteristic of anoxia, and there was no evidence of myocardial or cerebral damage. The authors attributed their results to the "Haldane effect" and calculated that the remaining functional hemoglobin was 11 and not 25%.

Fukuda and Kobayashi (1961) described a hyperglycemic response in rabbits whose COHb levels were increased to 40% after they inhaled 0.8% CO for 15 minutes. The hyperglycemic response was more marked than that seen in animals that had inhaled an O_2–N_2 mixture for 15 minutes and was blocked by bilateral adrenalectomy or splanchnotomy. Animals with 40% COHb levels produced by inhaling a low dose of CO (0.05% for 240 minutes), however, did not have a hyperglycemic response. The authors concluded that inhaling high levels of CO elicits a sympathetic response more marked than that seen with comparable hypoxias.

These results were not completely confirmed by Sokal (1975), who investigated biochemical disturbances arising from 50% COHb produced by two different CO dose regimes. The "longer-lesser" exposure (4000 ppm CO for 40 minutes) caused more pronounced changes than the "shorter-intense" (10,000 ppm CO for 4 minutes) one. Both exposures resulted in decreased blood pH, increased blood glucose, and increased lactate and pyruvate levels in blood and brain tissue; however, the longer-lesser exposure elicited greater changes than the shorter-intense exposure. The authors suggested that the longer-lesser exposure resulted in a more profound tissue hypoxia, but suggested also a possible direct effect of CO.

Chen *et al.* (1979) studied circulating blood glucose and lactate levels in rats receiving CO by two different routes of administration. One group inhaled 5000 ppm CO for 10 minutes (INH-I); a second group inhaled 2000 ppm CO for 40 minutes (INH-II); a third group received 100% CO i.p. (30 ml/kg). Blood glucose and lactate concentrations were determined when COHb concentrations reached the same level (64–67%). Both groups of animals receiving CO by inhalation became unconscious at COHb levels of 60%, whereas i.p. rats remained conscious throughout the experiment. Blood glucose levels were higher (239 and 360 mg/dl) in the INH-I and INH-II rats than in the i.p. rats (196 mg/dl). Blood lactate levels increased in both groups of rats inhaling CO yet remained virtually unchanged in the i.p. rats. Chen *et al.* (1981) also reported little change in heart rate in animals receiving CO by i.p. injection but an increased heart rate when CO was inhaled. These studies suggest that, despite similar COHb levels, CO inhalation is more stressful and elicits more powerful hyperglycemic and heart rate responses than CO injection.

Goldbaum *et al.* (1975) compared the effects of anemia to those of CO. Anemia was produced by bleeding and reinfusing either Ringer's lactate (hemoglobin reduced by 68%) or COHb red cells (COHb levels of 60%). Dogs breathing 13% CO for 15 minutes died in 15 minutes–1 hour with average COHb levels of 65%. The dogs bled and kept in an anemic state survived for an indefinite period of time. The dogs infused with COHb red cells also survived indefinitely with COHb levels of 60%.

In a second study (Goldbaum *et al.,* 1976), 100% CO was injected i.p. in dogs in doses varying from 20 to 200 ml/kg. No CO toxicity was observed, although COHb levels averaged 60% and reached as high as 80%.

Goldbaum *et al.* assert that even though high COHb levels are present in the dogs receiving CO by i.p. injection, the lack of toxicity suggests that there is little dissolved CO to compete with the relatively high O_2 tension for the cytochromes in the cells. Moreover, it is probable that after i.p. injection, all the dissolved CO is removed from the blood by the time the blood reaches the heart. In the animals inhaling CO experiencing high toxicity, the high CO tension in the lungs caused a high CO tension in the arterial blood. The findings of Goldbaum *et al.* are substantiated by the work of Wilks (1959) who injected CO into the peritoneal cavity of dogs and observed that none of the animals showed any ill effects even though COHb levels of 50% were obtained. Wilks noted that CO appeared in the alveolar air in 5 minutes and suggested that the pulmonary "blow-off" of CO prevented accumulations of COHb.

There is some disagreement over the rapidity with which CO and hemoglobin combine *in vivo*. Goldbaum *et al.* (1976) presented evidence that the *in vitro* combination of CO with Hb to form COHb is not an instantaneous reaction. They mixed blood with 100% CO and noted that complete saturation takes 20 minutes. On the other hand, Holland (1969), using a stopped-flow apparatus, demonstrated that the reaction between 100% CO and human blood is extremely rapid, occurring in seconds.

The studies by Goldbaum *et al.* (1975, 1976) have been criticized by Coburn (1979) as follows: (1) The concentration of inspired CO is expected to saturate blood leaving the lung, almost completely, for a few seconds after inspiration of the high concentrations of gas used. Thus, death in Goldbaum's dogs could have been the result of the sequalae of severe anoxia prior to mixing inspired CO in the entire body CO stores.

(2) There are data (Luomannaki and Coburn, 1969) that report survival of animals given CO via the lung very slowly at low concentrations in amounts sufficient to cause increases in COHb as high as reported by Goldbaum. (3) Death is a poor index and many other desirable measurements (i.e., blood pressure, blood gas tensions) that could have provided insight into the cause of death were not recorded in the study.

C. Tissue CO Tension

The question of whether CO has a direct effect *in vivo* has yet to be resolved, but several studies have suggested that there are significant partial pressures of CO in the tissues after CO inhalation. Campbell (1935) estimated CO pressures in the tissues of rabbits inhaling CO to be less than one-half that of the inspired air. Linderholm (1957) reported significant levels of CO in the brachial artery of patients breathing 500 and 1000 ppm CO during pulmonary function testing. Göthert *et al.* (1970), using nitrogen pneumoperitoneum techniques, found, in rabbits breathing 1000 ppm CO, CO concentrations of 154 ppm in the gas bubble after 2 hours and 460 ppm after 25 hours.

D. Isolated Tissues

The possibility that CO may have a direct toxic effect in addition to its hemoglobin mediated hypoxic effect has been investigated in several isolated organ preparations, also.

Duke and Killick (1952) have suggested that CO inhibits contractility of vascular smooth muscle. In their study, isolated cat lungs were perfused through the pulmonary artery while the lungs were ventilated with CO or nitrogen in air. Whereas ventilating the lungs with nitrogen in air produced an increase in pulmonary arterial pressure, ventilating with CO in air produced a decrease in pulmonary arterial pressure that was usually accompanied by bronchodilation. Pulmonary arterial pressure fell also when the lungs were ventilated with CO in nitrogen. The short, latent period between the administration of the CO and the decrease in pressure and the rapid return of pressure to its original level (when the lungs were ventilated with air alone) led the authors to conclude that the response could not be caused by the COHb concentration in the blood. Because the response could not be blocked by nerve-blocking agents, the authors speculated that CO may dilate some part of the pulmonary vascular bed directly.

Bassett and Fisher (1976) studied glucose metabolism in isolated rat lungs ventilated with 95% N_2–5% CO_2 and 95% CO–5% CO_2 mixtures.

Lactate plus pyruvate production was increased 187% with N_2 ventilation and 387% with CO ventilation. CO_2 production was decreased more markedly with CO ventilation. The authors concluded that CO ventilation is more effective than N_2 ventilation in inhibiting oxidative metabolism because CO not only displaces O_2 from the ventilating gas but also binds with cytochrome oxidase competitively. These investigators stated also that their results did not exclude the possibility that CO may have effects on lung tissues that are unrelated to cytochrome oxidase activity.

Scharf *et al.* (1975) studied the effects of CO and nitrogen hypoxia in the isolated, supported, dog heart preparation. They reported that at a constant perfusion pressure, as oxygen content was reduced from 20 to 5 vol %, coronary flow doubled with hypoxic hypoxia and tripled with CO hypoxia. Oxygen consumption and contractility increased with both types of hypoxia. Although the increase in oxygen consumption and contractility was eliminated by β-blockade, the difference in coronary flow between hypoxic hypoxia and CO hypoxia was not. In a second series, using β-blocked hearts, vascular resistance decreased more with CO than with hypoxic hypoxia. α- and β-Blockade eliminated the differences in vascular resistance completely, and the two types of hypoxia had the same effect. The authors concluded that CO has direct toxic effects in addition to those caused by lowering blood O_2 content. They suggested that the differences between CO and hypoxic hypoxia noted in their study could be attributed primarily to differences in adrenergic influences generated by each kind of hypoxia.

McGrath and Martin (1978) compared the effects of nitrogen-induced anoxia to CO-induced anoxia in the isolated, rat, right ventricle preparation stimulated to contract in hemoglobin-free Krebs–Ringer solution. Both gas stresses caused a decrease in developed tension, an increase in resting tension, and a reduction in percentage recovery when oxygen was reintroduced into the preparations. There was no significant difference in the decline in developed tension or the increase in resting tension between the two groups; however, after 10 min of reoxygenation, recovery was significantly greater in the CO-stressed muscle. Recovery from either anoxia was depressed severely when CO was present during the recovery phase. The results suggest also that the effects of CO on functioning tissue may be more complex than that caused solely by simple anoxia.

McGrath and Chen (1978) and Chen and McGrath (1978) studied the problem in both the spontaneously beating and stimulated Langendorf heart preparations. In the spontaneously beating preparation, heart rate and pulse pressure decreased more rapidly in the CO-challenged hearts. Recovery of heart rate and pulse pressure also were more rapid during reoxygenation after the CO challenge. Coronary flow increased initially with both challenges but then decreased. The initial increase was significantly greater and the subsequent decrease significantly less during nitrogen anoxia. During reoxygenation, coronary flow increased to control levels, and there was no significant difference between the two challenges. Lactate production increased with both CO and nitrogen stress; however, lactate production was significantly greater in the nitrogen-stressed hearts. During recovery, lactate production decreased to control levels in both the nitrogen-and CO-stressed hearts.

In stimulated Langendorf preparations both nitrogen and CO stress caused a rapid decrease in pulse pressure. There was no significant difference in the rate of pressure decline, and neither preparation recovered pulse pressure with reoxygenation. Coronary flow also decreased with both nitrogen and CO anoxia and recovered to 60% of the prestress value in both preparations. There was no significant difference in lactic acid production in both groups of hearts during either the stress or recovery periods. These studies suggest that CO does have a direct toxic effect on the heart in addition to its hypoxic effect, and, in the heart, the effect may be most pronounced on the conducting system.

This suggestion is supported by the observation that cardiac pacing during CO poisoning increases the chances of survival in dogs (Dergal *et al.*, 1976).

Coburn (1979) studied the effects of CO on isolated smooth-muscle strips. The author concluded that significant CO binding to cytochrome oxidase is not likely to be an *in vivo* mechanism of CO toxicity, at least in vascular tissue, because CO tensions more than 1000 times greater than those ever seen *in vivo* had only a small effect on oxygen uptake. A second finding, that CO relaxed aorta smooth muscle under conditions during which cyanide inhibited electron chain transport in mitochondria, suggests CO reacts with an intracellular enzyme other than cytochrome oxidase. The author discounted myoglobin, because so little of it was present in his preparation.

E. Altered Lethality

Several workers have reported experiments in which CO lethality was altered by factors that did not effect COHb levels.

Winston *et al.* (1974) described the effects of pretreatment with several drugs on the lethality of inhaled CO. Mice were pretreated with ethanol, chlorpromazine, or phenobarbital and then exposed for 4 hours to 1900 ppm CO or hypoxia (7.5% O_2). Pretreatment with either chlorpromazine or ethanol increased both CO and hypoxia lethality, whereas pretreatment with phenobarbital had no effect on CO lethality but increased hypoxia lethality. Because Jaeger and McGrath (1973) had shown that the development of hypothermia is associated with protection from CO lethality, Winston *et al.* tested the hypothesis that chlorpromazine and ethanol treatments may increase lethality by maintaining body temperature higher than it is in the nontreated controls. Before CO or drug treatment rectal temperatures were 38°C. After 4 hours exposure to CO, mice pretreated with chlorpromazine or ethanol had rectal temperatures that were equal to or lower than controls'. Thus, an increased CO lethality occurred in spite of a potential protective effect provided by drug-induced hypothermia. These workers studied the possibility of the drug pretreatments altering 2,3-DPG levels or the affinity of hemoglobin for oxygen and/or CO. *In vitro* studies demonstrated that drug pretreatment did not significantly change 2,3-DPG concentrations and had no effect on the rate of carboxyhemoglobin formation.

The authors investigated also the relationship between mortality and drug-induced changes in carboxyhemoglobin levels. Pretreatment with chlopromazine, while resulting in a significant increase in CO lethality, was associated with significantly lower carboxyhemoglobin concentrations. On the other hand, pretreatment with phenobarbital significantly increased carboxyhemoglobin levels, but had no significant effect on CO lethality. One-hour ethanol pretreatment significantly increased CO lethality without affecting carboxyhemoglobin content.

Winston and Roberts (1977) demonstrated that pretreating mice with potassium cyanate for 15 days resulted in a significant increase in hemoglobin–oxygen affinity, but had no significant effect on either CO-induced lethality or carboxyhemoglobin levels after the CO exposure. The authors pointed out that this observation was unexpected because it is generally accepted that an increased affinity of hemoglobin for oxygen should impede the unloading of oxygen to the tissues and aggravate the existing hypoxemia.

F.   Conclusions

Thus, there are studies that suggest CO has direct toxic effects as well as a hemoglobin-mediated effect. The effects may be caused, in part, by reactions with intracellular enzymes other than cytochrome oxidase or may be mediated by the sympathetic nervous system. In any event, the direct effects are evident only at high CO exposures.

III. CARDIOVASCULAR EFFECTS

A. Association with Cardiovascular Disease

The effects of CO on the heart and its association with cardiovascular disease have been reported in both human and animal studies. Anderson et al. (1973) demonstrated a shortening in the time to onset of angina in exercising human subjects exposed to CO. These workers showed that the mean duration of exercise before the onset of pain was reduced in patients breathing 50 or 100 ppm CO. Similar findings were reported in patients breathing 50 ppm CO (Aronow and Isbell, 1973) and in patients smoking low nicotine (Aronow and Swanson, 1969) and nonnicotine (Aronow and Rokaw, 1971) cigarettes.

That carbon monoxide and not nicotine may be the etiological agent responsible for the excess mortality seen in cigarette smokers is suggested by the studies of Wald et al. (1981). These workers determined that serum levels of cotinine (a principal metabolite of nicotine) were significantly higher in pipe smokers than in cigarette smokers, whereas CO levels were significantly higher in cigarette smokers. Wald et al. related these results to studies demonstrating that pipe smokers, in contrast with cigarette smokers, have no material excess risk of coronary disease and concluded that nicotine is not likely to be the major cause of excess coronary heart disease and mortality in smokers.

B. Cardiac Catheterization

The cardiovascular responses of humans to CO inhalation were studied during diagnostic cardiac catheterization (Ayres et al., 1969, 1970). Human subjects with and without coronary artery disease inhaled either 1000 ppm CO for 8–15 minutes or 50,000 ppm for 30–120 seconds. These procedures produced COHb levels of 5–25%. Both cardiac output and minute ventilation increased when the higher CO concentration was breathed; however, breathing 1000 ppm CO while increasing minute ventilation did not increase cardiac output. These results are in agreement with the earlier report by Chiodi et al. (1941) that gradual administration of lower levels of CO increases minute ventilation but not cardiac output.

Ayres et al. (1969) reported also that during CO breathing, both arterial P_{O_2} and mixed venous P_{O_2} decreased and that O_2 extraction from arterial blood increased. Coronary blood flow increased significantly in patients with noncoronary heart disease who breathed 50,000 ppm CO, but the increase was not significant in patients with coronary heart disease. The decreased coronary sinus P_{O_2} and the decreased myocardial lactate extraction indicate that the heart was subject to hypoxic insult.

The results of these studies suggest a threat to patients with coronary heart disease who are exposed to CO because of their inability to increase coronary blood flow to compensate for the COHb-induced hypoxia.

Ayres et al. observed similar decreases in coronary sinus P_{O_2} and lactate extraction in their concurrent studies of dogs. The authors concluded that CO inhalation decreased myocardial O_2 tension by three possible mechanisms:

1. Stimulation of the adrenergic system, causing increased ventricular work and oxygen demand.

2. Decreased oxygen extraction.

3. A leftward shift of the oxyhemoglobin dissociation curve, causing a decreased capillary oxygen tension.

The changes seen by Ayres et al. were consistent with the increase in cardiac output and pulse rate reported by Thomas and Murphy (1960) in healthy subjects after cigarette smoking and attributed, in part, to an acute elevation of COHb. These changes are similar also to those seen with adrenergic stimulation and lead Ayres to speculate that the circulatory response to cigarette smoking, carbon monoxide, and hypoxia may be mediated through the sympathetic nervous system.

C.   Ventricular Fibrillation Studies

The possibility that CO exposure may be associated with cardiac arrhythmias, particularly ventricular fibrillation, is suggested by several human and animal studies. In healthy firemen, CO exposure may be followed by disturbances in cardiac rhythm (Turino, 1981).

Other studies have demonstrated a reduced threshold for ventricular fibrillation in CO-exposed animals. After exposing monkeys with myocardial infarcts to 100 ppm CO, Debias et al. (1973) described electrophysiological changes suggestive of myocardial ischemia. Later, in 1976, Debias et al. reported a reduced threshold for ventricular fibrillation in monkeys after exposure to 100 ppm CO for 6 hours. Aronow et al. (1979) confirmed these observations in normal and infarcted dogs breathing 100 ppm CO for 2 hours.

D.   Hemodynamic Studies

The effects of CO on coronary flow, heart rate, blood pressure, cardiac output, and myocardial oxygen consumption have been described in a number of studies. The results are somewhat contradictory (partly because exposure regimes differed); however, most workers agree coronary flow increases with exposure to CO.

Increased coronary flow and heart rate and decreased myocardial oxygen consumption were described by Adams *et al.* (1973) in anesthetized dogs breathing 1500 ppm CO for 30 minutes. The decreased oxygen consumption indicated that the coronary flow response was not great enough to compensate for the decreased oxygen availability. The authors noted also that although there was a positive chronotropic response, there was no positive inotropic response. The authors speculated that (1) the CO may have caused an increase in the endogenous rhythm or blocked the positive inotropic response or (2) the response to CO was mediated reflexly through the cardiac afferent receptors to give a chronotropic response without the concomitant inotropic response. When they used β-adrenergic blocking agents, the heart rate response to CO disappeared. This suggests that the heart rate response may be mediated reflexly by the sympathetic nervous system.

In a later study, Young and Stone (1976) reported an increase in coronary blood flow with no change in myocardial oxygen consumption in chronically instrumented, awake dogs with COHb levels of 30%. The increased coronary blood flow occurred in animals with hearts paced at 150 beats/minute and as well as nonpaced animals and animals with propranolol and atropine blockade. Because the changes in coronary flow with arterial oxygen saturation were similar whether the animals were paced or not, the workers concluded that the increase in coronary flow is independent of changes in heart rate. Furthermore, the authors reasoned that if the coronary vasodilation were caused entirely by the release of a metabolic vasodilator material associated with decreased arterial oxygen saturation, the change in coronary flow in animals with both β-adrenergic and parasympathetic blockade would be the same as in the control dogs. The workers concluded that the coronary vasodilation observed with an arterial oxygen saturation reduced by CO is mediated partially through an active neurogenic process.

Increased myocardial blood flow after CO inhalation in dogs was confirmed by Einzig *et al.* (1980), who also demonstrated the regional nature of the blood flow response. Using labeled microspheres, these workers demonstrated that whereas both right and left ventricular beds were dilated maximally at COHb levels of 41%, subendocardial/subepicardial blood flow ratios were reduced. The authors concluded that in addition to the global hypoxia associated with carbon monoxide poisoning, there is also a relative underperfusion of the subendocardial layer, which is most pronounced in the left ventricle.

Horvath (1975) reported on the coronary flow response of dogs to COHb levels of 6.2–35.6% produced by continuous administration of precisely measured volumes of CO. Coronary blood flow increased progressively as blood COHb levels increased and were maintained for the duration of the experiment. However, when animals with complete atrioventricular block were maintained by cardiac pacemakers and exposed to COHb levels of 6 to 7%, there was no longer an increase in coronary blood flow. These results are most provocative, because they suggest an increased danger from low COHb levels in cardiac-disabled individuals.

Shephard (1972) exposed human subjects to CO in a double-blind experiment in which COHb levels of 2.4–5.4% were produced. Heart rates were measured for a 30-minute period before and a 30-minute period after gas exposure. CO had no effect on heart rate.

Weiss and Cohen (1974) exposed anesthetized rats to 80 and 160 ppm CO for 20-minute periods and measured tissue oxygen tension as well as heart rate. A statistically significant decrease in brain P_{O_2} occurred with inhalation of 160 ppm, but there was no change in heart rate.

Human volunteers were exposed by Stewart et al. (1973) to CO concentrations ranging from 1000 ppm for 10 minutes to 35,600 ppm for 45 seconds, producing COHb levels of 3–15%. There was no significant change in diastolic or systolic blood pressure, force of myocardial contractility, or heart rate during exposures to CO as high as 35,600 ppm.

Petajan et al. (1976) exposed unanesthetized rats to 1500 ppm CO to achieve blood COHb levels of 60–70%. After a slight transient increase, heart rate as well as blood pressure decreased throughout the exposure. The authors interpreted their data as indicating that the lowering of blood pressure was more important than the degree of hypoxia to the neurological impairment seen in their studies.

The effects of CO hypoxia and hypoxic hypoxia on arterial blood pressure and other vascular parameters were studied in carotid baroreceptor- and chemoreceptor-denervated dogs (Traystman and Fitzgerald, 1977). Arterial blood pressure was unchanged by CO hypoxia but increased with hypoxic hypoxia. Similar results were seen in carotid baroreceptor-denervated animals with intact chemoreceptors. Following carotid chemodenervation, arterial blood pressure decreased equally with both types of hypoxia.

In a subsequent report from the same laboratory (Sylvester et al., 1979), the effects of CO hypoxia and hypoxic hypoxia were compared in anesthetized, paralyzed dogs. Cardiac output and stroke volume increased during both CO and hypoxic hypoxia whereas heart rate was variable. Mean arterial pressure decreased during CO hypoxia, but increased during hypoxic hypoxia. Total peripheral resistance fell during both hypoxias, but the decrease was greater during the CO hypoxia. After resection of the carotid body, the circulatory effects of hypoxic and CO hypoxia were the same and were characterized by decreases in mean arterial pressure and total peripheral resistance. In a second series of closed-chest dogs, hypoxic and CO hypoxia caused equal catecholamine secretion before carotid body resection. After carotid body resection, the magnitude of the catecholamine response was doubled with both hypoxias. These workers conclude that the responses to hypoxic and CO hypoxia are different and that the difference is dependent on intact chemo- and baroreflexes and on differences in arterial oxygen tension, but not on differences in catecholamine secretion or ventilatory response.

E. Conclusions

Thus, the susceptibility of the cardiovascular system, and especially the heart, to CO exposure is reasonably well documented. The levels and the intensity of the exposure required to produce significant cardiovascular effects as well as the underlying mechanisms are areas of active research. Most studies indicate that CO increases coronary blood flow, has little effect on heart rate, and, depending on dosage, may decrease blood pressure. Cardiac output may be increased by exposure to high levels of CO and is unaffected by lower-level exposures. Myocardial oxygen consumption may be decreased by CO exposure, depending on the dosage.

IV. CO AT ALTITUDE

A. Emissions

The problems associated with CO exposure at altitude have not been studied extensively. Yet there are approximately 2.2 million people living at altitudes above 1524 m (5000 ft) in the United States. In addition, large numbers of tourists visit high altitude areas during the summer and winter months.

Increased use of heating devices (space heaters and fireplaces), for social effect as well as warmth, also increases CO emissions in mountain resort areas. Haagenson (1979) has reported that the NAAQS for CO of 9 ppm is frequently exceeded in Denver, Colorado (altitude 5280 ft), during the winter months.

Kirkpatrick and Reeser (1976) cited factors exacerbating ambient CO levels in mountain recreational communities of 8000 ft. If automobiles are tuned for mountain driving, they emit 1.8 times more CO at 8000 ft than at 5280 ft in Denver; they emit almost 4 times more if they are tuned for sea level driving conditions. The increased emissions from automobiles are compounded by driving at reduced speeds along steep grades under poor driving conditions. Because automobiles emit more CO and other pollutants at high altitude, if they are not tuned for such driving, influxes of tourists into high altitude resort areas may drastically increase pollution levels in general and especially the CO level (NRC, 1977). Finally, population growth concentrated along valley floors and the reduced volume of air for pollutant dispersal leads to accumulation of pollutants, including CO, in mountain valleys.

B. Compartmental Shifts

The possibility that CO may pose a special threat at altitude is suggested by the studies of Luomanmaki and Coburn (1969). These workers reported that during hypoxia, CO shifts out of the blood and into the tissues of anesthetized dogs. In experiments using ^{14}CO, they observed that radioactivity did not change when arterial oxygen tension was increased from 50 to 500 mm Hg; however, when arterial P_{O_2} was decreased to less than 40 mm Hg, ^{14}CO activity decreased to levels as low as 50% of control. When arterial P_{O_2} was returned to normal, the ^{14}CO reentered the blood. The possibility that the ^{14}CO had been sequestered in the spleen was excluded by determining that there was no significant difference between splenic and central venous ^{14}CO radioactivity, either before or after the ^{14}CO shift.

These workers studied also the problem of shift of CO out of the blood during hypoxia by measuring the rate of increase of blood COHb when CO was administered at a constant rate into a rebreathing system. They reasoned that if the partition of CO between vascular and extravascular stores remained constant, the increase in blood COHb should be proportional to the amount of CO administered. They found that COHb increased at a constant rate up to a saturation of 50%. With additional CO, there was a decrease in the rate at which COHb increased, suggesting that proportionately greater quantities of CO were going into the extravascular stores. The rate of COHb buildup became nonlinear at a COHb level of 50%, which corresponded to an arterial P_{O_2} of 80 mm Hg (Coburn, 1970). Agostoni et al. (1980) presented a theoretical model to support these observations: they developed a series of equations that predict that with decreased venous P_{O_2}, CO moves out of the vascular compartment and into skeletal and heart muscle, increasing the formation of carboxymyoglobin (COMB) in the tissues.

Available evidence suggests that during hypoxemia, CO moves into the extravascular compartment causing the COMB/COHb ratio to increase. During heavy smoking (COHb levels of 10%), as much as 30% of the cardiac myoglobin may be saturated with CO (Coburn, 1970). We may presume this situation would be worsened during altitude exposure because of the attendant arterial hypoxemia.

C. Cardiovascular Effects

Many studies compare the cardiovascular effects of CO to those of altitude, but there are relatively few studies of the effects of CO at altitude. Forbes *et al.* (1945) reported that CO uptake was increased during light activity at an altitude of 4877 m (16,000 ft). This was shown to be caused by altitude hyperventilation stimulated by a decreased arterial oxygen tension and not by a diminished barometric pressure per se.

An increased pulse rate in response to the combined stress of altitude and CO exposure was reported by Pitts and Pace (1947). The subjects were 10 healthy men who were exposed to simulated altitudes of 7,000, 10,000 and 15,000 ft with COHb levels of 0, 6, and 13%, respectively. The mean pulse rate during exercise and the mean pulse rate during the first 5 minutes after exercise had the greatest correlation with the change in the COHb level or inspired P_{O_2}. The authors stated that the response to a 1% increase in the COHb level was equivalent to that obtained by raising a normal group of men 335 ft in altitude. This relationship was stated only for a range of altitudes from 7000 to 10,000 ft and for increases in COHb up to 13%.

Weiser *et al.* (1978) studied the effects of low-level CO exposure on aerobic work at an altitude of 1810 m (Denver, Colorado). The young subjects inhaled a bolus of 100% CO to achieve COHb levels of 5%. CO impaired work performance to the same extent that it did at sea level. There were small but significant submaximal exercise changes in cardiorespiratory function during CO exposure. The working heart rate increased and the postexercise left ventricular ejection time shortened but not to the same extent as when filtered air was breathed. Exposure to CO resulted also in a lower anaerobic threshold and a greater minute ventilation at work rates heavier than the anaerobic threshold, because the blood lactate level was increased.

TABLE III

Approximate Physiologically Equivalent Altitudes at Equilibrium with Ambient Carbon Monoxide Concentrations (National Research Council, 1977)

Ambient carbon monoxide concentration (ppm)	Physiologically equivalent altitude at actual altitude of:					
	0 ft[a]	0 m[a]	5,000 ft	1,524 m	10,000 ft	3,048 m
0	0	0	5,000	1,524	10,000	3,048
25	6,000	1,829	8,300	2,530	13,000	3,962
50	10,000	3,048	12,000	3,658	15,000	4,572
100	12,300	3,749	15,300	4,663	18,000	5,486

[a] Sea level.

D. Psychophysiological Effects

Most physiological data that has been gathered on the effects of combined CO–altitude exposure comes from psychophysiological studies, and it is on the basis of these studies that the concept of physiologically equivalent altitudes has been determined (Table III).

McFarland et al. (1944) reported that changes in visual sensitivity took place at COHb concentrations of 5% or at a simulated altitude of approximately 2438 (8000 ft). Later, McFarland (1970) expanded on the original study and pointed out that a pilot flying at 6000 ft with 0.005% CO in the air is at an altitude physiologically equivalent to approximately 12,000 ft. McFarland states that the visual acuity test is so sensitive that even the effects of small quantities of CO absorbed from cigarette smoke are clearly demonstrable. In subjects inhaling smoke from three cigarettes, the saturation of blood with CO was equal to that at an altitude of approximately 7500 ft. The loss of arterial oxygen saturation resulting from the decreased P_{O_2} of the inspired air was approximately 4%. The absorption of a similar amount of CO at 7500 ft caused a combined loss of visual sensitivity equal to that which occurs at 10,000–11,000 ft.

The initial report (McFarland et al., 1944) was confirmed by Halperin et al. (1959), who determined also that recovery from the detrimental effects on visual sensitivity lagged behind carbon monoxide elimination from the blood.

Combined exposure to altitude and CO causes a decrease in flicker-fusion frequency (FFF), i.e., the crucial frequency in cycles per second at which a flickering light appears to be steady (Lilienthal and Fugitt, 1946). Whereas mild hypoxia (that occurring at 9000–12,000 ft) alone impairs FFF, COHb levels of 5–10% decreased the altitude threshold for onset of impairment to 5000–6000 ft.

The psychophysiological effects of CO at high altitude are a particular haz⟨...⟩ high performance aircraft (Denniston *et al.*, 1978). Acute as-⟨...⟩ titude causes increased ventilation caused by the stimulating ⟨...⟩ a reduced arterial P_{O_2} on the chemoreceptors. The increased ⟨...⟩ leads to a slight increase in blood pH and a leftward shift in the oxyhemoglobin dissociation curve. Although such a small shift would probably have no physiological consequences under normal conditions, it may take on physiological importance for aviators required to fly and perform tedious tasks involving a multitude of cognitive processes under a variety of operational conditions. The leftward shift of the oxyhemoglobin dissociation curve may be further aggravated by preexisting alkalosis caused by hyperventilation resulting from anxiety. The potential for this effect has been reported by Pettyjohn *et al.* (1977), who reported that respiratory minute volume may be increased by 110% during final landing approaches requiring night vision devices. Thus, the hypoxia-inducing effects of CO inhalation would accentuate the cellular hypoxia caused by stress and altitude-induced hyperventilation.

E. Conclusions

Thus, while the literature abounds with studies comparing and contrasting CO and altitude exposures, there are relatively few reports on the effects of CO at altitude. There are data to support the concept that the effects of these two hypoxia-inducing experiences are at least additive. The data presented by Luomanmaki and Coburn (1969) suggest that the effects of CO at altitude may be more than simply additive. When the Eisenhower Tunnel was built at an altitude of 11,000 ft, concern was expressed for the hypoxia that could potentially result from the decreased partial pressure of O_2 caused by the elevation combined with CO from smoking and automobile exhaust. Based on a study by Miranda *et al.* (1967), it was recommended that CO concentration in the tunnel be kept below 25 ppm.

V. CHRONIC EFFECTS

A. Early Studies

Numerous subgroups within the population are exposed chronically to low levels of CO from motor vehicles, fires, defective home heating equipment, and smoking (Goldsmith, 1970); however, the physiological consequences of prolonged exposure to low levels of CO remain largely unexplored.

Nasmith and Graham (1906), early workers in the field, reported an increased mass of circulating red blood cells in response to CO exposure. In later studies, Campbell (1932, 1935) demonstrated increased red blood cell counts, hemoglobin concentrations, and cardiac hypertrophy in mice and rabbits exposed to 0.3% for up to 308 days. Red blood cells in the mice increased from 9 to 17 million/mm^3, whereas the hemoglobin concentration increased up to 155% of control. Similar changes were noted for the rabbit. A curious observation was that mice were able to acclimate better to CO than to low oxygen in the air.

Clark and Otis (1957) exposed mice to gradually increasing concentrations of CO (maximum 0.15%) for a 14-day exposure. The blood oxygen-carrying capacity and hematocrit ratio increased 50 and 85%, respectively, above controls. These increases were similar to that seen in animals exposed to altitude (14,000–18,000 ft) for 14 days and were sufficient to provide some degree of protection against a subsequent altitude challenge of 34,000 ft.

B. Hematological Changes

The earlier reports of elevated hemoglobin concentrations, hematocrit ratios, and erythrocyte counts in response to CO exposure have been confirmed in more recent studies in which the time course as well as the intensity of the CO exposure have been considered (Wilks et al., 1959; Syvertsen and Harris, 1973; Jaeger and McGrath, 1975; Penney and Bishop, 1978; McGrath et al., 1979). Jaeger and McGrath (1975) reported elevated hematocrit ratios and hemoglobin concentrations in Japanese quail exposed to 300–350 ppm CO for 4 weeks to attain COHb levels of 30%.

Penney and Bishop (1978) reported that hemoglobin concentrations, hematocrit ratios, and red blood cell counts increased dramatically in animals exposed to 500 ppm CO. The initial increases in these parameters appeared within 4–7 days, plateaued after 25–30 days, and remained high as long as CO treatment continued. In their study, hemoglobin concentrations, hematocrit ratios, and erythrocyte counts increased to 50, 44, and 42% of control values, respectively. Reticulocyte counts increased sharply within 3 days of exposure. By day 26 of the exposure, reticulocyte counts returned to control levels and remained there as long as CO treatment continued. After the CO exposure was terminated, reticulocyte counts began to fall, coincident with decreases in hemoglobin concentration, hematocrit ratio, and red blood count.

A rapid increase in reticulocytes was reported also by McGrath *et al.* (1979), who noted that reticulocyte counts were significantly elevated after the first day of exposure to 450 ppm CO. Hemoglobin concentration and hematocrit ratios were elevated significantly by the third week and remained elevated for the remainder of the 5-week exposure.

The mechanism responsible for the hematological changes seen in chronic CO exposure is believed to be lowered tissue oxygen delivery resulting from elevated COHb levels. Reticulocyte release from the hemopoietic organs as well as other aspects of erythropoietic activity is stimulated by erythropoietin, a hormone released by the kidney in response to tissue hypoxia. Studies (Syvertsen and Harris, 1973; Guidi *et al.*, 1977) indicate that exposure to CO, as well as to high altitude, enhances erythropoietin production and that increased erythropoieses parallels blood erythropoietin levels. The enhanced erythropoieses increases the oxygen-transport capacity of the blood, which can be measured as an increase in hemoglobin concentration and hematocrit ratio.

Blood volume is increased by chronic CO exposure caused primarily by an increased circulating red blood cell mass; however, plasma volume, which may be depressed in the first few days of altitude exposure, is usually slightly elevated by CO exposure. Wilks *et al.* (1959) reported a 17% increase in plasma volume in dogs exposed intermittantly to 800–1100 ppm CO; however, blood volume increased 52% because of the increased red blood cell mass. Siggaard-Anderson *et al.* (1968) found little or no change in plasma volume with CO exposure. Jaeger and McGrath (1975) reported plasma volume increases of 5.3% and blood volume increases of 12.2% in CO-exposed Japanese quail.

Increased red blood volume may be at least partially responsible for the increased CO tolerance seen in animals chronically exposed or "acclimated" to CO. Wilks *et al.* (1959) point out that because the rate of CO uptake appears to be independent of the O_2 capacity of the blood, those animals with normal capacities will undergo more rapid rises in the percentage hemoglobin saturated with CO than acclimated animals with higher capacities. Furthermore, because acclimated animals have greater hemoglobin levels, even though they exhibit the same percentage of COHb in the steady state as normal animals, they have more hemoglobin available for O_2 combination when the steady state is reached. If, for example, the O_2 capacities of three dogs were 18, 26, and 30 ml/100 ml, the time to reach 50% COHb would then be x, $1.44x$, and $1.67x$ time units. At 50% COHb saturation, enough hemoglobin would be available to combine with 9, 13, and 15 ml O_2/100 ml blood, respectively, in the three dogs. Thus, the CO tolerance provided by an increased blood volume can be attributed to the longer time required to attain a given

CO saturation level and a greater hemoglobin reserve for oxygen transport at that CO saturation level.

C. Morphological Changes

The early studies reporting cardiac enlargement in response to chronic CO exposure have been verified by more recent studies in which different animal models as well as the time course of the change have been investigated. Theodore *et al.* (1971) reported cardiac hypertrophy in rats but not in dogs, baboons, or monkeys exposed to 400–500 ppm for 168 days. Penney *et al.* (1974a) noted that heart mass in rats exposed to 500 ppm increased rapidly, reaching within 14 days a weight 1/3 greater than that predicted. At the end of 42 days of exposure, heart weights ranged from 140 to 153% above those of unexposed controls. Changes in cardiac lactate dehydrogenase (LDH) were also reported by these workers and paralleled closely the heart weight changes. The changes in isoenzyme composition were similar to those seen in other conditions that cause cardiac hypertrophy, including aortic and pulmonary artery constriction, coronary artery disease, altitude acclimation, and severe anemia.

In an attempt to determine the threshold of cardiac hypertrophy and to characterize further the response thereto, Penney *et al.* (1974b) exposed rats continuously to 100, 200, and 500 ppm for varying lengths of time, which produced blood COHb levels of 9.26, 15.82, and 41.14%. Significant differences in heart weight were seen only after the 200 and 500 ppm exposures. The increases in heart weight were noted in the left ventricle plus septum, right ventricle, and especially the atria. Unlike cardiac enlargement caused by altitude, which primarily involves the right ventricle, cardiac enlargement caused by CO produces an overall increase in heart weight. The authors conclude that the threshold of the hemoglobin response is 100 ppm CO whereas that required for cardiac enlargement is near 200 ppm. Right and left ventricular hypertrophy was reported also in rats exposed intermittently (6 hours/day, 5 days/week, for 5 weeks) to 450 ppm CO (McGrath *et al.*, 1979).

Styka and Penney (1978) studied regression of cardiomegaly in rats exposed previously to moderate (400 ppm) or severe (500–1100 ppm) CO for 6 weeks. The ratio of heart weight to body weight (HW/BW) increased from 2.65 in controls to 3.52 and 4.01 with moderate and severe exposure, respectively. Myocardial lactate dehydrogenase M subunits (M LDH) were elevated 12–14% by severe and 5–6% by moderate CO exposure. Forty-one to 48 days after terminaton of the CO exposure, there were no significant differences in hemoglobin concentrations among treated and control groups. HW/BW ratios were similar in the control and moderately exposed animals, but remained significantly elevated in the severely exposed group. When the CO exposure was terminated, M LDH values decreased to control values in the moderately exposed animals within 44 days, but remained slightly elevated after 46 days in the severely exposed animals.

These data indicate that effects of the CO-induced change in myocardial mass and LDH isozyme composition remain after the CO stress is relieved, at least for the duration of these studies. Furthermore, the reversibility of the stress is determined in part by its severity. Although the effects of moderate CO stress may be readily reversible, recovery from severe stress may require a much longer recovery period.

D. Hemodynamic Changes

The hemodynamic consequences of chronic CO exposure have been examined by Penney and Bishop (1978) and James *et al.* (1979). Penney and Bishop (1978) exposed rats to 500 ppm CO for 1–42 days, achieving COHb levels of 38–42%. Hematocrit ratios increased from 49.8 to 69.7%, and cardiomegaly developed. The CO-exposed animals were then studied in an open chest, anesthetized preparation. Stroke index, mean stroke power, and mean cardiac output were reported to increase sharply with initial CO exposure and remain elevated for the duration of the exposure. Total systemic and pulmonary resistances were reported to fall sharply and remain depressed for the duration of the exposure. There were slight but nonsignificant changes in left and right ventricular systolic pressures, mean aortic pressure, and maximum left ventricular dp/dt. There was no consistent change in heart rate. The authors concluded that enhanced cardiac output via an increased stroke volume is a compensatory mechanism to provide tissue oxygenation during CO intoxication and that the increased work may be the major factor responsible for the development of cardiomegaly.

The effect of 2 weeks exposure to 160–220 ppm CO on cardiac function was studied in chronically instrumented goats by James *et al.* (1979). These workers reported that even though oxygen transport was reduced, there was no change in cardiac index or stroke volume. Left ventricular contractility and heart rate were also unchanged during the CO exposure, but were depressed significantly during the first week after termination of the exposure. The differences in the two studies may be caused in part by differences in the intensity of CO exposure, species differences, and, because the James study was conducted in Denver (altitude 1600 m), exposure to a moderately increased altitude as well as CO hypoxia.

E.   Critical Levels

Chronic exposure of mammals to CO concentrations less than 70 ppm has had little demonstrable effect on a wide variety of parameters (Stupfel and Bouley, 1970; Eckhardt et al., 1972); however, Musselman et al. (1959) did report slight but statistically significant increases in hemoglobin concentration, hematocrit ratios, and red blood cell counts in dogs exposed to 50 ppm for 3 months. There were no significant changes in reticulocytes, white blood cells, or differential count. There were also no differences in body, heart, adrenal, thymus, and spleen weights. These authors felt that this level of CO exposure did not produce harmful effects and stated that 50 ppm would be safe for continuous human exposure.

Exposure to CO concentrations of 100 ppm and higher elicits many of the physiological adjustments observed in chronic high altitude experiments, including increased hematocrit ratio and hemoglobin concentration, increased hypoxic resistance, and cardiac enlargement (Clark and Otis, 1957; Jones et al., 1971; Penney et al., 1974b).

Lewey and Drabkin (1944) reported that dogs exposed intermittently to 100 ppm CO for 11 weeks showed a consistent disturbance of postural and position reflexes and gait. Electrocardiograms of some of the animals showed pathological changes characteristic of anoxia. Furthermore, Lewey and Drabkin reported some necrosis of single heart muscle fibers and changes in the CNS that followed the course of blood vessels and were similar to those found in acute CO poisoning.

These investigators reported also on one dog in which the posterior coronary artery had been ligated before the CO exposure. This animal showed the earliest and most severe cardiac and cerebral responses to CO, which suggested that a compromised heart may increase the general risk in CO poisoning.

Thus, it appears that exposure to 50 ppm for prolonged periods does produce slight effects to which the healthy animal can usually adapt. It is not known, however, what constitutes the limits of safe exposure for animals with compromised circulation.

REFERENCES

Adams, J. D., Erickson, H. H., and Stone, H. L. (1973). Myocardiac metabolism during exposure to carbon monoxide in the conscious dog. *J. Appl. Physiol.* **34,** 238–242.
Agostoni, A., Stabilini, R., Viggiano, G., Luzzana, M., and Samaja, M. (1980). Influence of capillary and tissue PO_2 on carbon monoxide binding to myoglobin: A theoretical evaluation. *Microvasc. Res.* **20,** 81–87.

American Conference of Governmental and Industrial Hygienists (ACGIH) (1978). "Threshold Limit Values for Chemical Substances in the Work Room Air." ACGIH, Cincinnati, Ohio.

Anderson, D. E. (1971). Problems created for ice arenas by engine exhaust. *Am. Ind. Hyg. Assoc. J.* **32,** 790–801.

Anderson, E. W., Andelman, R. J., Strauch, J. M., Fortuin, N. J., and Knelson, J. H. (1973). Effects of low-level carbon monoxide exposure on onset and duration of angina pectoris. *Ann. Intern. Med.* **79,** 46–50.

Aronow, W. S., and Isbell, M. W. (1973). Carbon monoxide effects on exercise-induced angina pectoris. *Ann. Intern. Med.* **79,** 392–395.

Aronow, W. S., and Rokaw, S. N. (1971). Carboxyhemoglobin caused by smoking non-nicotine cigarettes. Effects in angina pectoris. *Circulation* **44,** 782–788.

Aronow, W. S., and Swanson, A. J. (1969). The effect of low-nicotine cigarettes on angina pectoris. *Ann. Intern. Med.* **71,** 599–601.

Aronow, W. S., Stemmer, E. A., and Zweig, S. (1979). Carbon monoxide and ventricular fibrillation threshold in normal dogs. *Arch. Environ. Health* **34,** 184–186.

Ayres, S. M., Mueller, H. S., Gregory, J. J., Gianelli, S., Jr., and Penny, J. L. (1969). Systemic and myocardial hemodynamic responses to relatively small concentrations of carboxyhemoglobin (COHb). *Arch. Environ. Health* **18,** 699–709.

Ayres, S. M., Gianelli, S., Jr., and Mueller, H. (1970). Myocardial and systemic responses to carboxyhemoglobin. *Ann. N.Y. Acad. Sci.* **174,** 268–293.

Basset, D. J., and Fisher, A. B. (1976). Metabolic response to carbon monoxide by isolated rat lungs. *Am. J. Physiol.* **230,** 658–663.

Bernard, C. (1857). Leçons surs les effets des substances toxiques et médicamenteuses. Baillière, Paris.

Bondi, K. R., Very, K. R., and Schaefer, K. E. (1978). Carboxyhemoglobin levels during a submarine patrol. *Aviat., Space Environ. Med.* **49,** 851–854.

Cambell, J. A. (1932). Hypertrophy of the heart in acclimatization to chronic carbon monoxide poisoning. *J. Physiol. (London)* **77,** 8P–9P.

Campbell, J. A. (1935). Growth, fertility, etc. in animals during attempted acclimatization to carbon monoxide. *Q. J. Exp. Physiol.* **24,** 271–81.

Chen, K. C., and McGrath, J. J. (1978). Effects of nitrogen and carbon monoxide on the stimulated rat heart. *Fed. Proc., Fed. Am. Soc. Exp. Biol.* **37,** 230.

Chen, K. C., McGrath, J. J., and Lee, P. S. (1979). Effects of intraperitoneal injection of carbon monoxide on circulating blood glucose and lactate. *Fed. Proc., Fed. Am. Soc. Exp. Biol.* **38,** 1313.

Chen, K. C., McGrath, J. J., and Vostal, J. J. (1981). Effects of intraperitoneal injection of carbon monoxide on heart and respiration. *Fed. Proc., Fed. Am. Soc. Exp. Biol.* **40,** 609.

Chiodi, H., Dill, D. B., Consolazio, F., and Horvath, S. M. (1941). Respiratory and circulatory responses to acute carbon monoxide poisoning. *Am. J. Physiol.* **134,** 683–693.

Clark, R. T., Jr., and Otis, A. B. (1957). Comparative studies on acclimatization of mice to carbon monoxide and to low oxygen. *Am. J. Physiol.* **169,** 285–294.

Coburn, R. F. (1970). The carbon monoxide body stores. *Ann. N.Y. Acad. Sci.* **174,** 11–22.

Coburn, R. F. (1979). Mechanisms of carbon monoxide toxicity. *Prev. Med.* **8,** 3310–3322.

Coburn, R. F., Blakemore, W. S., and Forster, R. E. (1963). Endogenous carbon monoxide production in man. *J. Clin. Invest.* **42,** 1172–1178.

Coburn, R. F., Forster, R. E., and Kane, P. B. (1964). Considerations of the physiological variables that determine the blood carboxyhemoglobin concentration in man. *J. Clin. Invest.* **44,** 1899–1910.

Cuddebach, J. E., Donovan, J. R., and Burg, W. R. (1976). Occupational aspects of passive smoking. *Am. Ind. Hyg. Assoc. J.* **37,** 263–267.

Davies, D. M. (1975). The application of threshold limit values for carbon monoxide under conditions of continuous exposure. *Ann. Occup. Hyg.* **18,** 21–28.

Debias, D. A., Banerjee, C. M., and Birkhead, N. C. (1973). Carbon monoxide inhalation effects following myocardial infarction in monkeys. *Arch. Environ. Health* **27,** 161–167.

Debias, D. A., Banerjee, C. M., Birkhead, N. C., Greene, C. H., Scott, S. D., and Harper, W. V. (1976). Effects of carbon monoxide inhalation on ventricular fibrillation. *Arch. Environ. Health* **31,** 42–46.

Denniston, J. C., Pettyjohn, F. S., Boyter, J. K., Kelliher, J. K., Hiott, B. K., and Piper, C. F. (1978). "The Interaction of Carbon Monoxide and Altitude on Aviator Performance: Pathophysiology of Exposure to Carbon Monoxide," U.S. Army Aeromed. Res. Lab. Rep. 78-7. Fort Rucker, Alabama.

Dergal, E., Hodjati, H., Goldbaum, L., and Absocon, K. (1976). Effects of cardiac pacing in acute carbon monoxide intoxication in dogs. *Chest* **70,** 424.

Drabkin, S. L., Lewey, F. W., Bellet, S., and Ehrich, W. H. (1943). The effect of replacement of normal blood erythrocytes saturated with carbon monoxide. *Am. J. Med. Sci.* **205,** 755–756.

Duke, H., and Killick, E. M. (1952). Pulmonary vasomotor responses of isolated, perfused cat lungs to anoxia. *J. Physiol. (London)* **117,** 303–316.

Eckhardt, R. E., McFarland, H. N., Alarie, Y. C., and Busey, W. M. (1972). The biological effect from long-term exposure of primates to carbon monoxide. *Arch. Environ. Health* **25,** 381–387.

Einzig, S., Nicoloff, D., and Lucas, R. J., Jr. (1980). Myocardial perfusion abnormalities in carbon monoxide poisoned dogs. *Can. J. Physiol. Pharmacol.* **58,** 396–405.

Environmental Protection Agency (EPA) (1979). "Air Quality Criteria for Carbon Monoxide," EPA-60018-79-022. EPA, Washington, D.C.

Forbes, W. H., Sargent, F., and Roughton, F. J. W. (1945). The rate of carbon monoxide uptake by normal man. *Am. J. Physiol.* **143,** 594–608.

Fukuda, T., and Kobayashi, T. (1961). On the relation of chemoreceptor stimulation to epinephrine secretion in anoxemia. *J. Jpn. Physiol.* **11,** 467–475.

Ginsburg, M. D. (1980). Carbon monoxide. *In* "Experimental and Clinical Neurotoxicology" (P. S. Spencer and H. H. Schaumberg, eds.), pp. 374–394. Williams & Wilkins, Baltimore, Maryland.

Goldbaum, L. R., Ramirez, R. G., and Absalon, K. (1975). Joint committee on aviation pathology. XIII. What is the mechanism of carbon monoxide toxicity? *Aviat., Space Environ. Med.* **46,** 1289–1291.

Goldbaum, L. R., Orellano, T., and Dergal, E. (1976). Mechanisms of the toxic action of carbon monoxide. *Ann. Clin. Lab. Sci.* **6,** 372–376.

Goldsmith, J. R. (1970). Contribution of motor vehicle exhaust, industry and cigarette smoking to community carbon monoxide exposure. *Ann. N.Y. Acad. Sci.* **174,** 122–134.

Göthert, M., Lutz, F., and Malorny, G. (1970). Carbon monoxide partial pressure in tissue of different animals. *Environ. Res.* **3,** 303–309.

Guidi, E. E., Buys, M. C., Madanes, A., Miranda, C., Scaro, J. L., and Carrera, M. A. (1977). Influence of ambient temperature on erythropoietin production in carbon monoxide-intoxicated mice. *Rev. Esp. Fisiol.* **33,** 67–68.

Haagenson, P. L. (1979). Meteorological and climatological factors affecting Denver air quality. *Atmos. Environ.* **13,** 79–85.

Haldane, J. S. (1895). The action of carbonic oxide on man. *J. Physiol. (London)* **45**, 430–462.

Halperin, M. H., McFarland, R. A., Niven, J., and Roughten, F. J. W. (1959). The effects of carbon monoxide and altitude on visual thresholds. *J. Aviat. Med.* **15**, 381–394.

Hart, M. H. (1979). Was the pre-biotic atmosphere of the earth heavily reducing? *Origins Life* **9**, 261–266.

Hoegg, V. R. (1972). Cigarette smoke in closed spaces. *Environ. Health Perspect.* **2**, 117–128.

Holland, R. A. B. (1969). Rate of dissociation from O_2Hb and relative combination rate of CO and O_2 in mammals at 37°C. *Respir. Physiol.* **7**, 30–42.

Horvath, S. M. (1975). "Influence of Carbon Monoxide on Cardiac Dynamics in Normal and Cardiovascular Stressed Animals." University of California Institute of Environmental Studies, Santa Barbara.

Jaeger, J., and McGrath, J. J. (1973). Effects of hypothermia on response of neonatal chicks to carbon monoxide. *J. Appl. Physiol.* **34**, 564–567.

Jaeger, J., and McGrath, J. J. (1975). Hematological and biochemical effects of chronic CO exposure on the Japanese quail. *Respir. Physiol.* **24**, 365–372.

Jaffee, L. S. (1970). Sources, characteristics and fate of atmospheric carbon monoxide. *Ann. N.Y. Acad. Sci.* **174**, 76–88.

James, W. E., Tucker, C. E., and Grover, R. F. (1979). Cardiac function in goats exposed to carbon monoxide. *J. Appl. Physiol.* **47**, 429–434.

Johnson, C. J., Moran, J. C., Paine, S. C., Anderson, H. W., and Breysse, P. A. (1965). Abatement of toxic levels of carbon monoxide in Seattle ice-skating rinks. *Am. J. Public Health* **65**, 1087–1090.

Jones, R. A., Strickland, J. A., Stunkard, J. A., and Siegel, J. (1971). Effects on experimental animals of long-term inhalation exposure to carbon monoxide. *Toxicol. Appl. Pharmacol.* **19**, 46–53.

Kirkpatrick, L. W., and Reeser, W. K., Jr. (1976). The air pollution carrying capacities of selected Colorado Mountain Valley ski communities. *J. Air Pollut. Control Assoc.* **26**, 992–994.

Lawther, P. J., and Commins, B. T. (1970). Cigarette smoking and exposure to carbon monoxide. *Ann. N.Y. Acad. Sci.* **174**, 135–147.

Lewey, F. H., and Drabkin, D. L. (1944). Experimental chronic carbon monoxide poisoning in dogs. *Am. J. Med. Sci.* **208**, 502–511.

Lilienthal, J. L., Jr., and Fugitt, C. H. (1946). The effect of low concentrations of carboxyhemoglobin on the "altitude tolerance" of man. *Am. J. Physiol.* **145**, 359–364.

Linderholm, H. (1957). On the significance of CO tension in the pulmonary capillary blood for determination of pulmonary diffusion capacity with the steady state CO method. *Acta Med. Scand.* **156**, 420–426.

Longo, L. (1977). The biological effect of carbon monoxide on the pregnant woman, fetus and newborn infant. *Am. J. Obstet. Gynecol.* **129**, 69–103.

Luomanmaki, K., and Coburn, R. F. (1969). Effects of metabolism and distribution of carbon monoxide on blood and body stores. *Am. J. Physiol.* **217**, 354–362.

MacEwen, J. D. (1973). Toxicology. *In* "Bioastronautics Data Book" (J. F. Parker and V. R. West, eds.), 2nd ed., pp. 455–487. Natl. Aeron. Space Admin., Washington, D.C.

McFarland, R. A. (1970). The effects of exposure to small quantities of carbon monoxide on vision. *Ann. N.Y. Acad. Sci.* **174**, 301–312.

McFarland, R. A., Roughton, F. J. W., Halperin, M. H., and Niven, J. (1944). The effects of carbon monoxide and altitude on visual thresholds. *J. Aviat. Med.* **15**, 381–394.

McGrath, J. J., and Chen, K. C. (1978). Comparative effects of nitrogen and carbon monoxide on the isolated rat heart. *Toxicol. Appl. Pharmacol.* **45**, 231.

McGrath, J. J., and Martin, Z. G. (1978). Effects of carbon monoxide on isolated heart muscle. *Proc. Soc. Exp. Biol. Med.* **157,** 681–683.

McGrath, J. J., Chen, K. C., and Vostal, J. J. (1979). Adaptive reaction to carbon monoxide-induced hypoxia. *Toxicol. Appl. Pharmacol.* **48,** A56.

Miranda, J. M., Konopinski, V. J., and Larsen, R. I. (1967). Carbon monoxide control in a highway tunnel. *Arch. Environ. Health* **15,** 16–25.

Musselman, N. P., Groff, W. A., Yevich, P. P., Wilinski, F. T., Weeks, M. H., and Oberst, F. W. (1959). Continuous exposure of laboratory animals to low concentrations of carbon monoxide. *Aerosp. Med.* **30,** 524–529.

Nasmith, G. G., and Graham, D. A. (1906). The haematology of carbon monoxide poisoning. *J. Physiol. (London)* **35,** 32–52.

National Research Council (NRC) (1977). "Carbon Monoxide." Committee on Medical and Biological Effects of Environmental Pollutants, Natl. Acad. Sci., Washington, D.C.

Peirce, E. C., Zacharias, A., Alday, J. M., Hoffman, B. A., and Jacobson, J. H. (1972). Carbon monoxide poisoning: Experimental hypothermic and hyperbaric studies. *Surgery* **72,** 229–232.

Penney, D., Dunham, E., and Benjamin, M. (1974a). Chronic carbon monoxide exposure: Time course of hemoglobin, heart weight and lactate dehydrogenase isozyme changes. *Toxicol. Appl. Pharmacol.* **28,** 493–497.

Penney, D., Benjamin, M., and Dunham, E. (1974b). Effects of carbon monoxide on cardiac weight as compared with altitude effects. *J. Appl. Physiol.* **37,** 80–84.

Penney, D. G., and Bishop, P. A. (1978). Hematological changes in the rat during and after exposure to carbon monoxide. *J. Environ. Pathol. Toxicol.* **2,** 407–415.

Petajan, J. H., Packham, S. C., Frens, D. B., and Dinger, B. G. (1976). Sequelae of carbon monoxide-induced hypoxia in the rat. *Arch. Neurol. (Chicago)* **33,** 152–157.

Pettyjohn, F. S., McNeil, R. J., Akers, L. A., and Faber, J. M. (1977). "Use of Inspiratory Minute Volumes in Evaluation of Rotary and Fixed Wing Pilot Workload," U.S. Army Aeromed. Res. Lab. Rep. 77-9. Fort Rucker, Alabama.

Pitts, G. C., and Pace, N. (1974). The effect of blood carboxyhemoglobin concentration on human hypoxia tolerance. *Am. J. Physiol.* **148,** 139–151.

Polabkin, D. L., Lewey, F. L., Bellet, S., and Ehrich, W. H. (1943). The effect of replacement of normal blood by erythrocytes saturated with carbon monoxide. *Am. J. Med. Sci.* **205,** 755–756.

Ratney, R. S., Wegman, D. H., and Elkins, H. B. (1974). In vivo conversion of methylene chloride to carbon monoxide. *Arch. Environ. Health* **28,** 223–226.

Root, W. S. (1965). Carbon monoxide. *In* "Handbook of Physiology" (W. O. Fenn and H. Rahn, eds.), Sect. 3, Vol. II, pp. 1087–1098. Am. Physiol. Soc., Washington, D.C.

Russell, M. A. H., Cole, P. V., and Brown, G. (1973). Absorption by non-smokers of carbon monoxide from room air polluted by tobacco smoke. *Lancet* **1,** 576–579.

Scharf, S. M., Permutt, S., and Bromberger-Barnea, B. (1975). Effects of hypoxic and CO hypoxia on isolated hearts. *J. Appl. Physiol.* **39,** 752–758.

Shephard, R. J. (1972). The influence of small doses of carbon monoxide upon heart rate. *Respiration* **29,** 516–521.

Siggaard-Andersen, J., Petersen, F. B., Hansen, T. I., and Mellemgaard, K. (1968). Plasma volume and vascular permeability during hypoxia and carbon monoxide exposure. *Scand. J. Clin. Lab. Invest., Suppl.* **103,** 39–48.

Sjöstrand, T. (1970). Early studies of CO production. *Ann. N.Y. Acad. Sci.* **174,** 5–10.

Sokal, J. A. (1975). Lack of correlation between biochemical effects on rats and blood carboxyhemoglobin concentrations in various conditions of single acute exposure to carbon monoxide. *Arch. Toxicol.* **34,** 331–334.

Stewart, R. D., Fischer, T. N., Hosko, M. J., Peterson, J. E., Baretta, E. D., and Dodd, H. G.

(1972). Experimental human exposure to methylene chloride. *Arch. Environ. Health* **25,** 342–348.

Stewart, R. D., Peterson, J. E., Fisher, T. N., Hosko, M. J., Baretta, E., Dodd, H. C., and Hermann, A. A. (1973). Experimental human exposure to high concentrations of carbon monoxide. *Arch. Environ. Health* **26,** 1–7.

Stupfel, M., and Bouley, G. (1970). Physiological and biochemical effects on rats and mice exposed to small concentrations of carbon monoxide for long periods. *Ann. N.Y. Acad. Sci.* **174,** 342–368.

Styka, P. E., and Penney, D. G. (1978). Regression of carbon monoxide induced cardiomegaly. *Am. J. Physiol.* **235,** H516–H522.

Sylvester, J. T., Scharf, S. M., Gilbert, R. D., Fitzgerald, R. S., and Traystman, R. J. (1979). Hypoxic and CO hypoxia in dogs: Hemodynamics, carotid reflexes, and catecholamines. *Am. J. Physiol.* **236,** H22–H28.

Syvertsen, G. R., and Harris, J. A. (1973). Erythropoietin production in dogs exposed to high altitude and carbon monoxide. *Am. J. Physiol.* **225,** 293–299.

Theodore, J., O'Donnell, R. D., and Back, K. C. (1971). Toxicological evaluation of carbon monoxide in humans and other mammalian species. *J. Occup. Med.* **13,** 242–255.

Thomas, C. G., and Murphy, E. A. (1960). Circulatory responses to smoking in healthy young men. *Ann. N.Y. Acad. Sci.* **90,** 186–189.

Traystman, R. J., and Fitzgerald, R. S. (1977). Cerebral circulatory responses to hypoxic hypoxia and carbon monoxide hypoxia in carotid baroreceptor and chemoreceptor denervated dogs. *Acta Neurol. Scand.* **56,** Suppl. 64, 294–295.

Turino, G. M. (1981). Effect of carbon monoxide on the cardiorespiratory system. *Circulation* **63,** 253a–259a.

U.S. Department of Health, Education and Welfare (USDHEW) (1979). "Smoking and Health. A Report of the Surgeon General," DHEW Publ. No. (PHS) 79-50066. USDHEW, Washington, D.C.

Wald, N. J., Idle, M., Boreham, J., Bailey, A., and Vanvunakis, H. (1981). Serum cotinine levels in pipe smokers: Evidence against nicotine as cause of coronary heart disease. *Lancet* **2,** 775–777.

Weiser, P. G., Morrill, C. G., Dickey, D. W., Kurt, T. L., and Cropp, G. J. A. (1978). Effects of low level carbon monoxide exposure on the adaptation of healthy young men to aerobic work at an altitude of 1610 meters. *In* "Environmental Stress, Individual Adaptations" (L. J. Folinsbee, J. A. Wagner, J. F. Borgia, B. L. Drinkwater, J. A. Gliner, and J. F. Bedi, eds.), pp. 101–110. Academic Press, New York.

Weiss, H. R., and Cohen, J. A. (1974). Effects of low levels of carbon monoxide on rat brain and muscle pCO. *Environ. Physiol. Biochem.* **4,** 31–39.

Wilks, S. S. (1959). Effects of pure carbon monoxide gas injection into the peritoneal cavity of dogs. *J. Appl. Physiol.* **14,** 311–312.

Wilks, S. S., Tomashefski, J. F., and Clark, R. T. (1959). Physiological effects of chronic exposure to carbon monoxide. *J. Appl. Physiol.* **14,** 305–310.

Winston, J. M., and Roberts, R. J. (1977). Effects of potassium cyanate on carbon monoxide and hypoxic hypoxia-induced lethality. *J. Toxicol. Environ. Health* **2,** 265–271.

Winston, J. M., Creighton, J. M., and Roberts, R. J. (1974). Alteration of carbon monoxide and hypoxia-induced lethality following phenobarbital, chlorpromazine or alcohol pretreatment. *Toxicol. Appl. Pharmacol.* **30,** 458–465.

Wright, G. R., and Shephard, R. J. (1979). Physiological effects of carbon monoxide. *Int. Rev. Physiol.* **20,** 311–368.

Young, S. H., and Stone, H. L. (1976). Effect of a reduction in arterial oxygen content (carbon monoxide) on coronary flow. *Aviat., Space Environ. Med.* **47,** 142–146.

7

Respiratory Airway Deposition of Aerosols

Richard M. Schreck

I. INTRODUCTION

Dusty and smoky environments have been recognized since ancient times as potentially harmful to those exposed to them; however, only recently have we begun to understand the interactions that result in the effects on health associated with such environments. During the last half-century, instrumentation that allows rapid characterization of the physical properties of aerosolized material has become available. These characterizations, coupled with concurrent improvements in our knowledge of lung anatomy and the physiology of mass transport within the lung airways, have provided insight into the way airborne particulate enters the body through the respiratory system, what characteristics of the aerosol can be used to predict its ability to enter the body, and how best to protect against harmful aerosols.

AIR POLLUTION

Aerosols have become a part of our daily lives, whether they be in the form of cigarette smoke, hairspray, paint aerosol, oil mists from machining operations, direct particulate emissions from combustion processes, or naturally occurring aerosols such as windblown dusts, mists, and fogs. An increased awareness of the presence of aerosols has brought with it increased concern that the materials constituting these aerosols may have a questionable impact on the health of individuals inhaling them and may, in some cases, constitute risks to health. Aerosolized compounds and compounds carried in the adsorbed or absorbed state on aerosols are of particular concern because they deposit according to the physical size and density of the aerosol particles rather than the physical characteristics of the delivered compound itself. Thus, gases such as nitrogen dioxide or formaldehyde, which may be effectively removed much higher in the respiratory tree due to their diffusivity in air and high water solubility in the mucosal lining of the nasopharynx and upper airways, may "piggyback" on a particle headed for deposition in the deepest and finest airways of the lung. There, the compound may be released by solubilization deep within the lung, reaching cells that do not normally deal with such foreign materials, and may also enter the circulatory system to be transported throughout the body.

This chapter presents those aspects of aerosol physics and fluid mechanics necessary to understanding the deposition of particles in the lung airways, together with *in vivo* inhalation experiments in aerosol deposition. These experiments demonstrate how the integration of physical deposition processes throughout the respiratory system results in an actual dose delivered to the respiratory system. This chapter also discusses how principles that underlie this research may be applied eventually to protect workers and the general population from unwanted exposures and to provide better health care. The literature cited is not meant to be an exhaustive review of the material available, but consists rather of readable, illustrative texts that can serve also as references for further in-depth studies of the various subject areas.

II. HISTORICAL OVERVIEW

Although a relationship between airborne dusts and respiratory and other forms of illness has long been grasped intuitively, it has been only recently that observations (Abraham, 1979) and measurements of deposition in the lungs have been made. Experiments on human subjects by Tyndall in the 1860s demonstrated the presence of particles in inspired air, and their absence in expired air, as visualized by scattering of light by the particles. More quantitative and definitive experiments were not forthcoming for another 100 years, because they depended on the development and application of aerosol sampling techniques, electron microscopy, aerodynamic and aerosol mobility sizing instruments, and the equipment to generate well characterized test aerosols (Tillery *et al.*, 1976; Raabe, 1976; Kotrappa and Moss, 1971).

Concurrent with the development of instrumentation to measure and classify aerosols came the simplified anatomical models of Weibel (1963) for airway lengths and diameters and the Horsfield and Cumming (1967) description of airway branching angle and its relationship to diameters. To this was added an improved understanding of the details of flow in the airways (Schroter and Sudlow, 1969; Schreck and Mockros, 1970). Next, the results of the anatomic and flow studies were combined with known information on sedimentation rates and Brownian diffusion of particles and new data on particle removal by impaction at bifurcations by Bell and Friedlander (1973) and Schlesinger and Lippmann (1972). Together, these three mechanisms can be used to explain the removal of the entire size spectrum of airborne particles in the various airway passages of the respiratory system and to predict generally the results of deposition studies of humans or animals inhaling known aerosol dispersions.

Recent applications of deposition studies have been in the field of air pollution to predict the dose to the body from a given ambient level of pollutant. Although aerosol deposition models are still developmental and not highly accurate predictors for all individuals undergoing all daily activity patterns, they can be used to predict within an order of magnitude or better the average mass of a pollutant that will deposit in human adults. The same methods can be used to predict the dose in animal exposures, compute dose levels in different species, and make comparisons with experimental studies (Soderholm, 1980). Aerosol deposition results have also found increasing application in the field of chemotherapy by allowing the lung to be used as an avenue for the delivery of therapeutic compounds in the form of microaerosols (Bell *et al.*, 1978; Swift, 1980). This methodology allows drugs for respiratory therapy to be applied directly to the affected region of the lung by properly sizing the aerosol to the breathing pattern and permits manifold lower doses than systemic administration of the same compound. This procedure also allows the possibility of continuous delivery of some materials via the lung, rather than intermittent medication by an oral or injection route.

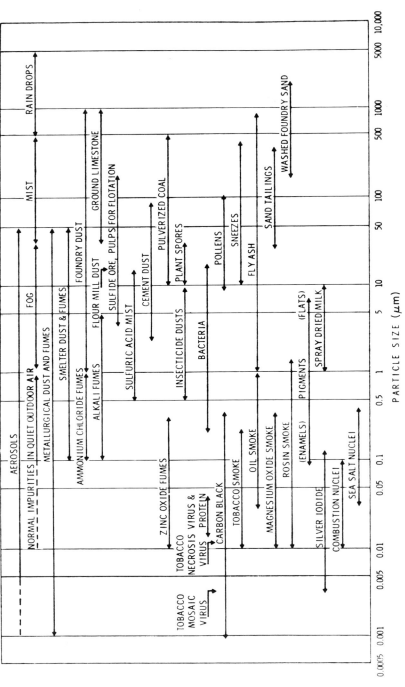

Fig. 1. Size ranges for solid and liquid airborne particulate materials that may be present in inhaled air. (Courtesy Mine Safety Appliances Co., Pittsburgh.)

III. AEROSOLS AND AEROSOL CHARACTERISTICS

The term "aerosol" is frequently used to describe dusts, smokes, mists, and fumes present in the atmosphere or in local environments. In this text, "aerosol" means a fine dispersion of material, solid or liquid, suspended in a gas such as air. Because large particles have an appreciable settling velocity and do not remain suspended for long periods of time in air, stable aerosols will generally be composed of particles less than 50 μm in diameter down to molecular clusters as small as 10 Å. Examples of materials that are encountered as aerosols and their size ranges are illustrated in Fig. 1. The figure demonstrates that innocuous materials such as fog and sea salt spray, as well as smokes and industrial wastes, may be present in the air and inhaled into the body. Classes of aerosolized materials that have a high potential for biological interaction are summarized in Table I. In each class and for each example listed, the mechanism of interaction with the body will depend on the physical and chemical properties of the material; however, a specific discussion of any of them (other than to note that the biological impact of each will vary with the site of deposition and the amount delivered) goes beyond the scope of this chapter.

Aerosols are found in all ambient atmospheres with mass concentrations normally ranging from a low of $10-20$ μg/m^3 after a rainshower to hundred of micrograms per cubic meter in many urban environments to well over 1000 μg/m^3 during high pollution episodes. Studies of the composition of the atmospheric aerosol in many cities have shown them to consist of primary particles, introduced directly into the atmosphere as particles, and secondary particles, resulting from gas-to-particle conversions in the atmosphere. The characteristic size distribution of these aerosols is illustrated in Fig. 2 from the work of Whitby and Cantrell

TABLE I

Classes of Aerosols with a Potential Biological Effect

Class	Examples
Organic	Cigarette smoke, soot, oil mist
Inorganic	Acid mists
Metallic	Metal oxides and fumes
Fiber	Asbestos, cotton
Radioactive	Radon daughters
Biological	Bacteria, fungi, viruses
Pharmacological	Bronchodilator drugs

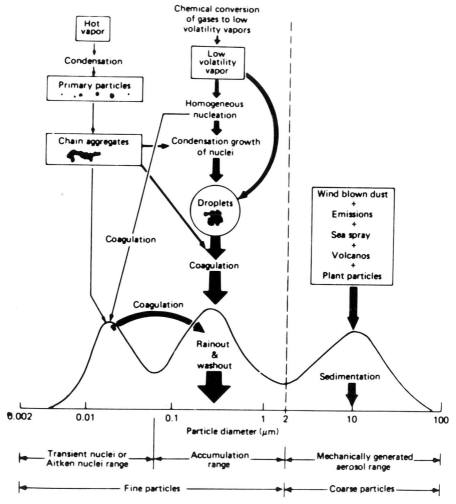

Fig. 2. Schematic representation of the atmospheric aerosol showing the three major modes, sources, and removal processes for the modes and mechanisms of mass transfer between modes. (Whitby and Cantrell, 1976.)

(1976). Generally, the aerosol size distribution is found to be trimodal, consisting of fine nuclei (smaller than 0.1 μm), an accumulation range from 0.1 to 2 μm, and a range of coarse particles larger than 2 μm. Furthermore, the chemical composition of the particles varies with the size range, the larger particles consisting of mechanically generated dusts, plant pollens, etc., whereas the two smaller ranges, often termed the "fine particles," consist primarily of organic materials. The mechanisms of generation and removal of these aerosols from the atmosphere are also illustrated in Fig. 2. Nuclei-range particles are produced by condensation processes and as primary particles in combustion. The particles in this size range have high diffusional mobility (caused by Brownian motion) and, consequently, collide with one another, adhere, and grow by coagulation with other fine and larger particles until they enter the second accumulation range. Here they may grow further by condensation of additional material or further coagulation into particles in the 0.1–2.0 μm size range and may remain airborne for hours or days. The chemical compositions of particles in these two size ranges are similar obviously because they are derived from the same sources and mass transfer takes place between them. Removal of material from the fine particle ranges occurs primarily by condensation of water and rain washout, whereas the coarse particles deposit by sedimentation in quiet air.

In health-related studies of all types of aerosols, the two most important characteristics of the aerosol are the airborne concentration and size, because these determine both the dose and site of delivery of the material. Aerosol concentration, the mass of material per volume of air, is easily understood and directly measurable in the environment, and the dose to the lung increases linearly with concentration. Determining aerosol size and characterizing it is a much more complicated matter for a number of reasons. In the first place, the physical processes that produce aerosols do not usually produce particles that are all exactly the same size. Thus, the monodisperse aerosol is unusual; most have a distribution of diameters, making them polydisperse. Aerosols formed by natural processes usually have a distribution of physical properties such as diameter, which can be represented by modeling this property with a log-normal distribution. The log-normal distribution can be characterized by its mean or median value and geometric standard deviation. The geometric standard deviation gives an indication of the degree of dispersity of the distribution with a value of 1 corresponding to a monodisperse property. Larger values up to 2 represent a more commonplace distribution, and values considerably greater than 2 indicate a very broad distribution.

Particles may be sized in a number of ways. The most fundamental is to capture them on a slide, observe their shape microscopically, tabulate a measured physical parameter, and compute the characteristics of the distribution. Although this procedure works well for spherical particles, experience has shown that many important aerosols may be ellipsoidal, cylindrical, or noncompact aggregates that have no useful characteristic physical dimension. An example of this is the diesel exhaust particle

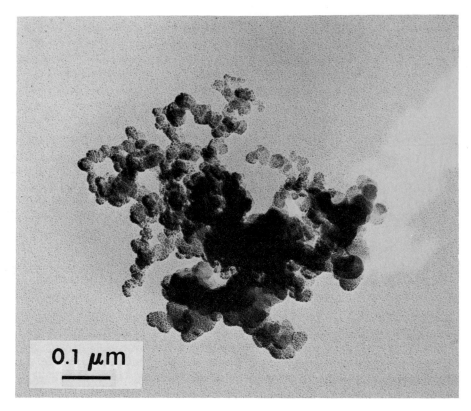

Fig. 3. Submicrometer particle of diesel soot, illustrating the complex shape of many airborne aerosols and the difficulty in determining an unambiguous diameter. (Electron micrograph courtesy of Dr. Sidney C. Soderholm.)

(shown in Fig. 3), which is an agglomerate of carbonaceous spherules. Not only does the particle lack a characteristic diameter, but other physical characteristics such as the bulk density vary also, depending on the degree of consolidation of primary spherical carbon units within the agglomerate. Despite these difficulties, the settling velocity and behavior of particles in a moving air stream can be inferred from aerodynamic shape factors such as those given by Mercer (1973, Chapter 3); once their shapes have been determined, these derived parameters can then be used to determine their deposition in the lung.

In biomedical research, the mass of material deposited in the lungs is of primary interest. Because most aerosols of practical importance are deposited by aerodynamic processes, it is useful to size particles in terms of their aerodynamic diameters. The aerodynamic diameter of a particle is the diameter of a sphere of unit density which has the same settling velocity. For any given aerosol, the aerodynamic diameter for which 50% of the aerosol mass resides in larger and smaller particles is called the mass median aerodynamic diameter (MMAD). In practice, it is reasonably straightforward to measure using aerodynamic devices such as the cascade impactor (see Mercer, 1973) to determine both the MMAD and geometric standard deviation of the aerosol. An excellent review of fine particles and their measurements and characterization can be found also in Hesketh (1977).

Other characteristics, such as the moments, M_i, of the particle size distribution are discussed by Friedlander (1977). In cases where the particle number distribution is known as a function of diameter, it is possible to derive the particle concentration (M_0), the number average particle diameter (M_1/M_0), and higher order moments proportional to the aerosol surface area and volume. These may be of value in determining the amount of surface-bound material adsorbed to an aerosol and delivered to the lung, but must be used in conjunction with the aerodynamic diameter information to determine the site of deposition.

IV. RESPIRATORY FLUID MECHANICS

For particles small enough to remain airborne for long periods of time and be inhaled, the flow of the air that suspends them is a critically important factor in their transport through the respiratory system. The airflow regime in which the particles travel through the interior airway passages may be turbulent and dominated by convective transport processes in some regions and laminar and orderly in others. It is therefore necessary to understand both the general anatomy of the lung airways and several fundamental concepts in fluid mechanics in order to appreciate the interplay of factors governing the transport and possible deposition of particles carried by the inspired air.

A most useful model of lung anatomy and the one having the greatest applicability in analytical studies is the uniform dichotomy or "A" model described by Weibel in 1963. In this study, Weibel describes the human lung as a dichotomous branching system of 24 generations of airways, beginning with the trachea, or zeroth generation, splitting into right and left mainstem bronchi of equal length and diameter, and proceeding in a similar fashion over progressively smaller and shorter airway generations to produce a treelike structure. The lengths and diameters of the airways of each generation can be described most simply by their mean values, although Weibel also presents data on their size distributions. In general, the model describes airways to be approximately three diameters long and airway diameters to decrease in successive generations by approximately 20% throughout the conducting airways. No data are presented on the angle of daughter airway branching from the parent airway or on the angle between the planes through successive generations of daughter airways. Although the model has cumulative airway volumes and dimensions equal to those of a human lung, anatomical accuracy is obviously lost in a simplified dichotomous model, because segmental bronchi are not represented accurately, and all airways do not proceed for this many generations before terminating in an alveolus or alveolar sac. This model, however, has proved to be very useful in computing averaged effects on the lung and has been employed in numerous analyses.

Studies that included branching angle were done by Horsfield and Cumming (1967) and showed the mean branching angle to vary somewhat with generation, but in general to increase in the smaller airways. Thus, the mean branching angle varied from approximately 32° for 4-mm airways up to 50° for airway diameters in the range 0.7–0.9 mm. This information was developed further in a subsequent publication by Horsfield *et al.* (1971), which described the anatomic structure of a single bifurcation. The work showed that an essentially circular parent airway becomes elliptical with the major elliptical axis in the plane of the daughter airways in a transition zone upstream from a bifurcation. The minor axis of the ellipse then becomes shortened, whereas the major axis enlarges with distance through the transition zone until the cusp formed by the inner surfaces of the two daughter airways is encountered. The two daughter airways then become nearly circular in cross section and, in turn, become parent airways for the next generation. The dimensions of the bifurcation were studied quantitatively by Schreck (1972), who demonstrated that the total cross-sectional area available for airflow increases by approximately 40% in a very gradual manner through a bifurcation and that the daughter airway walls changed direction from the parent airway as smooth, curved surfaces with radii of curvature that were approximately twice the parent airway diameter.

The nature of the airflow through this complex air distribution system will depend on a number of factors including the dimensions and shape of the airways and the pattern of breathing. Factors that alter airway shape include smooth muscle contraction, inflammation, and thickened mucous accumulations, which reduce the caliber of the airways, as well as disease states such as emphysema, which destroy much of the fine-airway structure. Characteristics of the breathing pattern such as tidal volume and breathing frequency affect the instantaneous flow velocities through the respiratory airways and the time for velocity patterns to develop. The most fundamental descriptive parameter of air flow through a conduit is the Reynolds number, R_e:

$$R_e = \frac{UD}{\nu} \tag{1}$$

where U is the average flow velocity; D, the airway diameter; and ν, the fluid kinematic viscosity. When the Reynold's number is small, say less than 2300, viscous effects tend to dominate in the flow, and, without additional excitation, the flow pattern tends to be laminar and orderly. For large values of the Reynolds number, inertial effects tend to influence the flow pattern, which will become turbulent with a great deal of convective mixing of the air in turbulent eddies and an overall velocity profile that is rather blunt compared to the laminar flow velocity profile in the same airway. Because the kinematic viscosity of air is constant for most lung inhalation work, the Reynolds number is a function of the airflow velocity and airway caliber. At a normal, resting breathing rate of 15 breaths/minute and a tidal volume of 600 ml, the peak flow rate through the lung is approximately 500 ml/second, and, at this flow, Reynolds numbers range from a little over 2000 in the trachea to below 100 by the eighth generation. With high inspiratory effort, these flows and Reynolds numbers can be increased by more than an order of magnitude. Throughout the breathing range, it follows that there is turbulent flow in some of the larger airways during each breath, but that flow in the smaller airways, and especially the pulmonary airways, is always quiescent and laminar. This is not to imply that the velocity profiles in the low Reynolds number airways are necessarily parabolic in shape as they are in fully developed laminar flow in infinite tubes. This happens because, even in laminar flows, a redistribution of the entering velocity profile occurs in the inlet region of a tube and, depending on the Reynolds number, may require many tube diameters to settle into the parabolic shape. It follows that because most airways are only approximately three diameters long, velocity profiles are not parabolic in shape, despite their being laminar. This was demonstrated experimentally by Schreck and Mockros (1970), whose velocity maps in models of daughter airways were not axisymmetric and showed markedly different flow distribution from fully developed flow.

Schroter and Sudlow (1969) injected smoke into plastic tube models of lung bifurcations to visualize the flow patterns during steady, laminar inspiratory or expiratory flow. The patterns indicated that during inspiration, high-speed flow moved from the center of the parent airway

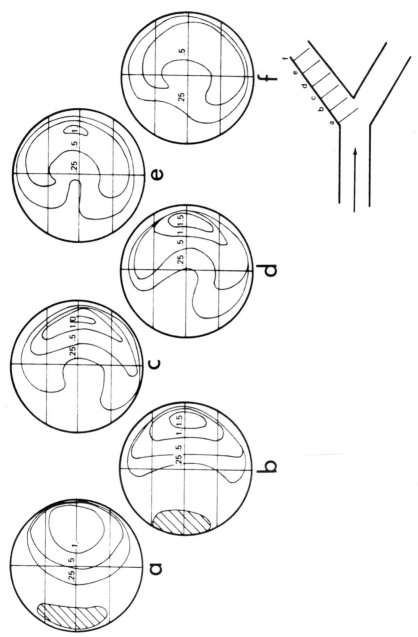

Fig. 4. Distal axial velocity field maps at six locations in an ideal bifurcation model (Schreck, 1972) for a parabolic inlet velocity profile having an average velocity of 0.34 m sec^{-1} and a Reynolds number of 710. Isovelocity lines are shown relative to the average inlet velocity, and hatching represents reverse flow.

to the inside walls of the daughter airways and gradually rolled around the airway in a double vortex. The studies also showed a zone of flow separation on the outside walls of the daughter airways. During exhalation, smoke introduced into both daughter airways produced a quadruple vortex around the axis of the parent airway, indicating again that besides the predominant axial flow, highly complex secondary flows are present at bifurcations. Subsequent studies by Schreck (1972) measured the axial velocity distribution map in the daughter airways in considerable detail as shown in Fig. 4. This work shows that with a symmetric parabolic velocity profile in the parent airway, the higher speed flow enters off axis (Fig. 4a) and gradually redistributes as it moves down the daughter airway until it is well distributed at the distal end of the airway (Fig. 4f), but is certainly not axisymmetric at the location where the flow is again split into two smaller daughter airways. Subsequent studies using casts of actual airways showed that the regions of separation and backflow (Fig. 4a and b) and probably those found by Schroter and Sudlow were caused by the geometry of the tubular model and did not occur in actual lung bifurcations.

A further aerodynamic consideration is the validity of modeling the cyclic flows of the real lung with steady flow model measurements or calculations of airflow and resultant aerosol transport based on steady flow. When the frequency of breathing is slow enough, airflow patterns in airways rearrange as if they were a succession of steady-state patterns because the inertia of the air does not influence the succession of velocity profiles. At higher frequencies, however, this is not the case, and steady flow characteristics would not apply. This effect can be estimated, using the frequency parameter of Womersley (1955), R_w:

$$R_w = R(\omega/\nu)^{1/2} \qquad (2)$$

where R is the airway radius; ω, the breathing frequency; and ν, the kinematic viscosity of air. For R_w less than 1.0, steady flow patterns are a valid approximation of the unsteady case, but for larger values they may not be. For a breathing frequency of 15 breath/minute and the Weibel airway diameters, the parameter R_w is less than 1.0 for all but the first three generations. From this it follows that steady flow conditions are a valid way to study the model lung airflows for aerosol transport and deposition.

An understanding of aerosol transport and mixing by aerodynamic processes will be helpful in studying the literature and the various models that have been developed to determine aerosol deposition in the lung. In turbulent flow with vigorous mixing, the incoming aerosol as well as the axial velocity are well distributed across the airway diameter by the turbulent eddies, and aerosol particles of all sizes are literally swept downward through the larger airways toward the deep lung. When Reynolds numbers decrease and flow settles into a laminar pattern, however, it is less clear how the aerosol rides the incoming air stream into the smaller lung airways. Recent work by Haselton and Scherer (1980) using flow visualization techniques to study the transport of a tracer through a bifurcation during oscillatory flow has shown that the different velocity profiles that occur in inhalation and exhalation results in a net forward transport of tracer through the bifucation. Thus, if particles are too large to have an appreciable diffusion mobility, convective processes facilitate their mixing and transport in the middle and finer airways where turbulent flows are no longer found. In addition to this, the Brownian diffusion of the very fine particles results in radial mixing and altered transport in the fine airways by the mechanism described by Taylor (1953). In this case, forward transport of aerosol is retarded as material on the faster-moving air fronts diffuses radially into slower-moving air and lags behind the average flow through the airway. Both experimental and analytical studies presented later in this chapter indicate that mixing in the laminar flow generations of the lung has an important effect that must be considered in understanding the process of fine aerosol deposition.

V. AEROSOL DEPOSITION

The health effects of any airborne particle depend, in part, on its ability to penetrate the respiratory defenses and deposit and on the body's ability to remove or neutralize it. The primary route of entry into the body is through the respiratory tract, which draws in 10,000–20,000 liters of air every day. For the purposes of analysis, the respiratory system can be thought of as comprising three sequential regions: the nasopharyngeal region, the tracheobronchial tree or conducting airways, and the pulmonary or gas-exchange zone deep in the lung (Fig. 5).

As inhaled air enters the respiratory tract with each inspiration, it passes sequentially through each of the three anatomical zones. The interior surfaces of the passages in each zone are coated with a sticky mucous layer to which particles of any size become immediately affixed upon contact. Thus, if it contacts the mucous blanket during either inspiration or expiration, a particle is filtered from the air by the respiratory system. The mechanics of particle deposition are a function of the

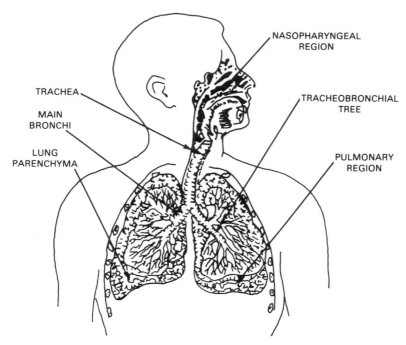

Fig. 5. Major features of the respiratory airways including the three regions of particle deposition: the nasopharynx, tracheobronchial tree, and the pulmonary region.

particles' size, density, and shape. They do not depend directly on particle composition or on whether the particles are in a solid or liquid state. Each particle behaves as an independent unit and, within limits, the actual number concentration is unimportant. Several useful survey articles are available in the literature on aerosol deposition for further reference (Stuart, 1973; Brain and Valberg, 1979).

A. Deposition of Ordinary Particles

There are four mechanisms by which ordinary particles are brought into contact with the mucous blanket: sedimentation, inertial impaction, interception, and diffusion (Fig. 6). The deposition efficiency of each method depends on the region of the respiratory system and on the particle size.

1. Sedimentation

All particles larger than 0.1 μm suspended in a gas sediment slowly under the influence of gravity. The effect of the viscosity of the sur-

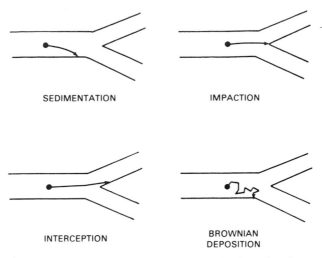

SEDIMENTATION   IMPACTION

INTERCEPTION   BROWNIAN
                                       DEPOSITION

Fig. 6. Schematic representation of the four mechanisms for the removal of un-
charged inhaled aerosols in the respiratory tract airways.

rounding gas is such that each particle falls at a constant speed (the
terminal velocity) approximated by Stokes' law. The effect of sedimentation is superimposed on inertial impaction and diffusion throughout the
respiratory system; however, in the upper regions at high airflow rates,
its effect is outweighed by inertial impaction. As air velocity slows and
airway diameter decreases, however, a particle's settling velocity becomes significant, and, because of the close proximity of the airway walls
in the smaller airways, many of the remaining large and medium-sized
particles (2–8 µm) deposit by sedimentation.

Sedimentation occurs because of the gravitational attraction on an
airborne particle, which is represented by the force:

$$F_g = \frac{\pi d^3}{6} (\rho_p - \rho_a)g \tag{3}$$

in which F_g is the gravitational force on the particle; d, the particle
diameter; ρ_p, the particle density; ρ_a, the air density; and g, the gravitational constant.

Using Stokes' law for a spherical particle, the drag force on this falling
particle can be written:

$$F_d = 3\pi \, d\mu v \tag{4}$$

in which F_d is the drag force; μ, the absolute viscosity of air; and v, the
particle velocity.

At the terminal velocity, the particle is no longer accelerating; hence, the forces on it are equal, and we may equate these two expressions and solve for the terminal velocity, v_t, obtaining:

$$v_t = \frac{gd^2}{18\mu} (\rho_p - \rho_a) \qquad (5)$$

This is a valid relationship for particles in which $1.0 \leq d \leq 40$ μm. For smaller particle diameters, this can be corrected by multiplying by the Cunningham slip correction factor (Cunningham, 1910) to obtain:

$$v_t(\text{actual}) = v_t\left(1 + \frac{2A\lambda}{d}\right) \qquad (6)$$

where λ is the mean free path of a molecule in air; and $A = 1.26 + 0.4e^{-1.1d/2\lambda}$.

For reference, Table II compares the terminal velocities for diameters of respirable-sized aerosols over several orders of magnitude, as well as their root mean square displacements in one second resulting from Brownian motion.

2. Inertial Impaction

Inhaled particles follow the moving airstream to a greater or lesser extent, depending on their size and mass. In the nasopharyngeal zone and upper tracheobronchial tree where airflow velocities and Reynolds numbers are high, the turbulent and/or laminar bolus flow sweeps practically all of the smaller particles with it; only the larger ones have difficulty readjusting their velocities and sideslip across the flow streamlines. Thus, when the airstream changes direction abruptly, as in the turbi-

TABLE II

Particle Displacement for Unit Density Spheres[a]

Particle diameter, d (μm)	Terminal velocity, v_t (μm/second)	Brownian displacement per second (μm)
10	2900	3.8
5	740	5.5
1	33	13
0.5	9.5	20
0.1	0.81	64
0.05	0.35	120

[a] Computed for air at 760 mm Hg and 27°C.

nated bony structures or one of the many bifurcations or due to turbulence, the larger particles are often unable to follow it and collide with the mucous-lined walls. Thus, the hundreds of bifurcations of the upper tracheobronchial tree serve not only as a distribution manifold for the respired air but also as sites for inertial impaction.

Deposition by inertial impaction may be understood by noting that a particle moving in the center of a parent airway at speed v in which the streamline changes direction by an angle θ, experiences a lateral displacement, L, across the streamlines of the form:

$$L = \frac{v_t v \sin \theta}{g} \tag{7}$$

in which v_t, the terminal velocity, is computed from Eq. (5) or (6). This displacement can be compared to the airway radius, R, to derive an expression equivalent to the Stokes' number, St:

$$St = \frac{\rho_p d^2 v}{18 \mu R} \tag{8}$$

which can be used to predict the tendency for inertial deposition in airways based on the physics of the flow (e.g., Chan *et al.*, 1978b).

3. Interception

Particles of any size may be removed in the airways by contacting the wall incidentally as the streamline on which they are traveling passes near an airway surface. This contact is not produced through inertia, but is rather a result of the particle's shape, orientation, or rotation while in close proximity to the surface. In effect, the center of the particle moves to within its own radius of the wall and is intercepted physically by the wall. Consequently, interception most often applies to nonspherical particles such as fibers having appreciable aspect ratios.

4. Diffusion

Airborne particles are subjected continuously to bombardment by the surrounding gas molecules. This induces a random movement of the particles called Brownian motion, which results in their transfer or diffusion from one region of a gas volume to another. The diffusivity, D_c, of a particle is inversely proportional to its diameter, but is independent of its density:

$$D_c = \frac{kT}{3 \pi \mu d} \tag{9}$$

In this expression, k is the Boltzmann constant, and T is the absolute temperature. In terms of the diffusivity, the particle's root mean square displacement, δ, after a time interval, t, may be written:

$$\delta = \sqrt{6D_c t} = \sqrt{\frac{2kTt}{\pi \mu d}} \tag{10}$$

From this it can be noted that the only important variables affecting the Brownian displacement of airborne particles in the respiratory system are the particle's diameter and the time it remains in the body. Decreasing particle diameter or increasing time by maneuvers such as breath-holding will increase the amount of Brownian deposition because of larger displacements and greater probability of contact with the respiratory epithelium. Table II illustrates typical particle displacements in 1 second for a range of diameters of unit density spheres.

Very small particles diffuse rapidly, although the rate of their diffusion is orders of magnitude below that of gases. Both diffusion and impaction are inefficient for particles with diameters of approximately 0.5 μm, and, as a result, they follow any movement of the surrounding air very closely and are deposited minimally in the respiratory system.

Diffusion, like sedimentation, works in all three respiratory zones; however, it is the dominant mechanism in the alveolar region, because airflow velocities therein are much too low for impaction to occur and most of the large particles have already been removed by sedimentation in the larger airways.

B. Deposition of Hygroscopic Particles

The relative humidity in the respiratory airways below the pharynx is estimated to be 96–100% (Scherer *et al.*, 1979) and can have a marked effect on the deposition of hygroscopic aerosols (Ferron, 1977). Mercer (1973) has demonstrated that, for small aerosols of sodium chloride, the growth to a larger equilibrium diameter can take place in a fraction of a second. This means that the aerodynamic characteristics determining particle settling and removal can change abruptly over time intervals that are short compared to the breathing cycle. This is caused by the rapid uptake by the particle of water from the surface of the respiratory airways, whose vapor pressure is approximately that of 0.9% sodium chloride at 37°C. A similar effect would be expected also for aerosols that are mixtures of salts, although the equilibrium sizes would be more complicated to calculate (Hänel and Zankl, 1979).

Experimental studies of enhanced deposition of hygroscopic aerosols have been conducted by Bell *et al.* (1978) and Scherer *et al.* (1979) in models of bifurcating airways to verify the enhancement of deposition and quantify the magnitude of the effect for application in the delivery of therapeutic drugs. Experiments on human subjects, described by Stuart (1973), also illustrate this effect and indicate that hygroscopic aerosols deposit in the lung according to their final diameters.

C. Deposition of Charged Particles

Aerosol particles carrying electrical charges are also known to deposit differently in the airways. When a charge-carrying particle approaches the electroconductive epithelial tissue of the airway wall, an "image charge" of opposite polarity is induced on the surface and attracts the particle (Fuchs, 1965). This effect has been demonstrated experimentally by Chan et al. (1978a) in model studies using casts of the human upper airways and by Melandri et al. (1977) in human subjects. Brock and Marlow (1975) have theoretically analyzed the image-force effect on fine particles in the respiratory system and conclude that the effects of particle charge on deposition should not be neglected when the net number of charges of either sign per particle exceeds 10. This is supported by the tracheal model deposition experiments of Chan et al. (1978a) for particle-charge levels of 360–1100 negative charges per particle with mass median aerodynamic diameters of 2–7 μm. Particle deposition was enhanced in the trachea, particularly at the lower steady flow rates of 15 and 30 liters/minute. At these charge levels, deposition appeared to correlate well with an electrostatic parameter:

$$\frac{CQ^2}{dU} \tag{11}$$

where C is the Cunningham slip correction factor; Q, the charge per particle; d, the particle diameter; and U, the local airflow velocity.

These results verify that particle charge enhances deposition by amounts larger than the minimum suggested by Brock and Marlow's analysis and suggest that the effect may be quite important for submicron charged particles. Work by Heyder et al. (1978) also confirms the effect on human subjects using 0.6-μm particles carrying mean charges as low as 27 or 55 per particle. The experiments by Melandri et al. (1977) demonstrate the effect at particle diameters of 0.3 and 1.1 μm and for unipolar charging of 30–110 elemental charges per particle. The net effect of the charging was to increase deposition for mouth breathing by 15–30%. Fry (1970) measured nasal deposition in humans, however, and noted that at charge levels found in ambient and most workplace atmospheres, no deposition enhancement would occur.

D.   Human and Animal Deposition Experiments

Measurements to quantitatively describe the deposition of aerosols in the respiratory system were begun in the first half of this century, and experimental research in this area developed in close association with theoretical studies designed to predict deposition based on physical mechanisms. These studies, particularly the earlier work by Findeisen (1935), Landahl (1950), and Beeckmans (1965), were instrumental in defining the form of the deposition relationship for the experimentalists and will be discussed in greater detail in the following section. Because aerosol deposition is a complex process dependent on the physical characteristics of the particles, lung anatomy, and the characteristics of the breathing pattern, clinical studies were essential in developing an early empirical understanding of the deposition process and useful later in testing the results of model predictions. In some cases, the procedures and apparatus used in this early work evolved later into the standard practices and equipment used in more routine clinical work, such as drug administration via nebulizers.

Early studies in humans, by Drinker *et al.* (1928), of inhalation of high concentrations of aerosol were used to estimate the dose of coal dust to the lung. The data showed an average deposition of 55% for aerosols ranging in diameter from submicron to 6 μm. More detailed particle deposition versus size studies were done by Landahl *et al.* (1951, 1952), using various sizes of nonhygroscopic triphenyl phosphate aerosol (0.25, 0.55, 1.4, 2.9, 3.8, and 6.3 μm) produced by a Sinclair condensation aerosol generator. Healthy male and female subjects inhaled the aerosols, using three different breathing patterns: flows of 300 ml/second for 4, 8, and 12 seconds, resulting in tidal volumes of 450, 900, and 1350 ml. Four equal expired fractions were sequentially collected via impingers and compared with the inspired aerosol concentration to obtain the percentage deposited versus inspired particle size in each of the fractions. These fractions were thought to correspond roughly to the respiratory system regions from upper (tracheobronchial) to lower (pulmonary). The resulting plots of percentage deposited versus particle diameter showed a general minimum in total deposition for particles approximately 0.3–0.5 μm diameter, a continual increase in deposition with increase in particle diameter for particles greater than 1 μm, and a more gradual increase in percentage deposition with decrease in particle size smaller than 0.3 μm. The trend with increased tidal volume was to increase deposition in all four expired fractions and, therefore, in total deposition at all particle diameters. Landahl *et al.* (1952) extended these experiments to a second set of breathing patterns in which each subject inspired for 3/8 of the cycle, held his breath for 1/8 cycle, expired for 3/8 cycle, and held his breath for 1/8 cycle. This procedure was followed for tidal volumes of 1500 and 3000 ml and, in comparison to the early study, produced increased deposition, with the highest values corresponding to the 3000 ml tidal volume. This is explained as resulting from higher impaction deposition of larger particles caused by higher inspired flow rates and greater Brownian deposition of the smaller sizes due to the longer breath-holding period.

In similar experiments, Altshuler *et al.* (1957) aerosolized triphenyl phosphate in sizes ranging from 0.14 to 0.32 μm, but employed a Tyndallometer to continuously measure the aerosol concentration in the expired air and a pneumotachograph to record the flow of air at the mouth. Their subjects respired six sizes of monodisperse aerosols by mouth, and the data confirm the minimum deposition at approximately 0.4 μm at breathing frequencies of 9, 15, and 21 breaths/minute as well as a decrease in total deposition with increased breathing frequency.

Studies reported by Lourenco (1969) used a monodisperse iron oxide aerosol tagged with ^{198}Au produced by a spinning disc aerosol generator and administered to normal subjects, bronchitic smokers, and patients with increasingly severe bronchitis. Deposition of the 2-μm aerosol was found to be more central in the lungs with increased airway resistance and was accompanied by slower clearance rates, leading the author to conclude that depressed clearance rates in diseased lungs may lead to the genesis and perpetuation of infections. Further inhalation studies by Lourenco *et al.* (1971) using the same aerosol administered to groups of healthy nonsmokers and smokers showed that the material deposited identically in each group, approximately half in the ciliated and half in the nonciliated (alveolar) regions. Particle clearance differences were noted between the two groups with nonsmokers clearing most of the large airway-deposited material immediately, whereas smokers' clearance was retarded by from 1 to 4 hours. In 1972, Lourenco *et al.* used the 2-μm gold-tagged aerosol again in experimental studies comparing deposition in healthy subjects and patients with bronchitis or bronchiectasis. In comparison to the healthy subjects (but not to patients who had bronchitis only), patients with bronchiectasis showed abnormally high central deposits of particulate material.

Davies (1974) reviewed the published literature on the subject and concluded that there was a large scatter in the reported experimental measurements caused by a variety of uncontrolled experimental variables, such as breathing pattern, difference in aerosols, nose versus mouth breathing, and anatomic variability in subjects. In reviewing more recent experimental data from well-controlled experiments, he noted that several investigators obtained very similar results that indicate the percentage deposition for particles in the 0.5–1.0 μm size range is actually only 10–15% during steady breathing, a value much lower than that given by the theory of Beeckmans (1965), or the Task Group on Lung Dynamics (1966). Davies argued that the lower percentage is logical, primarily because of the ability of particles in this narrow size range to ride the high Reynolds number flows in the large airways with little axial mixing and then, despite being convectively drawn into the fine airways, to resist mixing with the residual deep lung air effectively as does oxygen because its diffusion coefficient is four orders of magnitude lower.

Heyder *et al.* (1975), using a condensation-produced monodisperse aerosol of di-2-ethylhexyl sebacate ranging in size from 0.1 to 3.2 μm, investigated the effects of functional residual capacity, tidal volume, and inspiration velocity on deposition in five male subjects. Subjects inhaled a tidal volume of 1000 ml from residual volume by mouth and, for comparison, by nose via a mask. By altering tidal volume and breathing frequency, the flow was held constant for each test. The resulting deposition curves show an increase in total deposition at low breathing frequencies, but the deposition minimum remains in the 0.4–0.7 μm size range, despite a factor of 10 difference in breathing frequency (3.75–30 breaths/minute). They conclude that the average residence time in the respiratory system is the dominant parameter controlling deposition, because diffusion and sedimentation are the dominant deposition mechanisms for particles of this size. Velocity-dependent mechanisms were not found to acccount for significant effects associated with this aerosol size range. They also concluded that deposition increases with increased tidal volume and confirm the need to measure end tidal volume of the respiratory tract, because deposition decreases with increased lung volume. For nose breathing, a remarkable difference was noted only for high flow rates. This indicates that the nose can be an effective filter, albeit one that, as the data show, has little effect on the diameter of minimum deposition, which remained approximately 0.4 μm. By varying the ratio of time for inspiration to time for expiration for a given tidal volume of 1000 ml at a frequency of 15 breaths/minute, the authors were further able to demonstrate the importance of breathing pattern, because deposition increased with the time fraction devoted to inspiration. Generally, this data showed lower deposition than the earlier model predictions by Landahl (1950) and Beeckmans (1965); however, it showed rather good agreement with the more complex model proposed by Taulbee and Yu (1975). When compared to the experimental studies of others, these deposition curves were generally lower throughout the size range than those of some investigators (e.g., Altshuler *et al.*, 1957), but quite similar to the findings of Davies (1974), especially near the deposition minimum at 0.5 μm. These differences are attributed by the authors to the failure of investigators in earlier experiments to control parameters such as lung initial volume, respiratory pauses, breathing frequencies, inspiratory and expiratory flow patterns, and the degree of aerosol charging.

The effects of breathing pattern on deposition were further studied by Taulbee *et al.* (1978), using 0.5 and 1.0 μm particle diameter aerosols of di-2-ethylhexyl sebacate, a nonhygroscopic aerosol having a low vapor pressure. Their single-breath inhalation studies in healthy subjects employed a laser photometer to measure concentration and a pneumotachograph to measure flow. Breathing was synchronized with a metronome, and flow was displayed on an oscilloscope to aid the volunteer in following the desired pattern. Results were recorded and analyzed via computer to determine the deposition versus fraction of exhaled air. For a variety of tidal volumes ranging from 250 to 2000 ml and inhalation times ranging from 1 to 8 seconds, deposition of the 0.5-μm particles increased almost linearly with tidal volume. Also, at a given tidal volume of 1000 ml, the deposition was shown to decrease with increasing inspiratory flow rates. In both cases, the effect is attributable to the longer particle-residence time in the lung. The trends were also noted for the 1.0-μm particles; however, the amount deposited was generally greater for the larger particle. These results are generally in agreement with the work reported by others and the theoretical model of Taulbee and Yu (1975). They appear to support the assumption used in the author's model that relatively little aerosol mixing occurs in the large airways because of the blunt velocity profiles, but that mixing with clean air does occur in the smaller airways because of the nonuniform velocity distributions, nonuniform ventilation of regions, secondary flows, and even the mechanical agitation of the heartbeat.

In an effort to resolve discrepancies in the deposition data of two laboratories, joint studies were performed by Heyder *et al.* (1978) of the important, uncontrolled parameters that may lead to variations in results from different investigators. Aerosols of Carnauba wax (solid) and di-2-ethylhexyl sebacate (liquid) were generated in sizes of 0.6, 1.0, and 1.5 μm and inhaled by mouth by six male subjects. Careful control of subject residual volume and aerosol electric charge was maintained during inhalation studies with both aerosols. The resulting data showed that after losses in lines, transducer calibration, and other technical details are accounted for, similar deposition results were obtained. Artifacts in measurements resulted from factors such as charging of the aerosol and failure to humidify the inhaled aerosol cloud. When these were accounted for, intrasubject variability remained as the major factor in the scatter of the data. This variability may be caused by anatomic factors or as yet undefined characteristics of the breathing pattern of subjects.

Regional deposition studies of larger particles of ^{99m}Tc were conducted by Foord *et al.* (1978) in healthy male nonsmokers. The monodisperse aerosols, having diameters of 2.5 to 7.5 μm, were inhaled by mouth at tidal volumes of 0.5–2.7 liters and frequencies of 10–25 breaths/minute; regional deposition was measured using external detectors. Deposition of these larger-sized particles was found to correlate well with the so-called impaction parameter:

$$\text{Impaction parameter} = d^2F \qquad (12)$$

where d is the particle diameter in μm and F, the average inspiratory flow rate in liters/minute. This study not only verified that inertial impaction is the primary deposition mechanism for this aerosol size range, but also correlated well with the deposition predicted by the Task Group on Lung Dynamics (1966).

In further studies of deposition in the larger airways, Chan and Lippmann (1980) used a hollow cast of the human larynx and first six generations of the tracheobronchial tree to gather aerosol deposition data for comparison with the results from nonsmoking human volunteers, smokers, and patients with lung disease. Iron oxide aerosols tagged with ^{198}Au or ^{99m}Tc were generated in sizes from 0.2 to 7 μm and inhaled by mouth at 14 breaths/minute and a tidal volume of 1000 ml. External scintillation detectors were used to quantify the amount of material deposited 5 minutes and 24 hours after exposure.

Results of the model studies indicated that the deposition of larger particles (>2 μm) was dependent on inspiratory velocity and correlated well with the Stokes number:

$$\text{Deposition efficiency} = 0.803 \times \text{St} + 0.0023 \qquad (13)$$

where St is the Stokes number defined by Eq. (8).

Deposition in volunteer subjects varied in accordance with an anatomical Stokes number, St*, defined as:

$$\text{St}^* = \frac{4d^2F}{18\mu\pi(\text{BDS})^3} \qquad (14)$$

In this expression, BDS refers to a bronchial deposition size and may be thought of as an equivalent bronchial radius, ranging from 1.2 cm in healthy individuals to 1.02 cm in smokers to as low as 0.60 cm in patients with severe bronchitis. It can readily be seen that the anatomic Stokes number resembles the impaction parameter of Foord *et al.*, and indicates the importance of impaction in the removal of this particle size. Particles in the smaller size range (0.1–0.5 μm) were deposited at only half the efficiency predicted by the Task Group (1966) model, adding further evidence to the observation of Davies (1974) that the model is in error for submicrometer particles.

More detailed studies of the effect of the laryngeal jet and secondary flows in the larger airways were described by Chan *et al.* (1980), who used a similar aerosol with particle sizes of approximately 2.8, 4.8, and 7.5 μm. Again, deposition in the tracheobronchial region represented by the cast could be well represented by an empirical equation incorporating the first power of the Stokes number for inspiratory flow rates of 15, 30, and 60 liters/minute. An area of enhanced deposition was noted in the trachea approximately 2 cm distal to the larynx and appears to result from significant secondary flows caused by the larynx. Comparison of these model results to those obtained from a similar test using the uniform dichotomy model of Bell and Friedlander (1973) without the larynx showed considerably different deposition in the symmetric ideal model with considerable enhancement of deposition at local hot spots. The differences in deposition were found to result from differences in secondary flows in the large airways, which result from the local anatomy as shown by Schreck (1972). It was concluded that the enhancement of deposition in the upper trachea reduced particle penetration into the lower lung and facilitated their rapid clearance by the mucociliary escalator.

Because many studies of aerosol inhalation are for toxicological, radiological, or pharmacological research, animal models are frequently used to measure the effects resulting from a given dose by inhalation. Lung anatomy and breathing patterns are different in different species, and the dose per gram of lung or body weight may be very different in animals than in man; nonetheless, there is a surprising similarity in the general deposition pattern versus particle size (Thomas, 1972). Cuddihy *et al.* (1973) studied the deposition of radioactive lanthanum oxide aerosols having diameters from 0.4 to 6.6 μm in beagle dogs during nose breathing and found the total and regional deposition patterns were reasonably similar to those the Task Group (1966) projected for humans. Studies of rats and Syrian hamsters by Raabe *et al.* (1977) used monodisperse fused clay aerosols with diameters ranging from 0.1 to 3.2 μm. Deposition in rodents differs from man in that particles with diame-

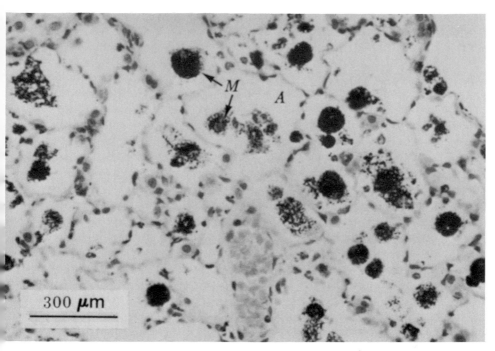

Fig. 7. Submicrometer soot particles accumulate in the lung of a rat exposed to diesel exhaust diluted to a concentration of 6000 $\mu g/m^3$ for 9 weeks. Note that many alveoli (A) contain macrophages (M) laden with diesel particles, and that little or no soot is found "free" in the airspaces. Hematoxylin—eosin stain, magnification 100×. (Photo courtesy of Dr. Harold White.)

ters much larger than 1.0 μm are removed in the head very efficiently, resulting in lower pulmonary deposition in the rodent (6–9%). (Pulmonary deposition in humans ranges from 20% for nose breathing to as high as 50% for mouth breathing.) For particles less than 1.0 μm, the deposition efficiencies become comparable; therefore, extensive studies of deposition of 0.1- to 0.2-μm diesel-exhaust soot particles have been conducted in rodents. Figure 7 illustrates that aerosol penetration and deposition in the alveolar region of the rat is considerable, but that the particles, which were deposited randomly by Brownian diffusion, are gathered by large scavenger cells called macrophages. A general discussion of the deposition, internal transport, and clearance of submicron carbonaceous particles in the rat and guinea pig is presented by Vostal *et al.* (1982).

E. Predictive Models of Respiratory Deposition

Deposition of aerosol particles in the respiratory airways is governed by a number of physical factors, including the properties of the aerosol, the airway anatomy, and the velocity fields in the airways that result from a given pattern of breathing. In principle, this information can be applied to compute the probability of deposition of a particle through the mechanisms of impaction, sedimentation, and diffusion in any airway of the lung, as well as the effect summed over the respiratory airways to determine total as well as regional deposition. This was first attempted by Findeisen (1935) using a simplified lung anatomy consisting of a trachea, four generations of bronchi, two generations of bronchioles, and alveolar ducts and sacs. He modeled the breathing pattern to be steady inspiration or expiration flow at 200 ml/second and a frequency of 14 breaths/minute. Using this model, he was able to show that impaction was the dominant removal mechanism in the largest airways, sedimentation in the middle region, and diffusion in the pulmonary zone. Findeisen also derived a deposition curve that was similar to that shown in Fig. 8 and predicted the deposition patterns to be found by later investigators.

Landahl (1950) improved on this model by increasing the number of airway generations of the model and adding the mechanism of interception for particle removal. For comparison with experimental measurements, he calculated the aerosol that would remain in four fractions of exhaled air as a function of aerosol size. Curves derived from his computations predict increased deposition of all sizes of aerosol with increased tidal volume, especially for those particles larger than 1.0 μm. Altshuler (1959) designed his model output to produce predictions of aerosol concentration in expired air that could be compared directly with the output of a light-scattering aerosol monitor used in experiments on human subjects. He exercised his model with and without convective axial mixing and showed that a contribution from axial mixing of aerosol gave improved model accuracy.

Landahl's geometric model was later used by Beeckmans (1965), who also used a constant inhalation and exhalation flow, but with a pause between each. His equations allowed for both axial and radial diffusion as well as mixing to occur. The model computed the probability per generation for deposition during inspiration and expiration, arranged the probabilities into arrays and matrices, and summed over all airway generations using a computer. Results showed that variations in tidal volume did not influence the locations of the maximum and minimum alveolar deposition for a given minute volume, but did influence the

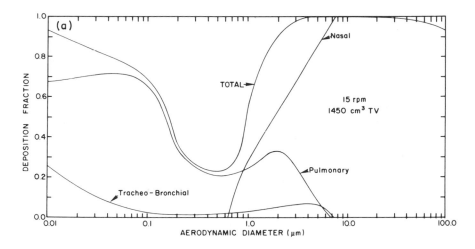

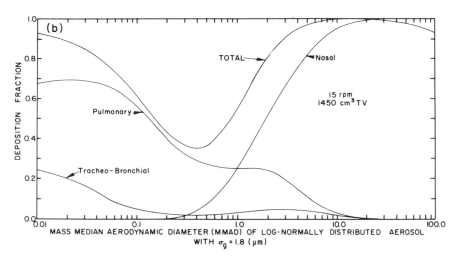

Fig. 8. (a) The deposition of monodisperse aerosols of various aerodynamic diameters in the respiratory tract of man (assuming a respiratory rate of 15 rpm and a tidal volume of 1450 cm³) as suggested by the Task Group on Lung Dynamics of the ICRP (1966). Compare these curves to the particle sizes illustrated in Figs. 1 and 2. (Figure courtesy of Dr. Otto G. Raabe.) (b) The deposition of log-normally distributed aerosols characterized by their mass median aerodynamic diameters (all with geometric standard deviation = 1.8) in the respiratory tract of man (assuming a respiratory rate of 15 rpm and a tidal volume of 1450 cm³), calculated by integrating over the whole distribution using the deposition fractions for individual sizes suggested by the Task Group on Lung Dynamics of the ICRP (1966). (Figure courtesy of Dr. Otto G. Raabe.)

total amount deposited, with slower breathing generally causing greater deposition for particles smaller than 6 μm. Also, the particle density was shown to strongly influence the diameter of maximum alveolar deposition, but not the value of the maximum, because the particle's mass median aerodynamic diameter, not its physical diameter, is important in this determination.

In 1964, the International Commission on Radiological Protection created a special Task Group to review existing lung models and develop a set of model curves that would predict the dose of dust to the lungs and aid in estimating the deposition of fine radioactive aerosols in the lung. The report of the Task Group on Lung Dynamics (1966) defines three zones of the respiratory system for consideration in the model. The first is the nasopharyngeal area, which includes the nares back to the larynx and is often referred to as the upper respiratory tract. The lower respiratory tract is divided into two regions, the tracheobronchial region, which includes the airways from the top of the trachea to the terminal bronchioles, and the pulmonary region, which includes the respiratory bronchioles, alveolar ducts, alveolar sacs, and alveoli. The Task Group emphasized (1) the importance of working with aerosol aerodynamic diameters in lung deposition modeling, and (2) that aerosols of practical concern always contain a distribution of sizes and are not monodisperse. Both of these concepts are important in using the measured characteristics of real aerosols to predict their site and amount of deposition.

The computation model used by the Task Group is a slight modification of Findeisen's, with a 4-second breathing cycle divided into 1.74 seconds inspiration, 0.20 second pause, and 2.06 seconds expiration. Nasal deposition was determined from the model of Pattle (1961). Lung tidal volumes, rather than frequency of breathing, were determined to change with the level of work performed; therefore, volumes of 750, 1450, and 2150 ml at 15 breaths/minute were used in the calculation to simulate various active work states. The deposition calculations were performed on a computer for aerosol particle diameters ranging from 0.01 to 100 μm. Deposition efficiency curves for the three lung compartments, plus the total respiratory system deposition, are shown for a tidal volume of 1450 ml in Fig. 8. Figures 8a and 8b facilitate a comparison between the deposition of monodisperse aerosols (Fig. 8a) and more realistic, log-normally distributed aerosols having a geometric standard deviation of 1.8 (Fig. 8b) and show the smoothing effect that a range of aerosol sizes has on the deposition of material. Because most naturally occurring aerosols have distributions of many particle sizes, the curves of Fig. 8b are qualitatively similar in shape to those that would be computed for a given test aerosol and breathing pattern; however, the exact shape and location of the curves will vary. The curves show that nasal deposition removes large particles primarily, but is ineffective for aerodynamic diameters less than 1.0 μm. Tracheobronchial deposition removes only a small fraction of micrometer-sized material and does not really become effective until diffusional deposition becomes important, below 0.05 μm diameter. Pulmonary deposition accounts for the major fraction of deposited material with diameters less than 1.0 μm and may remove more than 60% of the inhaled material smaller than 0.1 μm due to sedimentation, the large surface area of the pulmonary region, and the high Brownian mobility of these particles. It is interesting to note the deposition minimum at approximately 0.5 μm, which occurs because the various deposition mechanisms discussed earlier are all relatively inefficient for these particle sizes. Although this minimum shifts slightly with breathing pattern, it is a standard feature of all deposition curves, whether determined computationally, or experimentally.

Mercer (1975) reviewed the Task Group's model in the light of new experimental work. He concluded that the model predicted nasopharyngeal deposition adequately, but, considering new measurements by Lippmann (1977), appeared to overestimate pulmonary deposition at the expense of tracheobronchial deposition. From the standpoint of protecting health, this is a conservative estimate; however, if the models are to be used for other purposes, such as estimating the delivery of aerosolized medications, accuracy is of primary importance.

In the late 1970s, more elaborate predictive models that account better for the role of lung aerodynamics in particle transport and deposition were developed. Taulbee and Yu (1975) employed the Findeisen–Landahl–Beeckmans model using Weibel's (1963) anatomy model, impaction estimates derived from 90° bend data for smooth tubes, diffusion modeled from fully developed flow analysis in circular tubes, and sedimentation estimates derived from flow in horizontal circular tubes. Their analysis assumed that the three mechanisms were supcrimposed at each generation and the equations were applied for mouth breathing only. For each deposition mechanism, they estimated a time constant for the airways of the different respiratory regions and compared them to the period of the breathing cycle to determine whether the steady flow deposition estimates are valid. They found that the quasi-steady assumption is valid for impaction everywhere in the respiratory system, but that the time constant for diffusion and sedimentation may be too large for small particles in some parts of the lung; thus, solutions of the equations are necessary to predict deposition in some cases. Using this information, Taulbee and Yu performed the computations and derived a model that can determine the time-dependent particle concentration at the mouth during breathing, as well as the buildup of aerosol with airway generation during wash-in of a test aerosol into a clean lung.

This model's accuracy was demonstrated in comparisons with experimental data in Yu and Taulbee (1977), which showed total deposition estimates for a variety of tidal volumes at a flow rate of 250 ml/second and various flow rates at a tidal volume of 1000 ml. The data for particle diameters of 0.2–3.0 μm agree well with the experimental data from five subjects, showing that deposition efficiency increases with tidal volume at a constant flow rate and with decreasing flow rate for a given tidal volume. The model was also exercised to compute regional pulmonary and tracheobronchial deposition of aerosol at these conditions and to illustrate changes in the distribution of deposited aerosol mass over the 23 Weibel model generations for different aerosol sizes. For the size range considered, the prediction was that the largest diameters would produce the highest mass fractions deposited, the bulk of the aerosol mass would go to the pulmonary region below generation 17, and the peak in the mass deposition curve would move deeper into the pulmonary region with decreasing particle size.

Taulbee et al. (1978) modified the earlier model of Taulbee and Yu (1975) to improve the deposition computation using Weibel's lung model A (Weibel, 1963). Impaction deposition efficiency was computed as 1.3 times the local Stokes number in the airway generation; diffusion deposition efficiency was given by:

$$\text{Diffusion deposition efficiency} = 4.07h^{2/3} - 2.4h - 0.446h^{4/3 + \cdots}$$

$$(15)$$

where the dimensionless parameter, h, is $2 D_C z/U d^2$; D_C is the Brownian diffusion coefficient; z, the axial distance down the airway; and U, the flow velocity in the airway.

The removal efficiency by settling simplifies for small particles to:

$$\text{Settling deposition efficiency} = \frac{v_t z}{U d}$$

where v_t is the Stokes settling velocity, Eq. (5). This model ignored direct computation of factors such as mixing by secondary flows at each bifurcation and mixing because of regional nonuniformities in ventilation; however, the mixing effect was reintroduced via the Taylor (1953) diffusion coefficient and an empirical mixing parameter for the mixing effect of flow through the bifurcation. The resulting model gave very accurate single-breath aerosol wash-in/wash-out computations of concentration versus time or volume respired for 0.5- and 1.0-μm droplets of di-2-ethylhexyl sebacate. The agreement with the experimental data appears to verify the author's contentions that mechanical mixing in the upper airways plays a minor role in aerosol transport and that most mixing occurs in the alveolar region.

Gerrity *et al.* (1979) used Weibel's lung anatomy model and Landahl's model methods to develop a model similar to the previous ones, but verified it using iron oxide aerosols with MMAD's of 1.6–16 μm. Their model also computed the regional surface concentration of delivered aerosol material for use in medical studies. The model confirms Yu and Taulbee's (1977) findings that the percentage deposition per generation reaches its peak value higher in the lung for large particles and that the trend continues up to 16-μm diameter particles, which peak in their deposition in generations three or four. For the most part, the model data matches the experimental results of Heyder *et al.* (1975), also, and shows the usual minimum in total deposition for diameters of 0.5 μm. For a test computation using 7.9-μm particles, the highest epithelial surface concentrations were found to occur near generation four, a region of enhanced bronchial carcinomas in many human patients.

Austin *et al.* (1979) again use Landahl's equations, but add to them a more realistic mouth and tracheobronchial anatomy, as well as the airway branching angle data from Horsfield and Cumming (1967), to compute the aerosol deposition losses of a bolus of particulate-containing air in the respiratory system. Also, the model computes aerosol growth due to humidity at each generation to account for hygroscopic aerosol deposition. The minimum in the total deposition curve is shown to occur at 0.5 μm; however, the model tends to underestimate deposition compared to the most recent experimental deposition data. The model is, however, instructive in showing the effect of changes in aerosol density, airway diameters, and breathing pattern in regional deposition.

Deposition models have become increasingly more complex and accurate in predicting experimental deposition values of aerosol in the lung and, in turn, have been instructive in defining which variables of the breathing pattern are significant in affecting aerosol deposition. Their usefulness to an investigator lies in their ability to predict approximate mass doses of aerosolized material to the body once the aerosol's size has been determined. Because some of the more complex models also compute mass dose to the various anatomical regions of the respiratory system, these models are useful in understanding the etiology of diseases related to the deposition of harmful materials.

VI. FUTURE RESEARCH PERSPECTIVE

An increased awareness of the potential health hazards posed by aerosols and their effects on illnesses of respiratory system origin has developed in the past two decades. This has resulted from our improved ability to measure these materials in the environment and our knowledge that some aerosols such as asbestos fibers, cigarette smoke, and radioactive particles are a proved health hazard to man. Other aerosols of anthropogenic origin such as fly ash, sulfate and nitrate compounds, and combustion-derived soots appear to be well tolerated by man at reasonable dose levels encountered in everyday life. Still other natural aerosols such as pollens, spores, fungi, bacteria, and viruses cause host-mediated effects that are highly dependent on the individual exposed. Given the tools already developed to generate, characterize, and measure the deposition of aerosols of various materials, health-related questions that remain unanswered concern the potential synergism of particles and gases, the synergistic effect of multiple compounds present in a single aerosol, and the future development of aerosols for the delivery of therapeutic drugs and general pulmonary therapy.

High concentrations of aerosols were shown by LaBelle *et al.* (1955) to potentiate the effect of formaldehyde vapor but decrease the toxicity of nitric acid fumes. Other studies by Wagner *et al.* (1961) showed that fine oil-droplet aerosols, administered in high concentrations, significantly altered the effect of the atmospheric pollutants ozone and nitrogen dioxide under some conditions but not under others. The interplay of factors such as particle size, aerosol concentrations, solubility of gases, and particle surface area and absorptivity in determining the synergistic effects of particles and gases are currently not well understood. Additional work in this area is necessary, especially at concentrations relevant to ambient and workplace exposure levels. In addition, many commercial products, such as hairspray, are designed to be dispensed as aerosols, and studies have shown that their use leads to depression in mucociliary transport (Borum *et al.*, 1979). This is an area of potential concern in view of the large number of consumer products and commercial processes that employ aerosolized materials, and additional work is necessary to ensure that aerosols used in these ways do not compromise the health of the user.

Because aerosols from both atmospheric and local sources are often mixtures of two or more compounds, the potential synergism of multiple compounds on the same particle or different compounds on different particles in the same aerosol comes into question. In studies of many combustion-related aerosolized sulfate and nitrate compounds, little or no effect was noted on pulmonary function in animals (Sackner *et al.*, 1978b 1979), or in man (Sackner *et al.*, 1978a, 1979; Utell *et al.*, 1979) from inhalation of submicron particles of pure compounds at concentrations more than two orders of magnitude above ambient levels. A single inhalation study in animals using a mixture of sulfate compounds administered as a submicron aerosol showed no synergistic effect on pulmonary function or tracheal mucous velocity (Sackner *et al.*, 1981). Still more chemically complex mixtures are found in combustion-derived soots (Schreck *et al.*, 1978; McGrath *et al.*, 1978), and inhalation studies of these aerosols to determine their respiratory system effects (Abraham *et al.*, 1980) and potential impact on health are of current research interest (Schreck *et al.*, 1980); however, the lack of a more general understanding of the synergistic effects of complex mixtures prevents conclusions from being drawn on the potential impact of any given combination of compounds on health.

Perhaps the most constructive use to be made from aerosol deposition research will be in the area of aerosol-administered pharmacological and therapeutic agents as described by Swift (1980). Since aerosols deposit in the respiratory system according to size, it may be possible, by adjusting particle size and aerosol concentration, to deliver predetermined doses of material to regions of the lung itself, or via the lung to the circulatory system. In this manner it would be possible to deliver some therapeutic compounds on a continuous basis without the limitations imposed by other routes of administration.

REFERENCES

Abraham, J. L. (1979). Documentation of environmental particulate exposures in humans using SEM and EDXA. *Scanning Electron Microsc.* **2**, 751–766.

Abraham, W. M., Kim, C. S., Januszkiewicz, A. J., Welker, M., Mingle, M., and Schreck, R. M. (1980). Effects of a brief low-level exposure to the particulate fraction of diesel exhaust on pulmonary function of conscious sheep. *Arch. Environ. Health* **35**(2), 77–80.

Altshuler, B. (1959). Calculation of regional deposition of aerosol in the respiratory tract. *Bull. Math. Biophys.* **21**, 257–270.

Altshuler, B., Yarmus, L., Palmes, E. D., and Nelson, N. (1957). Aerosol deposition in the human respiratory tract. I. Experimental procedures and total deposition. *AMA Arch. Ind. Hyg.* **15**, 293–303.

Austin, E., Brock, J., and Wissler, E. (1979). A model for deposition of stable and unstable aerosols in the human respiratory tract. *Am. Ind. Hyg. Assoc. J.* **40**, 1055–1066.

Beeckmans, J. M. (1965). The deposition of aerosols in the respiratory tract, I. Mathematical analysis and comparison with experimental data. *Can. J. Physiol. Pharmacol.* **43**, 157–172.

Bell, K. A., and Friedlander, S. K. (1973). Aerosol deposition in models of a human lung bifurcation. *Staub—Reinhalt Luft*, **33**, 183–187.

Bell, K. A., Martonen, T. L., and Ho, A. (1978). Growth rate measurements and deposition modeling of hygroscopic medicinal aerosols in the respiratory tract. *Pap., 71st AIChE Meet.*

Borum, P., Holten, A., and Loekkegaard, N. (1979). Depression of nasal mucociliary transport by an aerosol hairspray. *Scand. J. Respir. Dis.* **60**, 253–259.

Brain, J. D., and Valberg, P. A. (1979). Deposition of aerosol in the respiratory tract. *Am. Rev. Respir. Dis.* **120**(6), 1325–1376.

Brock, J. R., and Marlow, W. H. (1975). Charged aerosol particles and air pollution. *Environ. Lett.* **10**(1), 53–67.

Chan, T. L., and Lippmann, M. (1980). Experimental measurements and empirical modeling of the regional deposition of inhaled particles in humans. *Am. Ind. Hyg. Assoc. J.* **41**, 399–409.

Chan, T. L., Lippmann, M., Cohen, V. R., and Schlesinger, R. B. (1978a). Effect of electrostatic charges on particle deposition in a hollow cast of the human larynx-tracheobronchial tree. *J. Aerosol Sci.* **9**, 463–468.

Chan, T. L., Schreck, R. M., and Lippmann, M. (1978b). Effect of turbulence on particle deposition in the human trachea and bronchial airways. *Symp. Transp. Processes Biol. Syst., AIChE Meet. 1978* Paper No. 124d.

Chan, T. L., Schreck, R. M., and Lippmann, M. (1980). Effect of the laryngeal jet on particle deposition in the human trachea and upper bronchial airways. *J. Aerosol Sci.* **11**, 447–459.

Cuddihy, R. G., Brownstein, D. G., Raabe, O. G., and Kanapilly, G. M. (1973). Respiratory tract deposition of inhaled polydisperse aerosols in beagle dogs. *Aerosol Sci.* **4**, 35.

Cunningham, E. (1910). On the velocity of steady fall of spherical particles through fluid medium. *Proc. R. Soc. London, Ser. A* **83**, 357–365.

Davies, C. N. (1974). Deposition of inhaled particles in man. *Chem. Ind. (London)* pp. 441–444.

Drinker, P., Thompson, R. M., and Finn, J. L. (1928). Quantitative measurements of the inhalation, retention and exhalation of dust and fumes by man. I. Concentration of 50 to 450 mg per cubic meter. *J. Ind. Hyg. Toxicol.* **10**, 13–25.

Ferron, G. A. (1977). The size of soluble aerosol particles as a function of the humidity of the air: Application to the human respiratory tract. *J. Aerosol Sci.* **8**, 251.

Findeisen, W. (1935). Über das Absetzen kleiner, in der Luft suspendierten Teilchen in der menchlichen Lunge bei der Atmung. *Pfluegers Arch. Gesamte Physiol.* **236**, 367–379.

Foord, N., Black, A., and Walsh, M. (1978). Regional deposition of 2.5–7.5 μm diameter inhaled particles in healthy male nonsmokers. *J. Aerosol Sci.* **9**, 343–357.

Friedlander, S. K. (1977). "Smoke, Dust and Haze: Fundamentals of Aerosol Behavior." Wiley, New York.

Fry, F. A. (1970). Charge distribution on polystyrene aerosols and deposition in the human nose. *J. Aerosol Sci.* **1**, 135–146.

Fuchs, N. A. (1965). "The Mechanics of Aerosols." Pergamon, Oxford.

Gerrity, T. R., Lee, P. S., Hass, F. J., Marinelli, A., Werner, P., and Lourenco, R. V. (1979). Calculated deposition of inhaled particles in the airway generations of normal subjects. *J. Appl. Physiol.: Respir., Environ. Exercise Physiol.* **47**(4), 867–873.

Hänel, G., and Zankl, B. (1979). Aerosol size and relative humidity: Water uptake by mixtures of salts. *Tellus* **31**, 478–486.

Haselton, F. R., and Scherer, P. W. (1980). Bronchial bifurcations and respiratory mass transport. *Science* **208**, 69–71.

Hesketh, H. E. (1977). "Fine Particles in Gaseous Media." Ann Arbor Sci. Publ., Ann Arbor, Michigan.

Heyder, J., Armbruster, L., Gebhart, J., Grein, E., and Stahlhofen, W. (1975). Total deposition of aerosol particles in the human respiratory tract for nose and mouth breathing. *J. Aerosol Sci.* **6**, 311–328.

Heyder, J., Gebhart, J., Roth, C., Stahlhofen, W., Stuck, B., Tarroni, G., DeZaiacomo, T., Formignani, M., Melandri, C., and Prodi, U. (1978). Intercomparison of lung deposition data for aerosol particles. *J. Aerosol Sci.* **9**, 147–155.

Horsfield, K., and Cumming, G. (1967). Angles of branching and diameters of branches in the human bronchial tree. *Bull. Math. Biophys.* **29**, 245–259.

Horsfield, K., Dart, G., Olson, D. E., Filley, G. F., and Cumming, G. (1971). Models of the human bronchial tree, *J. Appl. Physiol.* **31**(2), 207–217.

Kotrappa, P., and Moss, O. R. (1971). Production of relatively monodisperse aerosols for inhalation experiments by aerosol centrifugation. *Health Phys.* **21**, 531–535.

LaBelle, C. W., Long, J. E., and Christofano, E. E. (1955). Synergistic effects of aerosols. *AMA Arch. Ind. Health* **11**, 297–304.

Landahl, H. D. (1950). On the removal of airborne droplets by the human respiratory tract. I. The lung. *Bull. Math. Biophys.* **12**, 43–56.

Landahl, H. D., Tracewell, T. N., and Lassen, W. H. (1951). On the retention of airborne particulates in the human lung. II. *AMA Arch. Ind. Hyg.* **3**, 359–366.

Landahl, H. D., Tracewell, T. N., and Lassen, W. H. (1952). Retention of airborne particulates in the human lung. III. *AMA Arch. Ind. Hyg.* **6**, 508–511.

Lippmann, M. (1977). Regional deposition of particles in the human respiratory tract. *In* "Handbook of Physiology" (D. H. K. Lee and S. Murphy, eds.), Chap. 14, pp. 213–232.

Lourenco, R. V. (1969). Distribution and clearance of aerosols. *Am. Rev. Respir. Dis.* **101**, 460–461.

Lourenco, R. V., Klimek, M. F., and Borowski, C. J. (1971). Deposition and clearance of 2 μ particles in the tracheobronchial tree of normal subjects—smokers and non-smokers, *J. Clin. Invest.* **50**, 1411–1420.

Lourenco, R. V., Loddenkemper, R., and Carton, R. W. (1972). Patterns of distribution and clearance of aerosols in patients with bronchiectasis. *Am. Rev. Respir. Dis.* **106**, 857–866.

McGrath, J. J., Schreck, R. M., and Siak, J. J. (1978). Mutagenic screening of diesel exhaust particulate material. *Proc. Annu. Meet.—Air Pollut. Control Assoc.* **71**, No. 78-33.6.

Melandri, C., Prodi, V., Tarroni, G., Formignani, M., DeZaiacomo, T., Bompane, C. F., Maestri, G., and Giacomelli-Maltoni, G. (1977). *Inhaled Part. 4, Proc. Int. Symp., 1975* (W. H. Walton, ed.), pp. 193–200.

Mercer, T. T. (1973). "Aerosol Technology in Hazard Evaluation." Academic Press, New York.

Mercer, T. T. (1975). The desposition model of the task group on lung dynamics: A comparison with recent experimental data. *Health Phys.* **29**, 673–680.

Pattle, R. E. (1961). The retention of gases and particles in the human nose. *In* "Inhaled Particles and Vapors" (C. N. Davies, ed.), pp. 302–309. Pergamon, Oxford.

Raabe, O. G. (1976). The generation of aerosols of fine particles. *In* "Fine Particles: Aerosol Generation, Measurement, Sampling and Analysis" (B. Y. H. Liu, ed.), pp. 57–110. Academic Press, New York.

Raabe, O. G., Yeh, H. C., Newton, G. J., Phalen, R. F., and Velasquez, D. J. (1977). Deposition of inhaled monodisperse aerosols in small rodents. *Inhaled Part. 4, Proc. Int. Symp., 1975* (W. H. Walton, ed.), p. 3.

Sackner, M. A., Ford, D., Fernandez, R., Cipley, J., Perez, D., Kwoka, M., Reinhart, M., Michaelson, E. O., Schreck, R. M., and Wanner, A. (1978a). Effects of sulfuric acid aerosol on cardiopulmonary function of dogs, sheep and humans. *Am. Rev. Respir. Dis.* **118**(3), 497–510.

Sackner, M. A., Perez, D., Brito, M., and Schreck, R. M. (1978b). Effect of moderate duration exposures to sulfate and sulfuric acid aerosols on cardiopulmonary function of anesthetized dogs. *Am. Rev. Respir. Dis.* **117**(4) (abstr.).

Sackner, M. A., Dougherty, R. D., Chapman, G. A., Zarzecki, S., Zarzemski, L., and Schreck, R. M. (1979). Effects of sodium nitrate aerosol on cardiopulmonary function of dogs, sheep and man. *Environ. Res.* **18**, 421–436.

Sackner, M. A., Dougherty, R. L., Chapman, G. A., Cipley, J., Perez, D., Kwoka, M., Reinhart, M., Brito, M., and Schreck, R. M. (1981). Effects of brief and intermediate exposures to sulfate submicron aerosols and sulfate injections on cardiopulmonary function of dogs and tracheal mucous velocity of sheep. *J. Toxicol. Environ. Health* **74**, 951–972.

Scherer, P. W., Haselton, F. R., Hanna, L. M., and Stone, D. R. (1979). Growth of hygroscopic aerosols in a model of bronchial airways. *J. Appl. Physiol.: Respir., Environ. Exercise Physiol.* **47**(3), 544–550.

Schlesinger, R. B., and Lippmann, M. (1972). Particle deposition in casts of the human upper tracheobronchial tree. *Am. Ind. Hyg. Assoc. J.* **33**, 237–251.

Schreck, R. M. (1972). Laminar flow through bifurcations with applications to the human lung. Ph.D. thesis, Northwestern University, Evanston, Illinois.

Schreck, R. M., and Mockros, L. F. (1970). Fluid dynamics in the upper pulmonary airways. *AIAA Pap.* **70-788.**

Schreck, R. M., McGrath, J. J., Swarin, S. J., Hering, W. E., Groblicki, P. J., and MacDonald, J. S. (1978). Characterization of diesel exhaust particulate for mutagenic testing. *Proc. Annu. Meet.—Air Pollut. Control Assoc.* **71**, No. 78-33.5.

Schreck, R. M., Soderholm, S. C., Chan, T. L., Hering, W. E., D'Arcy, J. B., and Smiler, K. L. (1980). In "Health Effects of Diesel Engine Emissions: Proceedings of an International Symposium" (W. E. Pepelko, R. M. Danner, and N. A. Clarke, eds.), Vol. 2, EPA-600/9-80-057b, pp. 573–591. U.S. Environ. Prot. Agency, Cincinnati, Ohio.

Schroter, R. C., and Sudlow, M. F. (1969). Flow patterns in models of the human bronchial airways. *Respir. Physiol.* **7**, 341–355.

Soderholm, S. C. (1980). Physical characterization of diesel exhaust particles in exposure chambers. In "Health Effects of Diesel Engine Emissions: Proceedings of an International Symposium" (W. E. Pepelko, R. M. Danner, and N. A. Clarke, eds.), Vol. 2, EPA-600/9-80-057b, pp. 592–605. U.S. Environ. Prot. Agency, Cincinnati, Ohio.

Stuart, B. O. (1973). Deposition of inhaled aerosols. *Arch. Intern. Med.* **131**, 60–66.

Swift, D. L. (1980). Aerosols and humidity therapy: Generation and respiratory deposition of therapeutic aerosols. *Am. Rev. Respir. Dis.* **122** (5), Part 2, 71–77.

Task Group on Lung Dynamics (1966). Deposition and retention models for internal dosimetry of the human respiratory tract. *Health Phys.* **12**, 173–207.

Taulbee, D. B., and Yu, C. P. (1975). A theory of aerosol deposition in the human respiratory tract. *J. Appl. Phys.* **38**(1), 77–85.

Taulbee, D. B., Yu, C. P., and Heyder, J. (1978). Aerosol transport in the human lung from analysis of single breaths. *J. Appl. Physiol., Respir., Environ. Exercise Physiol.* **44**(5), 803–812.

Taylor, G. I. (1953). Dispersion of soluble matter in solvent flowing slowly through a tube. *Proc. R. Soc. London, Ser. A* **219,** 186.

Thomas, R. G. (1972). An interspecies model for retention of inhaled particles. *In* "Assessment of Airborne Particles" (T. T. Mercer, P. E. Morrow, and W. Stöber, eds.), pp. 405–455. Thomas, Springfield, Illinois.

Tillery, M. I., Wood, G. O., and Ettinger, H. J. (1976). Generation and characterization of aerosols and vapors for inhalation experiments. *Environ. Health Perspect.* **16,** 25–40.

Utell, M. J., Swinburne, A. J., Hyde, R. W., Speers, D. M., Gibb, F. R., and Morrow, P. E. (1979). Airway reactivity to nitrates in normal and mild asthmatic subjects. *J. Appl. Physiol.* **46**(1), 189–196.

Vostal, J. J., Schreck, R. M., Lee, P. S., Chan, T. L., and Soderholm, S. C. (1982). Deposition and clearance of diesel particles from the lung. *In* "Toxicological Effects of Emissions from Diesel Engines" (J. Lewtas, ed.), pp. 143–159. Am. Elsevier, New York.

Wagner, W. D., Dobrogorski, O. J., and Stokinger, H. E. (1961). Antagonistic action of oil mists on air pollutants. *Arch. Environ. Health* **2,** 523–534.

Weibel, E. R. (1963). "Morphometry of the Human Lung." Springer, Berlin and New York.

Whitby, K. T., and Cantrell, B. (1976). Atmospheric aerosols—Characteristics and measurements. *Int. Conf. Environ. Sens. Assess. [Proc.], 1976* pp. 1–6.

Womersley, J. R. (1955). Method for the calculation of velocity, rate of flow and viscous drag in arteries when the pressure gradient is known. *J. Physiol. (London)* **127,** 553.

Yu, C. P., and Taulbee, D. B. (1977). A theory of predicting respiratory tract deposition of inhaled particles in man. *Inhaled Part. 4, Proc. Int. Symp., 1975* (W. H. Walton, ed.), pp. 35–47.

8

Mechanisms of Silica and Diesel Dust Injury to the Lung

Milos Chvapil

I. INTRODUCTION

Direct communication of the lung with the atmosphere predisposes this organ to insults from environmental particulates or vapors. For many years, complex methodolgy has been used to ascertain the changes in various components of the lung exposed to a defined environment. Almost every scientific speciality has contributed to the vast knowledge on the morphology, biochemistry, biophysics, and function of the lung undergoing pathological changes caused by exposure to silica, asbestos, and diesel exhaust particulates. The research interest in various noxious agents has been shifting from quartz-induced silicosis to asbestos dust, to oxidants (ozone, nitrous oxide, paraquat), and, finally, to diesel particulates. It is obvious that industry's concern to protect the individuals

223

AIR POLLUTION

exposed to a particular noxious environment dictates the need for a better understanding of lung tissue reaction to that particular agent.

It is becoming apparent that the old classification of dust particles as inert, toxic, and fibrogenic will be difficult to defend, because new, highly sensitive methods of ascertaining tissue injury show definite abnormalities even after lung exposure to so-called inert dusts (e.g., TiO_2, coal). Furthermore, the toxic effect of a lengthy exposure to such dusts undoubtedly results in collagen accumulation, thus inducing a fibrotic reaction. For example, amorphous silica (aerosil Degussa, ~ 100 Å particle size) dissolves quickly to molecular silicic acid, which is toxic, and is quickly eliminated out of the lung. Thus, after a single exposure to the norm, transient acute lung damage occurs. Repeated exposure to somewhat lower doses of this dust results, however, in a fibrotic reaction (Chvapil and Holusa, 1965). Similar observations were reported with exposures to high levels of oxygen, ozone, and nitrogen dioxide—each of these exposures resulting under certain conditions in lung fibrosis in various species (Drozdz et al., 1977; Välimäki et al., 1975; Werthamer et al., 1974).

Formation of a fibrotic lesion, fibrosis, or abnormal accumulation of connective tissue consists of two major components, collagen and, in the final stage of fibroproductive (fibroproliferative) inflammation, glycosaminoglycans. As will be documented below, a certain increase in the synthesis of these structural macromolecules and their deposition (often only temporary) is part of any tissue injury and inflammatory reaction. It requires a sufficiently long exposure to change the interstitial tissue characteristics in order to obtain abnormal continuous collagen deposition. Although the topic of this review is the lung's reaction to inhaled dust particles, it must be kept in mind that only 1–2% of the inhaled respirable particles ($<$ μm size) are retained in the lung interstitial tissue. In other words, "about 98–99% of the primarily deposited dust mass is cleared from the respiratory system" (Privalova et al., 1980). This finding clearly indicates the important role of alveolar phagocytosis as the pulmonary clearance mechanism, a topic reviewed recently by Privalova et al. (1980).

II. HYPOXIA

Much evidence indicates that the actual amount and physical characteristics of connective tissue in the intercellular space control the cell nutrition and oxygen diffusion rate. An increase in collagen concentration with increased structural stability related to the degree of polymerization of collagen forms a barrier affecting the function of the inter-cellular-space interstitial tissue. Once the function of the intercellular space is altered and interferes with the cell metabolism, hypoxia develops and becomes another factor promoting more collagen deposition. This would occur even if the original noxious agent were not present.

Plenty has been written about the role of low oxygen tension in injured tissue in forming fibrotic tissue. As I indicated, hypoxia results from a change in the quality of interstitial tissue and also from the direct effect on the microcirculation. A hypoxic environment is, in fact, conducive to fibrosis: inactive fibroblasts (fibrocytes) are reactivated as fibroblasts, which are more active metabolically. Under hypoxic stress, other cells have the capacity to change into fibrogenic cells. In fact, next to macrophages, fibroblasts were shown to multiply and function at the lowest P_{O_2} (Hunt, 1964; Hunt et al., 1967). Also, some evidence indicates the enhanced growth of connective tissue under hypoxia. The budding of capillaries follows the direction of the hypoxic gradient, and the budding phenomenon is clearly associated with the production of collagen type IV in the basement membrane of the microvessels. It is not quite clear whether the accumulation of lactate in hypoxic tissue stimulates collagen synthesis by activating inactive forms of prolyl hydroxylase (Chvapil et al., 1970; Comstock and Udenfriend, 1970).

III.   SEQUENCE OF EVENTS LEADING TO FIBROSIS

Let us follow the sequence of reactions occurring in the lung tissue exposed to noxious particulates or vapors. It is obvious that the description of the inflammatory events constituting the fibroproductive reaction depends on the methodology used to analyze the process (Fig. 1). Thus, the nomenclature differs depending on whether the problem is approached by a morphologist, bioengineer, biochemist, or immunologist.

For the sake of simplicity, I propose to differentiate four closely related and overlapping phases in fibroproductive inflammation:

1. Cell mobilization with release of activators and mediators of inflammation affecting other cells and the microcirculation

2. Participation of glycosaminoglycans

3. Phase of collagen synthesis and deposition

4. Phase of final organization and remodeling with possible formation of degenerative structures (Chvapil, 1967).

Irrespective of the type of noxious agent (physical, e.g., radiation,

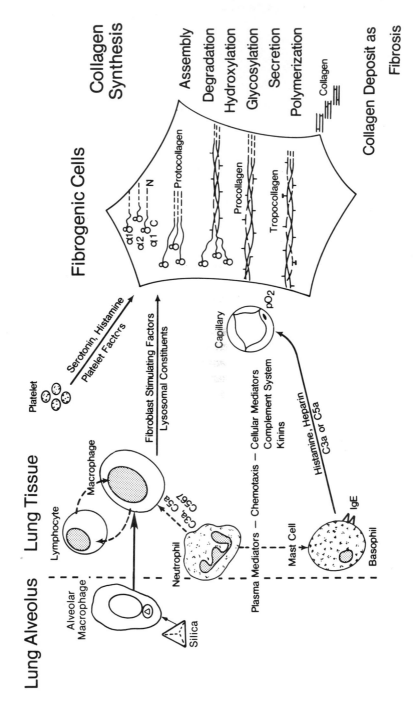

Fig. 1. Cellular aspect of fibroproductive reaction of the lung to inhaled dust.

TABLE I

Factors Modifying the Dynamics of Fibroproliferative
Inflammation

Endogenous	Exogenous
Race	Nutrition
Species	Temperature
Tissue	Mechanical factors
Sex	Oxygen
Nerve trophicity	Infection
Hormones	Immune state
Age	

incision, heat; chemical, e.g., CCl_4, paraquat; biological, e.g., bacterial, immune reaction), the sequence of the phases listed above is the same in all types of injuries. It is the duration of each phase and the magnitude of the reaction that may differ. In addition, several other factors (both exogenous and endogenous) modify the reaction as well (Table I).

Whereas in most tissues, the polymorphonuclear leukocyte is the first cell to react to an injurious agent, in the lung, it is the alveolar macrophage that responds first to an environmental stress. Rightfully, enormous research interest has been devoted to studying the multiple roles this cell plays in the inflammatory process, immune reactions, and the defense against malignancy. It was shown that the activated macrophage produces various substances that affect mitosis and the function of other cells. One of these products is a fibroblast-activating factor that stimulates both proliferation, as well as collagen and glycosaminoglycans synthesis, by these cells (Aalto et al., 1976).

One of the first reports indicating that silica-activated macrophages produce a factor that enhances the synthesis of collagenous hydroxyproline by fibroblasts was presented by Heppleston and Styles (1967) and reviewed later by Heppleston (1971). Their results prompted several studies that yielded rather conflicting results. Some of the studies reproduced and some failed to substantiate the original results. This is documented in a condensed form in Table II, which indicates the type of macrophage, type of stimulatory agent, and source of the target cell (or tissue) and its growth phase. The table also shows the indicator and the direction of the effect. Finally, the last column shows some characteristics of the fibroblast stimulating factor. As our knowledge increases, we begin to understand the reasons for the conflicting evidence: for instance, it has been shown that only activated macrophages are stimulating fibroblasts, and only fibroblasts in logarithmic phase are suscep-

TABLE II

Summary of Various Studies on Macrophage–Fibroblast Relationship

Investigators	Agent employed	Macrophage source	Target cells used	Stage of cell growth	Assay for	Effect of macrophage extract	Properties of factor from macrophage extract
Heppleston and Styles (1967)	Quartz (Doren-trop)	Rat peritoneal ex-udate cells (PEC)	Chick embryo	Confluent	Hydroxyproline	Stimulation in cell layer	Soluble
Burrell and An-derson (1973)	SiO$_2$ (crystalline)	Rabbit alveolar	Human WI-38		Hydroxyproline	Stimulation (4–16×) in me-dium; no effect in cell layer	
Nourse et al. (1975)	SiO$_2$ (in vivo)	Guinea pig alveolar	Guinea pig embry-onic lung	(a) Log (b) Confluent	Hydroxyproline Hydroxyproline	(a) 60% stimulation (b) No effect or inhibition	
Harington et al. (1973)	SiO$_2$ (fransil)	Hamster peritoneal	Hamster skin	Confluent	Hydroxyproline	80% inhibition	
Aalto et al. (1976)	SiO$_2$	Rat PEC	Granulation tissue slices		Collagen production Protein Nucleic acids Mucopolysaccha-rides	Stimulation Stimulation No effect No effect	Sensitive to sonication Sensitive to freeze–thaw

Lewis and Burrell (1976)	Connective tissue AB	Mouse PEC	Mouse fibroblasts	Confluent	Hydroxyproline	Stimulation	
Leibovich and Ross (1976)		Guinea pig peritoneal	Guinea pig wound fibroblast		Fibroblast proliferation	Stimulation	Nondialyzable; stable at 56°30'; stable with freeze–thaw
Calderone et al. (1974)		Mouse PEC	Mouse fibroblast		[³H]Thymidine incorporation	(a) one fraction stimulates (b) one fraction inhibits	Dialyzable, inhibits; nondialyzable, stimulates
Richards and Wusteman (1974)	SiO$_2$ (minusil)	Rabbit alveolar	Rabbit lung fibroblast	Log	Collagen	Stimulation	Resistant to trypsin; resistant to boiling; resistant to freezing
Kilroe-Smith et al. (1973)	Quartz (intratracheally or inhaled)	Guinea pig alveolar	Subcutaneous implantation of factor		Collagen	Stimulation	Resistant to acetic acid; TCA pH 4 nonlabile; stable for 4 years at 4°C; associated with insoluble portion of resistant to sonication

229

TABLE III

Effect of Extracts from Resting and Activated Macrophages on Synthesis of Sulfated GAG by Fibroblast[a] in Log Growth Phase

Treatment[b]	Synthesis of glycosaminoglycans $^{35}SO_4$ (dpm/μg protein)
None, control	72.9 ± 1.7
Resting macrophage	87.8 ± 1.2
Silica activated	125.3 ± 1.7
E. coli activated	66.5 ± 0.9

[a] WI-38 fibroblast in log growth; no change observed in confluent cells.

[b] Rabbit pulmonary alveolar macrophages were plated in tissue culture vessels, kept 2 hours at 37°C, then activators (silica and heat-inactivated E. coli) added for 60 minutes. Harvested cells were sonicated, spun at 2500 g, and supernate added to fibroblast cultures and incubated for 3 days. 24 hours before termination of experiment, $^{35}SO_4$ was added to the medium.

tible to macrophage factor (Table III). Furthermore, fibroblasts in the stationary phase are resistant and do not increase the activity after treatment with the extract from activated macrophages (J. Ulreich and M. Chvapil, unpublished results).

Kilroe-Smith et al. (1973) demonstrated that a fibrogenic factor in alveolar macrophages of guinea pigs exposed to quartz dust produces granuloma when implanted subcutaneously into guinea pigs. Thus, there seems to be no doubt that macrophages activated by either silica or any other activating agent produce a factor that triggers the fibroblastic activity that synthesizes more collagen and, eventually, glycosaminoglycans. The nature of the fibroblast-stimulating factor from macrophages has been the subject of several studies. Aalto et al. (1976) showed that among 30 fractions of macrophage media, one fraction inhibited and two stimulated DNA synthesis by fibroblasts and proline incorporation into both cells and collagen. One of the fractions also exhibited ribonuclease activity. In our experiments, the fibrogenic factor from macrophages was present in the extractable 15,000 g supernates. In experiments by Kilroe-Smith et al. (1973), the factor was found to be associated with the insoluble debris produced after sonic disintegration of the macrophages.

It is tempting to speculate on the possible nature of such a substance. It has been well established that activated macrophages release lysosomal enzymes. It may well be that modification of the fibroblast's surface membrane by lysosomal enzymes acting at a particular level would trigger higher activity in the fibroblast. This seems to be supported by the work of Rokosova and Bentley (1980), who found that fibroblast mitosis was enhanced when certain low concentrations of some lysosomal enzymes were added to the cultivating medium.

Another mechanism by which activated macrophages could stimulate the fibroblast activity may relate to increased synthesis of prostaglandins, especially PGE_1 and $PGF_{2\alpha}$. During macrophage activation the oxidation of unsaturated fatty acids (in particular, arachidonic acid) is enhanced, resulting in the synthesis of prostaglandins. Prostaglandins of the E and F types were found to stimulate adenyl cyclase activity in different fibroblasts' cultured cell lines, the most active being PGE_1. This suggests that prostaglandins synthesized by inflammatory cells (e.g., macrophage or platelet) may regulate, via cyclic AMP, the synthesis and, eventually, the secretion of glycosaminoglycans and of collagen by fibroblasts. Another experiment indicates the fibroblast-promoting activity of extract from whole organs, in this case, the lungs of rats injected intratracheally with a suspension of quartz and sacrificed at different time intervals from 3 to 24 days after the installation of the dust. The lungs were homogenized and extracted into the medium used for the cultivation of fibroblasts. The extract from various lungs was then equilibrated to the same content of proteins. Equal amounts of two concentrations of extract were added to the established line of 3T3 fibroblasts. The function of the fibroblasts was ascertained by measuring the synthesis of sulfated glycosaminoglycans. We found that within three days of the installation of the silica particles, the lung extract significantly enhanced

TABLE IV

Effect of Extracts from Silicotic Lungs of Rats on GAG Synthesis by 3T3 Fibroblasts[a]

Days after silica administration	GAG synthesis: $^{35}SO_4$ in GAG (dpm/μg cell protein)
0	281 ± 24
3	625 ± 24
10	525 ± 45
17	293 ± 45
24	290 ± 27

[a] Same lobe of the lungs at given times was extracted in MEM; protein content in 15,000 g supernate of the extract was adjusted to the same protein content. Data (means ± SEM) refer to the effect of 10 μg protein of extracts added to 3T3 fibroblast culture in log phase. Cells were incubated for 36 hours, then $^{35}SO_4$ was added (50 μCi/flask) and incubated for another 24 hours (M. Chvapil and G. Herring, unpublished results).

the activity of fibroblasts (Table IV). Significant activation occurred on the tenth day. In later stages of silicotic fibroproliferative inflammation, the fibroblast-stimulating activity in the injured lung decreased and returned to the control values. We concluded that the presence of the factor in the lung coincides with the accumulation of macrophages on the dust-exposed lung tissue.

We also found that other organs, such as the liver, in particular, produce the "fibroblast stimulating factor" when exposed to silica as well. Four and 7 days after an intravenous injection of 1 μm silica suspension, the extract from the homogenized liver stimulated the activity of prolyl hydroxylase in fibroblast exposed in tissue culture to the extract (Ulreich and Chvapil, unpublished results).

IV. MAST CELLS

The reaction of the lung tissue to noxious exogenous particles or vapors is characterized by some unique changes in cell content. It was reported that the number of mast cells in the lung increases after chronic exposure of rats or guinea pigs to diesel particulates (White and Garg, 1981).

The localization of mast cells in the bronchial lumen, in the bronchial mucosa, in the intraepithelial location, as well as in deeper perivenular collections, makes this cell susceptible to exogenous toxins. The important role of this cell in the inflammatory process and defense of the lung has been reviewed recently by Wasserman (1980). Evidence showing that serotonin, as well as histamine, stimulates collagen synthesis by fibroblasts (Boucek and Alvarez, 1970, 1971) and that the inhalation of histamine aerosol induces diffuse lung fibrosis (Uspenskiĭ, 1952) was presented, also.

V. ALVEOLAR TYPE II CELLS

Two characteristic features of the lung cell reactivity to toxic environments are the documented susceptibility of type I epithelial cells and the consequent hyperplastic response of the type II cells. In the intact lung, the alveoli are covered mainly by large squamous cells (type I cells). Only sporadically, smaller cubical cells (type II cells) are dispersed in the alveolar surface. The role of type I cells has been associated with gas exchange. The functions of type II cells are connected with production, storage, and secretion of the lung surfactant (Evans et al., 1973). In fact, the proliferation of type II cells with a characteristic localization has been found in lungs of rats exposed to diesel particulates even when no biochemical changes in the lung chemistry were detectable (White and Garg, 1981). The only other finding was an increase in the lung surfactant in the lavage of lung airways (Eskelson, 1981a). Although there might be several explanations for this finding, one possibility is that the proliferation of type II cells is responsible for an increase in the production of surfactant.

VI. FIBROBLASTS

The synthesis of glycosaminoglycans and collagen has been commonly ascribed to fibroblasts; however, some recent evidence indicates that many other cells (not exactly of mesenchymal origin) are capable of synthesizing these structural macromolecules and should be classified perhaps, as fibrogenic. In the lung, fibroblasts, endothelial cells, and, possibly, alveolar macrophages have been implicated in collagen biosynthesis (Kelleher et al., 1977; Hance and Crystal, 1975; Myllyla and Seppa, 1979). The environment of the injured lung contains various substances assumed to activate fibrogenic cells. It has been our experience that in tissue culture, the fibroblast reacts to various stimuli by the same pattern, i.e., by enhancing collagen (or glycosaminoglycan) synthesis. This coincides quite often but not regularly with the inhibition of DNA synthesis as measured by [^{3}H]thymidine incorporation (J. Ulreich and M. Chvapil, unpublished results).

VII. LIPIDS AND LUNG INJURY

In recent years, we were very impressed by the evidence that all the various species of lipids, but especially cholesterol and phospholipids, accumulate in the granuloma tissue or in the injured tissue before any increase in collagen is apparent.

The relationship and importance of lipids to the developing fibrotic lesion has not been determined; however, in most fibrogenic processes, there is a lipid loading (Pelliniemi, 1973; Boucek and Noble, 1955). Examples of these are (1) the lipid loading in silicotic lungs (Babushkina 1977; Katsnelson et al., 1964; Heppleston et al., 1974), (2) alcoholic fatty liver, which precedes alcoholic liver cirrhosis (Lieber, 1978), (3) lipid loading in atherogenesis (Chvapil et al., 1978), (4) lipid loading in subcutaneously implanted sponge (Boucek and Noble, 1955), and (5) nutritional muscular dystrophy (Shull et al., 1958).

An example of a fast accumulation of some lipid species in the rat lung

TABLE V

Pulmonary Lipids from Rats Exposed Intratracheally to 1 mg of Diesel Particulate[a]

Rat group	Phospholipids (mg/lung)	Cholesterol (mg/lung)	Triacylglycerols (mg/lung)
Experimental	51.7 ± 4.64^b	11.95 ± 1.11^c	20.00 ± 3.15^d
Control	30.8 ± 4.66^b	8.27 ± 0.63^c	24.52 ± 1.24^d

[a] Results expressed as mg of lipids per lung $\pm$ SD.
[b] $p < 0.001$.
[c] $p < 0.005$.
[d] Not significant.

given 1 mg diesel particulate intratracheally is shown in Table V. Five days after a single dose of this rather inert dust, the phospholipid and cholesterol contents were significantly increased. No change in the content of triacylglycerols was observed.

Speculations concerning the importance of lipid loading in prefibrotic lesions suggest that this energy-rich milieu may be an energy source that can be used by fibroblasts to form collagen. Some of the lipids in the prefibrotic lesion could be converted to toxic materials such as lipid peroxides, which stimulate fibroblasts for fibrogenesis. The increased peroxidation products and, in particular, the "free radicals" associated with lipid peroxidation were shown to bind with normal proteins. These altered proteins might be recognized by the living system as "foreign proteins," thus inducing an immunological response to the *de novo* formed "lipoprotein complex," resulting in an inflammation of the lung and an ensuing fibrotic lesion. Although we do not know where in the tissue the abnormally deposited lipids are located, we would like to speculate that a definite portion of the increased lipid pool affects cell membranes, possibly changing the proportions of membrane lipid constituents. This may result in a change of cell activity, reflecting the connection between membrane fluidity and cell function.

VIII. LUNG–LIVER LIPOGENESIS IN LUNG INJURY

Our recent studies showed that any exposure of lung to such "abnormal" stimuli, as silica, asbestos dust, or diesel particulates, or to intratracheal saline installation, results within a few hours in increased lipogenesis in the liver. Specific activity of phospholipids and cholesterol was significantly elevated. Although this stimulation lasted only 12 hours

TABLE VI

Hepatic Lipids from Rats Exposed Intratracheally to 1 mg of Diesel Particulate[a]

Rat group	Phospholipids (mg/g liver)	Cholesterol (mg/g liver)	Triacylglycerols (mg/g liver)
Experimental	27.9 ± 1.9[b]	3.36 ± 0.10[c]	8.57 ± 1.43
Control	31.3 ± 1.7[b]	4.02 ± 0.11[c]	9.07 ± 0.88

[a] Data presented as means ± SD.
[b] $p < 0.05$.
[c] $p < 0.001$.

after intratracheal saline installation, we found it increased for several weeks after insufflation of silica dust into the lung. Increased activity of both lipid species was identified in the serum and, finally, in the lung tissue.

As shown in Table V, the acute exposure of rat lung to a single dose of 1 mg diesel dust significantly increases the content of phospholipids and cholesterol. At the same time (Table VI), hepatic phospolipids and cholesterol content were decreased, whereas triacyglycerols were not significantly altered. The loss of hepatic lipids is paralleled by an increase in the specific activity of the lipid species mentioned above. We believe that this indicates stimulated hepatic lipogenesis. These briefly outlined experiments, as well as others, led us to the hypothesis that the "injured" lung produces a factor (or factors) that activates lipogenesis in the liver. The duration of the effect depends on the type of noxious agent to which the lung is exposed. Lipids are then secreted into the circulation, where they are picked up and processed into a specific form by "activated" lung tissue only. The testing of this hypothesis is under way.

IX. REACTION OF THE LUNG TO DUST PARTICLES

As indicated in the introduction, the concept of inert dust is questionable. Definite changes in cell population in the bronchial lavage fluid, increase of the content of lysosomal enzymes, and changes in surfactant composition and physical properties all indicate the activation of the lung defense mechanism (Seaton, 1975; C. Eskelson, unpublished data).

Using very sensitive methods to ascertain the synthesis of newly formed collagen (type III), even the reputed inert dusts, such as TiO_2, kaolin, korund, and aluminum oxide, stimulate collagen synthesis. To a certain extent, it seems that these changes are part of the "normal life of the lung tissue," reflecting the homeostatic response of this tissue to a frequently compromised environment. It is not surprising, therefore, to find that after chronic exposure of rats to different concentrations of diesel particulates, even when there are no changes in lung biochemistry (Misiorowski et al., 1980) or lung functional tests (Gross, 1981), there are significant increases in the number and size of macrophages. There is also higher activity among some lysosomal enzymes in the lavage fluid. It is interesting that the increase in macrophage counts precedes the oc-curence of polymorphonuclears and lymphocytes, which appear at the later stages of chronic exposure (Strom et al., 1981). Based on our expe-rience with the reaction of rat lungs to chronic exposures of up to 2 years to diesel particulate at concentrations ranging from 250 to 1500 $\mu m/m^3$, it seems that changes in bronchopulmonary fluid, including cellular population as well as modification in surfactant, are probably the most sensitive indicators of lung exposure. The reaction in lipid composition of surfactant is rather fast and temporary, its duration depending on the nature of the dust (Eskelson, 1981a). Five days after intratracheal ad-ministration of diesel particles at the dose 5mg/ml (rat), a significant increase of phospholipids, cholesterol, and protein in the lavaged fluid occurs (Eskelson, 1981a; Table VII). The increase of the surfactant lipids obviously accounts for the increased lipid content in the whole lung, as no difference in lipid content exists between exposed and con-trol lungs after lavaging the lungs (Table VIII).

This finding indicates the probability that lung stimulation by a nox-ious agent causes surfactant lipids to be secreted into the alveoli by type II alveolar epithelial cells, which are believed to synthesize and secrete the surfactant lipids. Our preliminary studies indicate, however, that lipids constituting the complex surfactant mixture are synthesized in the

TABLE VII

Analysis of Pulmonary Lavage Fluid from Rats Exposed Intratracheally to 5 mg of Diesel Particulate[a]

Rat group	Phospholipids (mg)	Cholesterol (mg)	Protein (mg)
Experimental	1.90 ± 0.48[b]	.539 ± .040[b]	8.47 ± 0.33[b]
Control	0.49 ± 0.18[b]	.151 ± .081[b]	2.09 ± 1.04[b]

[a] Results expressed as mg of lipids in the total lavage fluid. Variability is given as means ± SD.

[b] $p < 0.001$.

TABLE VIII

Analysis of Lavaged Lungs from Rats Exposed Intratracheally
to 5 mg of Diesel Particulate[a]

Rat group	Phospholipids (mg)	Cholesterol (mg)
Experimental	34.9 ± 1.34	15.9 ± 1.02
Control	31.3 ± 10.6	13.1 ± 3.23

[a] Results expressed as mg of lipid per lung ± SD.

liver and transported to the lungs. We hypothesize that alveolar II cells absorb and conform these lipids into the surfactant constituents. We also suggest that the increase in surfactant lipids and protein reflects both a higher influx of these components from the liver and an already established increase in the pool of alveolar type II cells due to lung injury (Eskelson, 1981b).

Another category of dusts, commonly classified as toxic, is represented by particulates that induce a fast, necrotic tissue reaction. These dusts are usually very soluble and are quickly removed from the tissue by simple diffusion. This is the case with amorphous silica Aerosil, which has a particle size between 100 and 400 Å and induces only temporal lung lesions. Still, the whole hierarchy of inflammatory reactions follows, involving even enhanced collagen synthesis and accumulation of this fibrillar protein.

Quartz and crystalline silica (and its modifications, e.g., crystobalit, tridymite) are probably the best-studied dusts. The mechanisms of silica's fibrogenic effect are not fully understood; however, most of the information points to the lysis of macrophages and the release of lysosomes caused possibly by H bonding of hydroxyl groups of the surface of silica crystals to membrane components. This results in the labilization of the membrane stability. Table IX lists various hypothesis on silica fibrogenicity. The solubility theory of King et al. (1953a,b) attracted enormous attention in the 1950s. The critiques of this theory cite a rather poor correlation between the solubility of various amorphorous and crystalline forms of silica and their fibrogenic capacity. It was suggested also that silicotic nodules are induced primarily by highly polymerized colloidal silicic acid, originating from oligomeric or molecular forms that permeate the tissue passing quickly through it and into the circulation. Polymerization of these low molecular weights in vivo is rather unlikely, because it is difficult to reach in vivo the critical con-

TABLE IX

Hypothesis on Silica Fibrogenicity

1. Mechanical
2. Solubility
3. Piezoelectricity
4. Epitaxis—surface
5. Hydrogen bonding
6. Immune mechanisms
   a. Lysosomal concept
      i. Degradation by enzymes
      ii. Stimulation by enzymes
   b. Lipid peroxidation
      i. As such
      ii. Prostaglandins

centration of the molecular silicic acid that would polymerize. (For a review, see Heppleston, 1969, and Ziskind *et al.*, 1976.)

The concept that the OH groups (silanol groups) of the silica crystal surface serve as a matrix, adsorbing various macromolecules (Stöber, 1966), is most attractive.

The surface of the quartz crystal is not a passive site, but serves as a catalyzer of various reactions, oxidation being the most important. The often-quoted work of Allison and Hart (1968) and Allison *et al.* (1966), implicating the hydrogen binding of the crystal to biological structures as a mechanism labilizing membranes (lysosomes, erythrocytes, etc.), is an extension of the above concept on the role of crystal surface. It could be assumed that due to the adsorbtion of proteins to the crystal, the protein structure is modified to such an extent that it becomes antigenic to the organism. This leads to considerations on the activation of the immune system of subjects with silicotic lesions (Vigliani and Pernis, 1958).

The brief outline, presented above, of the sequence of reactions that ultimately lead to the development of lung fibrosis allows one to select those steps that appear to be the most promising targets for a pharmacological attack aimed at preventing or curing respiratory diseases. On the basis of the scheme presented above, three approaches that appear to be fruitful are:

1. Prevention of the initial damage (lipid peroxidation)
2. Control of cellular and tissue response to this damage
3. Interference with the synthesis, accumulation, or maturation of collagen.

The stabilization of biomembranes by controlling lipid peroxidation is a relatively recent concept. One example of a practical application of this idea is the inhibition of lipoperoxidation by zinc: lysosomal and plasma membranes are stabilized, and the development of lung edema after exposure to NO_2 is prevented by low concentrations of zinc salts. The actual mechanism by which zinc prevents membrane damage is not yet known, but is being investigated (Chvapil, 1973). In the same line, any antioxidant should reduce the magnitude of the peroxidation related lung damage. Thus, α-tocopherol and selenium acting through glutathione peroxidase should be beneficial.

Contemporary knowledge of collagen metabolism may soon allow us to prevent fibrosis by selectively affecting specific steps in synthesis and degradation. This could be achieved in several ways (Fuller, 1981):

1. *Inhibiting hydroxylation of the collagenous polypeptide*

Of the several theoretical ways to interfere with collagen hydroxylation, only three approaches appear likely to succeed.

a. Removing the ferrous iron cofactor of protocollagen hydroxylase by chelating agents such as 1,10-phenanthroline or 2, 2'-dipyridyl.

b. Incorporating analogs of proline or lysine such as 3,4-dehydroproline, 4-fluoroproline, or *cis*-hydroxyproline, which cannot be hydroxylated into the collagen polypeptide.

c. Administering a synthetic polypeptide that has higher affinity (K) to protocollagen hydroxylase than does collagen, thus inhibiting the hydroxylation of the natural collagen polypeptide.

2. *Interfering with the maturation of collagen*

Collagen exerts its function only when polymerized into a fibrous structure, cross-linked by covalent bonds; thus, interfering with the cross-linking will change the mechanical properties of collagen. The enzyme inducing these links, lysyl oxidase, requires copper and oxygen for activity. It is inhibited by copper chelators such as penicillamine. Lathyrogens (β-aminopropionitrile and some antituberculosis drugs, e.g., isoniazid) competitively inhibit lysyl oxidase, resulting in an elevated level of soluble collagen. There is no information, however, as to whether this soluble collagen, formed in the presence of lathyrogens, is degraded faster than normal collagen. Mechanical properties of a tissue that has been treated with lathyrogens are significantly modified.

3. *Stimulating collagen degradation*

Because collagen deposition in fibrotic lesions is the result of an imbalance between synthesis and degradation, enhancing collagen digestion is tissues would be another method of contolling its accumulation. Unfortunately, little is known about this problem.

From the biochemical and metabolic viewpoints, regulating collagen metabolism in the lung is a difficult task, because its slow turnover rate is slower therein than collagen in the liver or intestine. However, because of the diseased state, collagen synthesis in the lung is enhanced, and a differential turnover rate of collagen in normal and affected tissue favors the therapeutic effect on newly synthesized collagen. The unique structure of the lung with its communication with the outer atmosphere makes it possible to administer therapeutic agents in the form of an aerosol, permitting a higher deposition of the drug in the lung.

Up to now, only those factors controlling some specific steps of collagen metabolism have been discussed; however, several other factors having rather nonspecific effects on the reactivity of connective tissue play an enormous role in the development of fibrotic lesions.

ACKNOWLEDGMENTS

The author was supported by a contract with General Motors Research Division. The author is grateful to Mrs. Judith B. Ulreich, M.S., for her help in preparing this manuscript.

REFERENCES

Aalto, M., Potila, M., and Kulonen, E. (1976). The effect of silica-treated macrophages on the synthesis of collagen and other proteins *in vitro*. *Exp. Cell Res.* **97,** 193–202.

Allison, A. C., and Hart, P. D. (1968). Potentiation by silica of the growth of *Mycobacterium tuberculosis* in macrophage cultures. *Br. J. Exp. Pathol.* **49,** 465–476.

Allison, A. C., Harington, J. S., and Birbeck, M. (1966). An examination of the cytotoxic effects of silica on macrophages. *J. Exp. Med.* **124,** 141–154.

Babushkina, L. G. (1977). Role of lipids in the pathogensis of silicosis. *In* "Industrial Diseases Due to Exposure to Dust" (S. G. Domin and B. A. Katsnelson, eds.), pp. 104–114. Ministry of Health, Moscow.

Boucek, R. J., and Alvarez, T. R. (1970). 5-Hydroxytryptamine: A cytospecific growth stimulator of cultured fibroblasts. *Science* **167,** 898–899.

Boucek, R. J., and Alvarez, T. R. (1971). Increase in survival of subcultured fibroblasts mediated by serotonin. *Nature (London)* **229,** 61–62.

Boucek, R. J., and Noble, N. L. (1955). Connective tissue: A technique for its isolation and study. *Arch. Pathol.* **59,** 553–558.

Burrell, R., and Anderson, M. (1973). The induction of fibrogensis by silica-treated alveolar macrophages. *Environ. Res.* **6,** 389.

Calderone, J., Williams, R. T., and Unaue, E. R. (1974). An inhibitor of cell proliferation released by cultures of macrophages. *Proc. Natl. Acad. Sci. U.S.A.* **71,** 4273–4277.

Chvapil, M. (1967). "Physiology of Connective Tissue." Butterworth, London.

Chvapil, M. (1973). New aspects in the biological role of zinc: A stabilizer of macromolecules and biological membranes. *Life Sci.* **13,** 1041–1049.

Chvapil, M., and Holusa, R. (1965). Zusammenhang der dosis von quarzstaub mit der grosse der entzundungsreaktion der lungen. *Int. Arch. Gewerbepathol. Gewerbehyg.* **21,** 369–378.

Chvapil, M., Boucek, M., and Ehrlich, E. (1970). Differences in the protocollagen hydroxy-lase activities from *Ascaris* muscle and hypodermis. *Arch. Biochem. Biophys.* **140,** 11–18.

Chvapil, M., Stith, P. L., Owen, J. A., and Eskelson, C. D. (1978). Biochemical reactivity of aortic connective tissue in chickens treated with cholesterol diet in combination with a nonspecific immunization or with epinephrine and thyroxine. *Life Sci.* **23,** 55–60.

Comstock, J. P., and Udenfriend, S. (1970). Effect of lactate of collagen proline hydroxy-lase activity in cultured L-929 fibroblasts. *Proc. Natl. Acad. Sci. U.S.A.* **66,** 552–557.

Drozdz, M., Kucharz, E., and Szyja, J. (1977). Effect of chronic exposure to nitrogen dioxide on collagen content in lung and skin of guinea pigs. *Environ. Res.* **13,** 369–377.

Eskelson, C. (1981a). Biochemical alterations in bronchopulmonary lavage fluids after intratracheal administration of diesel particulates to rats. *Environ. Prot. Agency Diesel Emissions Symp., 1981.*

Eskelson, C. (1981b). Lipid changes in lungs of rats after intratracheal administration of diesel particulates. *Environ. Prot. Agency Diesel Emissions Symp. 1981.*

Evans, M. J., Cabral, L. J., Stephens, R. J., and Freeman, G. (1973). Renewal of alveolar epithelium in the rat following exposure to NO_2. *Am. J. Pathol.* **70,** 175–190.

Fuller, G. C. (1981). Perspectives for the use of collagen synthesis inhibitors as antifibrotic agents. *J. Med. Chem.* **24,** 651–658.

Gross, K. (1980). Pulmonary function testing of rats chronically exposed to diluted diesel exhaust for 612 days. *In* "Health Effects of Diesel Engine Emissions: Proceedings of an International Symposium" (W. E. Pepelco, R. M. Danner, and N. A. Clarke, eds.). U.S. Environ. Prot. Agency, Cincinnati, Ohio.

Hance, A. J., and Crystal, R. G. (1975). The connective tissue of lung. *Am. Rev. Respir. Dis.* **112,** 657–711.

Harington, J. S., Ritchie, M., King, P. C., and Miller, K. (1973). The in vitro effects of silica-treated hamster macrophages on collagen production by hamster fibroblasts. *J. Pathol.* **109,** 21–37.

Heppleston, A. G. (1969). The fibrogenic action of silica. *Br. Med. Bull.* **25,** 282–287.

Heppleston, A. G. (1971). Observations on the mechanism of silicotic fibrogenesis. *Inhaled Part. 3, Proc. Int. Symp., 3rd, 1970* pp. 357–369.

Heppleston, A. G., and Styles, J. A. (1967). Activity of a macrophage factor in collagen formation by silica. *Nature (London)* **214,** 521–522.

Heppleston, A. G., Fletcher, K., and Wyatt, I. (1974). Changes in the composition of lung lipids and the "turnover" of dipalmitoyl lecithin in experimental alveolary lipo-pro-teinosis induced by inhaled quartz. *Br. J. Exp. Pathol.* **55,** 384–395.

Hunt, T. K. (1964). A new method of determining tissue oxygen tension. *Lancet* **2,** 1370.

Hunt, T. K., Twomey, P., Zederfeldt, B., and Dunphy, J. E. (1967). Respiratory gas tensions and pH in healing wounds. *Am. J. Surg.* **114,** 301.

Katsnelson, B. A., Babushkina, L. G., and Velixchdovskii, B. T. (1964). Changes in the total lipid content of the lungs of rats with experimental silicosis. *Bull. Exp. Biol. Med. (Engl. Transl.)* **57,** 699–702.

Kelleher, P. C., Thanassi, N. M., and Moehring, J. M. (1977). Prolyl hydroxylase in pul-monary alveolar macrophages. *FEBS Lett.* **81,** 125–128.

Kilroe-Smith, T. A., Webster, I., Van Drimmelen, M., and Marasas, L. (1973). An insoluble fibrogenic factor in macrophages from guinea pigs exposed to silica. *Environ. Res.* **6,** 298–305.

King, E. J., Mohanty, G. P., Harrison, C. V., and Nagelschmidt, G. (1953a). The action of different forms of pure silica on the lungs of rats. *Br. J. Ind. Med.* **10,** 9–17.

King, E. J., Mohanty, G. P., Harrison, C. V., and Nagelschmidt, G. (1953b). The action of

flint variable size injected at constant weight and constant surface into the lungs of
rats. *Br. J. Ind. Med.* **10,** 76–92.

Leibovich, S. J., and Ross, R. (1976). A macrophage-dependent factor that stimulates the
proliferation of fibroblasts *in vivo*. *Am. J. Pathol.* **84,** 501–513.

Lewis, D. M., and Burrell, R. (1976). Induction of fibrogenesis by lung antibody-treated
macrophages. *Br. J. Ind. Med.* **33,** 25–28.

Lieber, C. S. (1978). Pathogenesis and early diagnosis of alcoholic liver injury. *N. Engl. J.
Med.* **298,** 888–893.

Misiorowski, R. L., Strom, K. A., Vostal, J. J., and Chvapil, M. (1980). Lung biochemistry
of rats chronically exposed to diesel particulates. *In* "Health Effects of Diesel Engine
Emissions: Proceedings of an International Symposium" (W. E. Pepelko, R. M. Danner, and N. A. Clarke, eds.), pp. 465–480. U. S. Environ. Prot. Agency, Cincinnati,
Ohio.

Myllyla, R., and Seppa, H. (1979). Studies on enzymes of collagen biosynthesis and the
synthesis of hydroxyproline in macrophages and mast cells. *Biochem. J.* **182,**
311–316.

Nourse, L. D., Nourse, P. N., Botes, H., and Schwartz, H. M. (1975). The effects of
macrophages isolated from the lungs of guinea pigs dusted with silica on collagen
biosynthesis by guinea pig fibroblasts in cell culture. *Environ. Res.* **9,** 115–127.

Pelliniemi, T. (1973). Lipids of connective tissue with special reference to the effect of
dietary hyperlipidemia and oxygen deprivation on experimental granulation tissue
in the rat. Academic Dissertation, Polytypos, Turku.

Privalova, L. I., Katsnelson, B. A., Osipenko, A. B., Ysuhkov, B. H., and Babushkina, L. G.
(1980). Reponse of a phagocyte cell system to products of macrophage breakdown as
a probable mechanism of alveolar phagocytosis adaptation to deposition of particles
of different cytotoxicity. *Environ. Health Perspect.* **35,** 205–218.

Richards, R. J., and Wusteman, F. S. (1974). The effects of silica dust and alveolar macrophages on lung fibroblasts grown *in vitro*. *Life Sci.* **14,** 355–364.

Rokosova, B., and Bentley, J. P. (1980). The effect of lysosomal enzymes on the proliferation of aortic smooth muscle cells and fibroblasts. *Pathol Biol.* **28,** 493–500.

Seaton, A. (1975). Silicosis. *In* "Occupational Lung Disease" W. K. C. Morgan and A.
Seaton, eds.), pp. 80–111. Saunders, Philadelphia, Pennsylvania.

Shull, R. L., Ershoff, B. H., and Alfin-Slater, R. F. (1958). Effect of antioxidants on
muscle and plasma lipids of vitamin E deficient guinea pigs. *Proc. Soc. Exp. Biol. Med.*
98, 364–366.

Stöber, W. (1966) Silikose und Auflosung von Siliziumdioxid. *Med. Welt* **43,** 2313–2321.

Strom, K. A., Misiorowski, R. L., Vostal, J. J., and Chvapil, M. (1981). Collagen in lungs of
rats chronically exposed to dilute diesel emissions. *Toxicologist* **1,** 75.

Uspenskiĭ, V. I. (1952). Experimental histaminic pneumosclerosis. *Arkh. Patol.* **14,** 46–56.

Välimäki, M., Juva, K., Rantanen, J., Ekfors, T., and Niinikoski, J. (1975). Collagen metabolism in rat lungs during chronic intermittent exposure to oxygen. *Aviat., Space
Environ. Med.* **46,** 684–690.

Vigliani, E. C., and Pernis, B. (1958). Immunological factors in the pathogenesis of the
hyaline tissue of silicosis. *Br. J. Ind. Med.* **15,** 8–14.

Wasserman, S. I. (1980. The lung mast cell: Its physiology and potential relevance to
defense of the lung. *Environ. Health Perspect.* **35,** 153–164.

Werthamer, S., Penha, P. D., and Amaral, L. (1974) Pulmonary lesions induced by chronic
exposure to ozone. *Arch. Environ. Health* **29,** 164–166.

White, H. J., and Garg, B. D. (1981). Alveolar type II. Pneumocyte response in rats
exposed to high levels of diluted diesel exhaust. *Toxicologist* **1,** 72.

Ziskind, M., Jones, R. N., and Weill, H. (1976). Silicosis. *Am. Rev. Respir. Dis.* **113,** 643–665.

9

Physiological Effects of Cotton Dusts: Byssinosis*

Richard L. Ziprin, Stephen R. Fowler, and Gerald A. Greenblatt

I. INTRODUCTION

Byssinosis, commonly known as brown lung disease, is an occupational respiratory disease caused by inhalation of organic dusts generated from cotton, flax, soft hemp, and sisal, during the production of commercially

*Mention of a trade name, proprietary product, or specific equipment does not constitute a guarantee or warranty by the U.S. Department of Agriculture and does not imply its approval to the exclusion of other products that may be suitable.

243

AIR POLLUTION
ISBN 0-12-483880-4

useful fiber. Other botanically produced dusts may irritate the respiratory system, but do not produce the same characteristic disease symptoms.

Clinically, the disease is characterized by tightness in the chest, coughing, and subjective respiratory discomfort on Mondays or other days that follow a period of absence from work (Schilling *et al.*, 1955a,b). These symptoms may take several years to develop, but many people experience symptoms on their first exposure to dust (Bouhuys *et al.*, 1960). Although symptoms occur at first only on the first day of the week, they may come gradually to occur throughout the week. In most cases, measures of pulmonary function reveal a decrease in forced expiratory volume during acute episodes (Bouhuys *et al.*, 1969). Over a period of years, a chronic condition characterized by a sputum-productive, chronic cough and irreversible damage to lung function may develop (Morgan, 1975). The hypothesis that the acute response develops gradually into the chronic condition is logical but unsubstantiated.

Both internationally and domestically, the largest plant fiber industry, and consequently that with the greatest number of workers at risk, is cotton processing. In the United States, that working population is estimated at about 200,000 (Schilling, 1981).

A. History

The first description of respiratory difficulties associated with these botanically derived dusts was presented by Bernardino Ramazzini in 1713. He was the first to associate workers' continual coughing and eventual asthmatic complications with what must have been absurdly high dust levels in hemp and flax mills (Ramazzini, 1964).

In the early nineteenth century, similar situations and symptoms were reported in both England and France, in both flax and cotton processing mills (Anonymous, 1974). Even at that time, it was noted that the problem existed mainly in specific portions of the mills, the cardrooms and preceding work areas, the dustiest areas in the mills.

In 1831, J. P. Kay, a physician in Lancashire, described a respiratory disease, prevalent among cotton workers, that differed from chronic bronchitis in that patients experienced uneasiness caused by their inability to fully expand their chests. He called this disease Spinner's phthisis (Morgan, 1975).

Zenker, in 1866, established the term pneumonoconiosis to describe lung diseases caused by dust inhalation; the respiratory disease of cotton workers was added to the pneumonoconioses in 1871. Hirt named this disease lyssinosis pulmonum, which was changed to byssinosis by Proust in 1877 (Rooke, 1981a). During the early twentieth century, the British physician E. L. Collis noted the effects of cotton dust on health and related byssinosis prevalence to duration of employment and was also the first to describe the time course of the disease (Anonymous, 1974).

In 1933, the United States Public Health Service concluded that dust levels in cotton mills in the United States were too low to affect workers' health adversely. It was later maintained that byssinosis was not a problem in the United States, because of efficient dust control, high worker turnover, and cleaner grades of cotton (Schilling, 1981; Anonymous, 1945; Britten *et al.*, 1933).

After World War II there was a resurgence in activity concerning clinical, epidemiological, and etiological aspects of this occupational respiratory health problem. This resurgence was prompted primarily by Richard Schilling and colleagues in England. Their pioneering efforts have been supplemented by a growing number of scientists worldwide. It is now accepted that a significant prevalance of disease exists in all countries processing cotton, flax, or hemp. In all instances, the reported disease symptoms are identical.

B. Prevalence

The disease can occur in all areas of cotton processing, if the dust levels are high enough. When dust levels are high (above 3.0 mg/m^3), ginning and seed delinting operations are not exempt, and 10–20% of the work force may experience symptoms.

Estimates of the incidence of byssinosis and, more importantly, of chronic, disabling byssinosis vary widely. Bouhuys (1979) estimated 30,000 cases of disability among textile workers in the United States. The United States Department of Labor has estimated 85,000 cases among cotton workers. In contrast, there are very few instances of workers receiving compensation for byssinosis. For example, a study of 383 cotton workers in the United States indicated that 33 were considered totally disabled and 54 partially disabled, yet only one was receiving compensation (Brown, 1981). In Britain, approximately 5000 cases were diagnosed and found compensable over a 36-year period (Rooke, 1981b). A major mill in the United States has estimated the incidence of acute byssinosis at between 0.5 and 5% and claimed a three- to fourfold increased risk among smokers (Imbus, 1981). Such discrepancies and the divergence between incidence and compensability can be attributed to a lack of uniformly acceptable diagnostic criteria and such confounding complications as smoking and other nonoccupational causes of chronic bronchitis, asthma, and emphysema.

Results among various prevalence studies are not consistent when dust levels are substantially reduced via new engineering technology. Disease incidence may dramatically decline with lower dust levels or remain essentially the same (Roach and Schilling, 1960; Braun *et al.*, 1973). Accumulated evidence indicates that the use of tobacco exacerbates the problem, but there is not complete agreement on this issue (Bouhuys *et al.*, 1977, 1979; Fox *et al.*, 1973). There is anecdotal evidence that smoking eases symptoms (personal communication, M. G. Buck). Race and sex are probably inconsequential; length of exposure or years of employment is probably important (Fox *et al.*, 1973).

Generally, morbidity and mortality studies indicate that textile workers' longevity is the same as the general population's. Some studies indicate otherwise (Berry and Molyneux, 1981; Merchant and Ortmeyer, 1981). Longevity studies are inconclusive, because data on the causes of death are often incomplete.

Prevalence surveys have estimated the relative incidence of acute byssinosis among workers exposed to various textile dusts (Roach and Shilling, 1960; Merchant *et al.*, 1972; Berry *et al.*, 1973). These studies have shown also that certain segments of the industry may be exempt from the disease (Kondakis and Pournaras, 1965; Palmer *et al.*, 1978; Jones *et al.*, 1981). Occupational epidemiology is severely limited in demonstrating without ambiguity the incidence of acute transitory symptoms, because subtle complications by other initiators of pulmonary disease cloud interpretations of diagnostic data. Such complications and the "healthy worker" phenomenon (only those workers not made ill by their jobs tend to stay) make observed prevalence rates inevitably erroneous.

Considering the worldwide scope of the natural fibers industry and the numbers of workers involved, there has been a comparative lack of scientific effort to understand byssinosis. There are now probably fewer than 20 investigators concerned primarily with byssinosis epidemiology, pathophysiology, and causative agent research. Of these researchers, few are engaged in studies of byssinosis on a full-time basis. The bulk of current research is sponsored by the U.S. government and cotton industry in the United States and is directed toward dust-control engineering, the most logical approach to reducing incidence of disease. The allocation of resources to an engineering solution is nonetheless a costly expedient, which will prove difficult to implement on a worldwide scale and which has sorely hampered etiological research.

C. Botany

True cottons are malvaceous plants belonging to the genus *Gossypium*. Only four of the currently recognized 33 species have any commercial value (Fryxell, 1969). The four cultivated species are distinguishable from the wild types by their relatively long, spinnable seed hairs, which ultimately become the cotton of commerce.

Two of the cultivated species, *G. arborium* L. and *G. herbaceum* L., are indigenous to Asia and are referred to as Old World cottons. Cytologically, they are diploids with thirteen pairs of chromosomes (Hutchinson, 1959). The other two species, *G. barbadense* L. and *G. hirsutum* L., are indigenous to North and South America and are called New World cottons. These two species are allotetraploids with 26 pairs of chromosomes. They are considered hybrids between Asian diploid and wild American species (Beasley, 1940). Commercial cotton production relies on the American types, because of their inherently better fiber characteristics, e.g., strength and fiber length.

Although the plant is considered a perennial, it is usually grown as an annual. Depending on climatic types, the mature plant can be from 3 to 8 ft tall with numerous, fruiting branches. Usually the foliage is densely populated with subepidermal lysigenous glands (Fig. 1). These structures are known to be depositories of highly toxic secondary plant products (Bell and Stipanovic, 1978).

As the capsulate fruit or boll develops, fibers arise from specialized

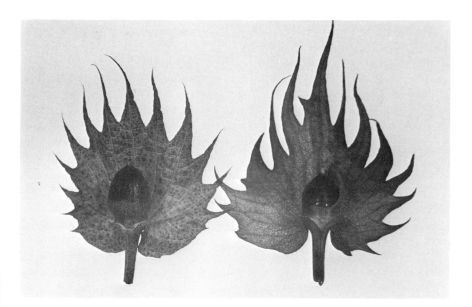

Fig. 1. Glanded and glandless cotton bracts. (Photo courtesy of Dr. John Halloin, National Cotton Pathology Laboratory, USDA.)

Fig. 2. (A) A cotton flower. (B) A closed boll. Glands are visible on the stem. (C) An open mature field-dried boll.

Fig. 2B and C.

epidermal cells on the seed coat. At maturity, the boll dries; the carpels split open along natural suture lines and flare out, exposing the mass of fibers attached to the seeds (Fig. 2). The common term applied to these fibers is lint; the total mass of seeds with attached lint is called seed-cotton.

D. Cotton Production

Seedcotton is a major crop in portions of the African tropics, Australia, China, Egypt, India, Mexico, Pakistan, the Soviet Union, the Sudan, the United States, and temperate regions of Central and South America and Asia minor. The 1979–1980 world crop yielded slightly more than 64 million bales (a standard bale contains 480 pounds) of fiber (Barlowe, 1980). This crop was grown on approximately 79 million acres of land and amounts to a production of approximately 30 billion pounds or 15 million tons of fiber. The world crop estimate for 1985 is 80 billion bales (Bowling, 1981).

The U. S. contribution to world production is approximately 20%. The 1979–1980 crop yielded 14.5 million bales, the 1981–1982 crops is estimated to yield 13.9 million bales. At an in-the-bale market price of about $0.70/pound, this amounts to $4.8 billion for the 1979–1980 crop. Mills in the United States utilized 6.5 million bales in 1979 and estimated use in 1980 at 7.0 million bales. Average exports were 5.8 million bales per year during the 1975–1980 period.

Depending on location, seedcotton is harvested in the late summer or fall. Harvesting is accomplished by hand labor in many third-world countries or by machine in developed countries. Hand-picked seedcotton is considered generally to be a cleaner product than that picked by machine, but hand picking does not eliminate the threat of byssinosis.

The cotton plant must be defoliated prior to mechanical harvesting. In most temperate areas of the world, this is accomplished by chemical desiccants or defoliants. In more northern climates, early frosts make chemical desiccation unnecessary. In both cases, the plant must be essentially denuded of leaves; those that remain must be sufficiently senescent that little chlorophyll remains, because fiber that is color tinged is downgraded and commands a lower price.

Regardless of how a cotton plant is manipulated prior to mechanical harvesting, the bract adheres tightly to the boll and much of it ends up admixed with fiber. It is the main botanical contaminant of baled cotton (Morey *et al.*, 1976a,b). There are attempts to reduce contamination by breeding lines that abscise or drop their bracts (caducous bract) prior to harvesting (Muramoto *et al.*, 1981). If it could be bred into all commercial cottons, caducous bract would reduce trash levels in the bale and, subsequently, the mill, significantly. Unfortunately, such a breeding program could take many years.

Seed and lint are mechanically separated in a process called ginning. The production phase ends with ginning and the two economic products of the cotton plant: lint and seed. The baled lint then undergoes a number of mechanical processes that generate yarn and, ultimately fabric.

Crop losses from all cotton diseases in the United States averaged 13.1% of production during the period 1953–1977 (Watkins, 1981). Losses from insect attacks are considered to be greater. No statistics on worldwide crop losses from disease or insect attack have been gathered, but there is reason to believe that they are much the same.

Breeding for resistance against plant pathogens and insects can influence greatly the type and concentration of toxic natural products in the plant. Current breeding programs stress the importance of increasing the level of certain constitutive terpenoids, which act as antimicrobial agents when the plant is infected (Bell *et al.*, 1975). Also, insect resistance is enhanced by various terpenoid aldehydes in the lysigenous pigment glands of the plant and by increasing the plant's tannin levels (Bell *et al.*, 1978; Lane and Schuster, 1981). In essence, programs to increase disease resistance are attempting to enhance the plant's defense mechanisms by increasing the levels of toxic and potentially toxic compounds in the plant. The role of these toxic compounds in byssinosis is as yet unknown.

Conversely, there are mutant recessive genes that completely eliminate the pigment glands (McMichael, 1960). Breeding lines with these genes have produced completely glandless, commercial-cotton cultivars, which yield nontoxic seed that can be used for inexpensive food for humans and animals. At present, these cultivars comprise only a small percentage of domestic production and even less of world production. Under many conditions, they are more susceptible to insect and plant pests. Commercial cotton cultivars have tremendous diversity with respect to toxic natural products. Except for limited work on our part, these various, chemically diverse cotton cultivars have never been studied systematically with regard to possible differences in byssinogenic activity.

At the mill, the initial processing steps prepare the baled cotton fiber for carding (a process that aligns the fibers and allows for further processing): these steps consist of opening, cleaning, and picking and result

Fig. 3. An opening carousel at Burlington's Flint Plant uses automation to open and separate large bales of cotton—a procedure that used to be performed by hand. As the carousel turns, it picks up cotton staple off the bottom of each bale and transports it through a system of chutes and pipes. This method is not only faster and more efficient, but much cleaner since the cotton is enclosed and does not enter the plant's atmosphere. (Photo and legend text courtesy of Burlington Industries, Inc., Greensboro, N.C.)

in the formation of a continuous, flat sheet of fiber called a lap. These operations open or loosen the compressed cotton and allow trash components to be removed by a series of pneumatic cleaners. The purpose is not only to clean the cotton but also to mix it uniformly before it is formed into laps (Fig. 3). This machinery is usually enclosed and may even be equipped with exhaust systems. Workers who clean this equipment should wear respirators; although fewer workers are involved, the incidence of byssinosis here is equal to that in the cardroom (Schilling *et al.*, 1955a,b).

In carding, a process fundamental to the use of any natural fiber, the randomly oriented fibers of the lap are aligned and cleaned still further (Fig. 4). First, the lap is forced under a rapidly rotating, toothed cylinder that opens the tufted fiber and allows foreign matter and short fibers to

Fig. 4. Burlington Industries has invested millions of dollars to protect the health and safety of employees. This new installation of high-speed, chute-feed carding machinery is equipped with modern dust filtration systems. Cotton fibers are entirely enclosed in chutes. (Photo and legend text courtesy of Burlington Industries Inc., Greensboro, N.C.)

be removed. The modified lap then passes between two toothed surfaces, one of which is a high-speed cylinder. The teeth on these surfaces are in opposition to each other and the resulting combing action further opens and straightens the fibers. The fibers are then condensed to form a rope or sliver approximately 1 inch in diameter. The cardroom has the highest dust levels in the mill (Berry *et al.*, 1974; Brown *et al.*, 1978). Those individuals who maintain the cards usually have the most severe byssinosis problem of any cardroom employees (Molyneux and Tombleson, 1970).

For the most part, the remaining operations are not believed to contribute significantly to the respirable dust load. The one exception may be the high-speed spinning and winding operation. Apparently, these high-speed operations can release a large proportion of the fine trash still entrained by the fibers.

E. Cotton Dust Composition

The dust released from cotton fibers during processing is a hetero-genous mixture composed primarily of micronized vegetable matter. The National Institute for Occupational Safety and Health (NIOSH) includes bacteria, fungi, soil, and agricultural chemicals as well as plant material in their definition of cotton dust (Anonymous, 1978).

A mechanial harvester must pick more than a ton of field material to yield a 480-pound bale of lint. More than 900 pounds of vegetable trash and 750 pounds of cottonseed are removed from the lint during gin-ning. Morey *et al.* (1976a,b) removed 84 g of trash from 1250-g sample of raw cotton taken after ginning; this amounts to 6.7% trash. The amount of visible trash isolated in baled cotton by a Shirley analyzer varies from 1 to 8%, averaging 1.6%. Approximately 75% of this trash is of cottonplant origin; the remainder is from common weeds (Morey *et al.*, 1976a,b).

The source of the active agents that cause byssinosis is the foreign material incorporated into raw cotton during harvesting and/or subse-quent storage. The cotton fiber probably has no role in disease etiology (Bouhuys, 1974). The inhalation of aerosolized, aqueous extracts of cot-ton dust or of bract elicits the acute symptoms of byssinosis in normal humans (M. G. Buck, personal communication; Buck and Bouhuys, 1981; Bouhuys and Nicholls, 1967); therefore, the vegetable trash con-taminating raw cotton is important in disease etiology.

Analysis of dust samples from 12 southeastern cotton mills revealed that cardroom dust contains, on the average, 55% cellulose, 23% non-cellulosic organic matter, and 12% ash. Dust from weaving rooms of cotton mills averaged 21% cellulose, 17% noncellulosic organic material, 12% ash, and approximately 41% sizings (material such as starch, car-boxymethyl cellulose, or polyvinyl alcohol, which are used in processing raw fibers). Nitrogen analysis showed that cardroom dust contains 12% protein, whereas weaving-room dust contains only 2.5% protein (Brown *et al.*, 1978). Because byssinosis is much more common in cardrooms than in weaving rooms, sizings probably play no role in the etiology of byssinosis. Noncellulosic organics, which include proteinaceous mate-rials, are more likely candidates.

Cotton-plant parts identified in this trash include bract, leaf, seed, endocarp, exocarp, mesocarp, bark, and stem. One study, involving se-lective removal of plant trash components from picked cotton, reported that leaf accounted for 20% and bract for 12% of the trash present in seed cotton (Corley, 1966). Another demonstrated convincingly that bract parts constitute the largest group of plant components contaminating raw cotton (Morey *et al.*, 1976b). As particle size decreases, bract becomes a more prominent contaminant. Bract is extremely friable; over 26% by weight was converted to particulate dust less than 10 μm in diameter in a laboratory milling test (Morey, 1979b). Bract was found to be significantly more friable than other raw-cotton trash materials. A second, laboratory milling test found that 98% of bract was converted to particulate dust of less than 250 μm and 33% to particulate dust of less than 10 μm (Morey, 1979a).

The percentage composition of six classes of botanical ingredients in less than 10 μm nonfibrous cotton dusts has been estimated: leaflike material, including bract, leaf, and weed leaf, accounted for 74% of these dusts; seed constituted 9.7%; stem, 7.3%; bark, 5.3%; exocarp–mesocarp, 3%; and endocarp, less than 1% (Morey, 1980). Because bract is significantly more friable than leaf or weed leaf, Morey concluded that bract may make up more than 90% of the leaflike material less than 10 μm in diameter.

Microorganisms are normally found on all plants. Although sterile in the unopen boll, the fibers of the cotton plant are contaminated soon after the boll opens. The majority of bacteria found on the plant and fiber are gram-negatives, chiefly *Enterobacter agglomerans*, but sometimes include *Klebsiella, Pseudomonas, Acinetobacter,* and *Agrobacterium* species (Rylander, 1981). Large numbers of these bacteria are found on the bract, particularly after senescence (Morey *et al.*, 1980).

A number of studies have found these same bacteria in the air of mills (Cinkotai *et al.*, 1977; Cinkotai and Whitaker, 1978; Fischer, 1979). The number of gram-negative bacteria in cardrooms was reported to be between 10^2 and 10^5 per cubic meter. A correlation was found between prevalence of byssinosis symptoms and amount of airborne gram-negative bacteria in workplace air of British textile mills (Cinkotai *et al.*, 1977).

Bacterial endotoxin was found in cotton dust and in cardroom air. Levels from 0.2 to 1.6 μg/m³ in cardroom air were reported by Cinkotai *et al.*, (1977). Several studies have attempted to correlate endotoxin in workplace air with prevalence of byssinotic symptoms (Fischer, 1980; Rylander and Snella, 1980; Rylander and Haglind, 1981). Although some studies seem to indicate a relationship, there is no convincing evidence that endotoxin is involved in the etiology of byssinosis.

Gram-positive bacteria found in cotton dust include *Bacillus subtilis, B. cereus, B. pumilis,* and *B. megatherium* (Fischer, 1979). Several genera of fungi have also been isolated from cotton dust. These include *Aspergillus, Penicillium, Hormodendrum, Fusarium, Alternaria,* and *Rhizopus* (Fischer, 1979). The causal relationships between these organisms and byssinosis is unknown and largely unexplored. One study has indicated that these organisms or their culture filtrates are not active in an *in vitro* assay for byssinogenic activity (Battigelli *et al.*, 1977).

II. PHYSIOLOGY

A. Respiratory Cell Physiology

The lung is histologically complex and contains more than 40 different cell types. The major function of most of these cell types is to allow gaseous exchange between air and blood yet maintain the integrity of the thin tissue barrier between air and blood despite the tissue's exposure to environmental pollutants and contaminants. The maintenence of lung integrity depends on the unique properties of these different cell types.

Mature, adult lung parenchyma consists of four types of cells: endothelial, mesenchymal, epithelial, and those of extrapulmonary origin. Endothelial cells lining the capillary network make up about 39% of the cell population. Mesenchymal cells, including fibroblasts, pericytes, and interstitial cells, fill three-quarters of the space between alveolar epithelium and capillary endothelium; they account also for nearly 39% of lung cells. Epithelial cells, which line the air space in the lungs, are divided into alveolar type I cells and alveolar type II cells. Type I cells are complex, branched cells, which have the largest surface area of any lung cell, a mean luminal surface area if 4000 μm^2. Type II cells, which cover 3% of the lumen surface, secrete surfactants that coat the lung epithelium (Weibel *et al.*, 1976). Cells of extrapulmonary origin include blood cells, lymphocytes, mast cells, and alveolar macrophages. These cells are responsible primarily for the defense of the lung by phagocytic, immunological, and inflammatory responses.

Cells in the lung parenchyma can respond to changes in their environment. Information about local changes is received via specific receptors on the cell surface or by phagocytosis and pinocytosis. Examples of these specific receptors are histamine receptors on alveolar macrophages and corticosteroid receptors on alveolar type II cells (Diaz *et al.*, 1979; Ballard *et al.*, 1978). Lung cells known to be phagocytic include the alveolar macrophage and the mast cell. The alveolar macrophages, the lymphocyte, and the mast cell can modify not only their own metabolisms but also those of nearby cells (Unanue, 1976, 1978; Davies and Bonney, 1979; Gupta and Good, 1980; Wasserman, 1979; David, 1975; Paterson *et al.*, 1976).

Alveolar macrophages function as the primary defense against particulate matter in the lungs: these phagocytes reside as free cells, at the air–tissue interface, a location that exposes them directly to inhaled environmental toxins and pollutants. Macrophages are responsible for clearance of particulates from the respiratory bronchioles and alveoli and from the ciliated portions of the tracheobronchial tree. Respiratory air space sterility is maintained by alveolar macrophages. Contact between alveolar macrophages and respirable particulates and toxicants may result in the release of soluble mediators of inflammation and immunity, which influence profoundly the metabolism of the surrounding tissue (Myrvik and Kohlweiss, 1975; Territo and Golde, 1979; Davies et al., 1979).

Several reviews discuss the secretory function(s) of mononuclear phagocytes (Unanue, 1976, 1978; Davies and Bonney, 1979). Among the products secreted by the alveolar macrophages are those than can stimulate or inhibit lymphocyte mitogenesis, stimulate antibody response, cause smooth muscle contraction, stimulate leukopoiesis, and attract blood granulocytes and mononuclear cells.

Lymphocytes, which make up approximately 7% of the free cells in the respiratory air space, also secrete mediators that affect the metabolic processes of the surrounding cells (Hunninghake et al., 1979; Johnson et al., 1979). Thymus-derived lymphocytes may be stimulated by inhaled toxins, resulting in the release of lymphokines. These lymphokines are capable of stimulating macrophages and/or B lymphocytes. Stimulated, human lung lymphocytes are known to produce the following lymphokines: migration-inhibition factor, leukocyte-inhibition factor, and monocyte-chemotactic factor. We are unaware of any research regarding the effect of cotton dust extract or inhalation of cotton dust on the production of these lymphokines.

Mast cells are potent mediators of immediate hypersensitivity. They are localized in bronchial submucosa and perivenular connective tissue. In human lung, they are present in concentrations of $1-10 \times 10^6$ cells/g of tissue (Paterson et al., 1976). The induction of hypersensitivity or inflammation by mast cells depends on specific antigen–antibody reactions. Immunoglobulin E is found on the surface of the mast cell. The Fc receptors for IgE are monomeric; the bridging of two or more receptor-bound IgE molecules by specific antigen initiates the noncytotoxic exocytosis of preformed mediators that are stored in cytoplasmic granules. Bridging results also in an influx of calcium ions, generation of arachidonic acid metabolites, and intracellular increases in cyclic AMP (Wasserman, 1979; Langunoff, 1976).

Alveolar type II cells play an important role in maintaining normal function and structure of the lung. These secretory cells are essential in the production of surfactants (Weibel *et al.*, 1976). Type II cells proliferate and differentiate to replace injured type I cells. The type I cell is very sensitive to injury, but is incapable of mitotic division. The stereotypic pattern of lung injury is invariable: first, the thin cytoplasmic flaps of type I cells exhibit damage; this is followed by proliferation of type II cells to form a cuboidal lining of the alveoli. Over time, these cuboidal cells extend to form the thin cytoplasmic flaps characteristic of type I cells. Failure of type II cells to perform properly may result in permanent lung damage. Although type II cells can be isolated and grown *in vitro*, no studies have been reported of the effect of cotton dust or extract on these cells (Mason *et al.*, 1976; Douglas *et al.*, 1976).

Although we can learn much from studies of the lung *in vivo* or in organ culture, the large number of cell types native to the lung create difficulties in understanding the role of any one type in both normal and pathological situations. For example, a lung parenchyma devoid of cells of the alveolar lumen and of blood elements still contains more than six cell types. Excellent brief reviews have been prepared by Gee and Smith (1981) and by Rylander (1980). It seems reasonable that attempts be made to study the effects of cotton dust on purified lung cell types in culture to learn the contributions of individual cell types to the disease.

B.   Symptoms and Histopathology

Chest tightness on the first morning of the work week (Monday morning syndrome) is the first complaint mentioned by affected workers. Some workers do not develop symptoms until several years of employment have elapsed, but many experience symptoms upon first exposure. These symptoms may continue to be mild or come eventually to persist throughout the work week. If this occurs, workers will probably develop a productive cough and shortness of breath. Later on, they may develop a permanently reduced ventilatory capacity and a chronic productive cough. Acute byssinosis may eventually become irreversible and lead to chronic airway obstruction. The final result is clinically indistinguishable from chronic bronchitis and emphysema.

One of the perplexing problems of byssinosis investigations is the relationship, if any, between acute and chronic byssinosis. The acute response could be considered nothing more than a passing annoyance. Those individuals who suffer severe attacks will seek other employment or be reassigned. Chronic byssinosis is a more vexing problem, because of the issues of liability and compensation. We do not know whether acute attacks are a forerunner of chronic byssinosis. Some groups deny the existence of chronic byssinosis, arguing that it is really an unrelated bronchitis–emphysema complex caused by excessive use of tobacco.

In experimental human byssinosis, there is an increase in polymorphonuclear leukocytes (PMN) on nasal airway muscosa (R. Rylander, personal communication; Merchant *et al.*, 1974, 1975). Neutrophile influx occurs in experimental animals exposed to cotton dusts (Kilburn *et al.*, 1973; Lunn *et al.*, 1974; Walker *et al.*, 1974). We are, however, unaware of any evidence of neutrophile influx into the bronchi and small airways of human byssinotics. Chemotactic phenomena can be studied *in vitro*. The papers by Ainsworth and by Kilburn are of interest (Kilburn *et al.*, 1977, 1979; Ainsworth and Neuman, 1977, 1980; Ainsworth *et al.*, 1979a).

Histopathology of lungs from mill workers or former mill workers has produced inconsistent findings. The pseudonym for byssinosis, brown lung disease, is undoubtedly a misnomer. The lungs of nonsmoking byssinotics are usually unremarkable. So-called "brown bodies" are far from an inevitable finding, and there is confusion in the literature concerning emphysema (Rooke, 1981a,c; Heyden and Pratt, 1980; Pratt, 1981).

Pratt (1981) has done an extensive histopathological study of the lungs of former mill workers and has made some interesting observations from which he has developed a reasonable hypothesis for the development of chronic byssinosis: chronic byssinosis resembles that complex of problems called chronic obstructive pulmonary disease (COPD) and is characterized by a reduced FEV no longer related temporally to the inhalation of dusts.

Histopathological examinations of the lung sections reveal the frequent presence of small-airway disease characterized by gland-cell hyperplasia. Small-airway disease is not usually associated with COPD or reduced FEV, and Pratt has shown that gland-cell hyperplasia is more prevalent among nonsmoking cotton workers. Many people have asymptomatic emphysema. P. C. Pratt (personal communication) has shown with *in vitro* experiments using excised lung that there is a reduced FEV in lungs of individuals having *both* inapparent centrilobular emphysema and small-airway disease. These observations can be related to the low incidence of chronic byssinosis.

It seems likely that small-airway disease results from dust inhalation. Emphysema, known to occur in both smoking and nonsmoking populations, may be a secondary problem. The relationship of cotton-dust inhalation to the development of emphysema is not clear. Pratt's work indicates that emphysema present in chronic byssinosis is not caused by cotton-dust inhalation. Nevertheless, earlier work by Rooke noted an emphysema rate of 36% in the lungs of cotton-mill workers at autopsy (Edwards *et al.*, 1975).

C. Mechanisms

Numerous hypotheses of physiological mechanisms that explain how exposure to cotton dusts can result in bronchial constriction have been proposed. Below, we list six of these:

1. Mechanical Irritation

Cotton dust is assumed to contain no pharmacologically active material. The physical irritation of the inhaled dust on the airway epithelium somehow causes a reflexive narrowing of the airways. Inert dust, e.g., coal dust, in concentrations much higher than that required of cotton dust may also cause such a reaction. Cotton dust causes a much longer-lasting response at $\frac{1}{30}$ the concentration at which coal dust causes a reaction (Edwards, 1981). Finally, dust from washed cotton does not cause byssinotic symptoms. Mechanical irritation may contribute to the symptoms of byssinosis, but it is certainly not the primary mechanism of disease induction.

2. Smooth Muscle-Stimulating Compounds

Several pharmacologically active agents have been identified in cotton dust, including at least two known to cause contraction of smooth muscle. The presence of histamine in cotton dust was reported by Haworth and Macdonald (1937). Serotonin has also been identified (Davenport and Paton, 1962). The levels of these compounds in dust are far too low to be responsible for the effects produced by inhalation of cotton dust (Nicholls, 1962). Russell *et al.* (1980) described a bronchoactive component from cotton bract that causes contraction of isolated canine trachealis strips.

3. Histamine-Releasing Factors

The acute, reversible broncho-constriction could be caused, at least in part, by histamine released locally by nonantigenic agents in the dust. This hypothesis is supported by three lines of evidence: (a) dust and bract extracts elicit histamine release from minced lung tissue (Evans and Nicholls, 1974a); (b) workers exposed to cotton dust excrete increased amounts of the histamine metabolite methylimidazole acetic acid (Evans and Nicholls, 1973); and (c) antihistamines such as isoproterenol and disodium cromoglycate prevent the Monday fall in FEV_1 (Valic and Zuskin, 1973).

At least four natural products from the cotton plant have been implicated as histamine-releasing agents. Hitchcock *et al* (1973) proposed that the component of dust capable of releasing histamine from human lung tissue was methyl piperonylate. Evans and Nicholls (1947b) proposed rutin and trimethylamine as histamine-releasing compounds in aqueous dust extracts. The occurrence of these in cotton dust has not been verified. The fourth and most active compound was described as an aminopolysaccharide. This compound has not been characterized fully and its histamine-releasing properties have not been verified. It is probably not the aminopolysaccharide byssinosan (Mohammed *et al.*, 1971).

There are indications that lacinilene C 7-methyl ether, another natural product from the cotton plant, is capable of releasing histamine from mast cells (Northrup *et al.*, 1976) and from porcine platelets (Ainsworth *et al.*, 1979b). The lacinilenes are a group of naturally occurring sesquiterpenoids of cotton (Stipanovic *et al.*, 1975). Only recently, have critical aspects of their chemistry relevant to their biological activity been assessed (Stipanovic *et al.*, 1981). The lacinilenes are formed from their precursors by autoxidation. Lacinilene C 7-methyl ether can now be prepared in amounts reasonable for testing, and there is the distinct possibility testing can include the separate use of the optical isomers. Also, lacinilene C 7-methyl ether was found to act as a chemotactic agent *in vitro* (Kilburn *et al.*, 1977). At somewhat higher concentrations, it was cytotoxic. Another report has indicated that this compound is not chemotactic; the activity is caused solely by the wetting agent used to solubilize the compound (Ainsworth *et al.*, 1979a). As yet, there are no confirmations of either report. The activity of this natural product must be reassessed carefully and completely.

Experiments on the histamine-releasing capacity of cotton-dust and bract extracts are almost entirely dependent on the use of minced, excised human and other mammalian lungs. Aside from being very tedious, minced lung assays do not always yield reproducible data (Greenblatt, 1977; Ainsworth *et al.*, 1979a). The results from histamine-release experiments indicate that many synthetic natural products and extracts from cotton and other plants can cause histamine release (Ainsworth *et al.*, 1979b). Definitively, however, we know only that aqueous extracts of field-dried bract cause histamine release from pig peripheral blood and that there are differences in the abilities of glanded and glandless bract to cause histamine release (Greenblatt, 1978). Recently, methods for the purifying mast cells *in vitro* and for measuring degranulation of mast cells in response to toxin exposure have progressed to where a practical bioassay may be developed (Elissalde *et al.*, 1981).

Histamine may not be the predominant bronchoconstrictive agent in acute byssinosis: only weak evidence suggests that it is (Schachter *et al.*, 1981). Byssinosis is a complex problem, and histamine-induced smooth muscle contraction cannot explain all the salient observations.

4. Immediate-Hypersensitivity Reactions

This hypothesis proposes that an antigen in cotton bract causes the release of histamine and possibly other inflammatory mediators by reacting with IgE on mast cell surfaces in a classical type I reaction; however, we do not believe that byssinosis is a type I reaction. There is often a 1 to 4-hr lag between exposure and onset of symptoms; type I reactions are more immediate. Furthermore, there is an even distribution of byssinotics among nonatopics and atopics. Evidence against a type I reaction includes also the failure to detect elevated IgE levels in byssinotics or to detect IgE specific for cotton-dust components (Edwards, 1981; Ainsworth *et al.*, 1981).

5. Antigen–Antibody-Mediated Reactions

Massoud and Tyalor (1964) first demonstrated that extracts of cotton dust or bract reacted with sera from byssinotics; however, sera from normal subjects also reacted with their extract. The "antigen" purified from cotton by Taylor *et al.* (1971) was a condensed polyphenol called leukoanthocyanidin. Some question exists whether the precipitin lines they observed in their immunodiffusion assay were the result of a specific antigen–antibody reaction or a nonspecific precipitation of protein (Massoud and Taylor, 1964; Edwards and Jones, 1974: Taylor *et al.*, 1971; Edwards, 1981; Kutz *et al.*, 1981). Polyphenols present in cotton dust extracts react with serum proteins in a "pseudo-immune" response, giving a false impression that an antigen–antibody reaction has occurred (Edward, 1981; Kutz *et al.*, 1980).

Mares and Sekul (1980) have isolated a cotton-dust antigen that is primarily a polysaccharide with a molecular weight of approximately 300,000. This antigen will react with approximately 10–20% of serum samples collected from both cotton-mill employees identified as byssinotics and naive subjects. When endotoxin and house dust are checked for cross-reactivity with the serum samples that react with the cotton-dust antigens, there is no reaction. Furthermore, Sekul does not see bands of identity when reactive serum from normal and byssinotic subjects are reacted with his cotton antigen (A. A. Sekul, personal communication). Therefore, it is not yet possible to say that there is an antigen–antibody reaction involved in triggering a byssinotic episode.

Ainsworth *et al.* (1981) have studied serum immunoglobulins among byssinotic and nonbyssinotic mill employees. Measurements of IgE, IgG, IgM, IgA, and complement were determined for samples collected before work started on Monday mornings and on Fridays. Immunoglobulin and complement levels were not found to be abnormal and the overall results tend to diminish the possibility that the type I or type II immunological reactions are involved in byssinosis.

While considering "immunological phenomena," we might consider also complement and complement activation. The complexities of the complement system are beyond the scope of this work. Suffice it to say that activation of complement by either the classic or alternate pathway can result in airway changes through histamine release from mast cells. Because some complement components are chemotactic for neutrophils, the activation of complement could result in amplification of the response and account for the phenomenon of neutrophile influx observed in animals exposed to cotton dust. (There is no evidence that neutrophils move into the human, lower respiratory airways during an acute byssinotic episode.)

Kutz et al. (1979) have shown that cotton-dust extracts can cause complement activation. Endotoxin, a component of cotton dust, is a known activator of the alternate pathway. Endotoxin levels in a cardroom dust sample were reported as 438.4 ng/g. In complement conversion experiments, a dust suspension of 0.2 mg/ml was positive. Kutz et al. (1979) state that the endotoxin present in their reactions could not account for the quantity of complement consumed, because the endotoxin concentration was below the nanogram range (438 ng/g $\times$ 0.0002 g/ml = 0.0876 ng/ml), and microgram quantities of endotoxin are needed for the amount of complement conversion they observed. Thus, they speculated that other unidentified substances were responsible. If this information is correct, the amount of endotoxin reaching the lungs of mill workers is insufficient to cause byssinosis by the activation of complement.

The cotton dusts used by Kutz et al. contained less than 500 ng endotoxin/g dust. Using this figure, OSHA-approved dust levels of 200 μg dust/m^3 air (Anonymous, 1978) translates to 0.1 ng endotoxin/m^3 air. Because some workers experience symptoms even at OSHA-approved levels, it appears unlikely that endotoxin–complement interactions are responsible for byssinosis. It is unlikely that enough endotoxin could be respired either at 2.5 mg/m^3 of air or OSHA-approved levels to cause a significant consumption of complement.

Our position is that complement-mediated histamine release caused by the inhalation of endotoxin is probably not a tenable hypothesis; however, a less direct pathway, perhaps involving an amplified response to endotoxin, might be plausible. As yet, there is no evidence for such a scheme.

6. Endotoxin

Some researchers have promoted the idea that endotoxin in cotton dust is responsible for the development of byssinosis (Rylander, 1981; Pernis *et al.*, 1961; Fischer, 1980; Rylander and Snella, 1980). Endotoxin inhalation results in an increase in the number of free cells in the pulmonary air space. Macrophages, normally present, increase in number, whereas neutrophils, which usually make up less than 1% of pulmonary air space free cells, migrate into the lungs and become a signficant fraction of the total population (Hudson *et al.*, 1977; Hunninghake *et al.*, 1979). This effect is a general inflammatory mechanism, not a specific response to endotoxin. Other irritants such as manganese dioxide, cigarette smoke, and silica particles also elicit influx of cells into the lungs. Relevant information on the physiological effects of endotoxin inhalation is derived mainly from animal experiments by Snell (1966), Cavagna *et al.* (1969), De Maria and Burrell (1980), Rylander *et al.*, (1975a,b), Helander *et al.* (1980), and Morrison and Ulevitch, (1978).

Work by Asobe-Hansen and Glick (1958), Sandusky *et al.* (1973), and by Morrison and Betz (1977) demonstrated that rat peritoneal mast cells will not degranulate when treated *in vitro* with bacterial endotoxins at concentrations as high as 1 mg/ml. If we may fairly transfer this information to the human byssinosis problem, it would appear unlikely that a direct endotoxin–mast cell interaction is the primary cause of histamine release in the byssinotic lung.

Attempts to correlate endotoxin in workplace air with prevalence of byssinosis symptoms among workers have been unsuccessful. Cinkotai *et al.* (1977) found that the relationship between airborne endotoxin in cotton-mill cardrooms and prevalence of byssinotic symptoms was not statistically significant. Cavagna *et al.* (1969) estimated that a mill worker inhales 10–15 liters air/minute, and calculated that in workplace air containing 8–9 $\mu g/m^3$ endotoxin, a worker would be exposed to 40–50 μg endotoxin during an 8-hour shift. They found that exposure of normal human volunteers to 40 μg aerosolized endotoxin did not result in decreased lung function as measured by forced expiratory volume. Exposing volunteers to 80 μg of aerosolized endotoxin resulted in significant FEV reduction in only two of eight workers tested. In patients with chronic bronchitis, one of four suffered decreased FEV when exposed to 40 μg endotoxin and one of three when exposed to 80 μg.

There have been experiments to test the effects of washing cotton prior to carding on byssinosis development. These attempts are based in part on the observations that mills that process medical-grade cotton have an extremely low incidence of byssinosis (Batawi *et al.*, 1962). Treatment of raw cotton by washing appears to decrease byssinosis symptoms among processors (Merchant *et al.*, 1973). Rylander (1981, and personal communication) reported that washing also lowers the endotoxin levels in dust; however, he reported also that mill workers still respond to the dusts of washed cotton, in which endotoxin levels are low (R. Rylander, personal communication). The United States Department of Agriculture and the National Institute of Occupational Safety and Health are conducting experiments to determine whether it is practical to use washed cotton for processing. B. Boehlecke (personal communication) could not detect a statistically significant change in FEV when reactor subjects were exposed to cardroom air in which washed cottons were processed.

It appears probable that endotoxin or Lipid A (a hydroxy fatty acid that can be isolated from endotoxin) has a role in the etiology of byssinosis. Endotoxin can cause the release of inflammatory mediators from alveolar macrophages, blood platelets, and neutrophils, and these mediators, amplified by the body's inflammatory mechanisms, may cause many of the symptoms associated with byssinosis. Lipid A has been shown to possess most of the inflammatory activity of endotoxin. Rather than a single compound, Lipid A has been shown to be composed of a class of hydroxy fatty acids, most commonly β-hydroxymyristic acid. A point many researchers seem to have overlooked is that the cotton plant itself may be a source of Lipid A-like material. Intact bract may contain up to 2% lipid. The cuticle, external to the cellulose cell wall of the epidermis, is about 80% lipid. Common constituents of this material include long-chain hydrocarbons, alcohols, ketones, fatty acids, and hydroxy fatty acids. These are not water soluble in the living plant, but may increase in polarity after senescence. Water extracts of bract contain some highly polar lipid material (Wakelyn *et al.*, 1976).

D. Endogenous Mediators of Inflammation

We believe that many, if not all symptoms of byssinosis may be explained by the release of endogenous mediators of inflammation by cells in the lung. The alveolar macrophage and the mast cell are primary candidates for such activity. Other inflammatory cells that may be involved include the neutrophile, platelet, and basophil. Release of mediators by one or more of these cell types results frequently in amplification of inflammation caused by recruitment of other cell types. For example, release of prostaglandins by alveolar macrophages may result in potentiation of mast cell degranulation and aggregation of platelets.

Of special interest in the study of byssinosis are mediators that results in the contraction of airway smooth muscle. Histamine release has been a main theory of the mode of action of cotton dust in the lungs; however, leukotrienes produce a more sustained contraction than histamine (Hanna *et al.*, 1981), and thromboxane A_2 is a more potent bronchoconstrictor than histamine (Svenson *et al.*, 1975). Platelet activating factor may be the most potent bronchoconstrictor yet discovered (Vargaftig *et al.*, 1980). Platelet-activating factor also causes platelet aggregation and secretion of platelet histamine and serotonin. Platelet aggregates are sequestered in the lung capillary network, resulting in decreased efficiency of gas exchange. Release of these mediators may explain the subjective symptom of chest tightness and observation of decrease P_{O_2} in blood (Lopez-Merino *et al.*, 1973).

A large portion of membrane-bound fatty acids in the alveolar macrophage is arachidonic acid (Mason *et al.*, 1972). Consequently, a major source of arachidonic acid metabolites in the lungs is the alveolar macrophage (Stenson and Parker, 1980). The products of arachidonic acid oxygenation released by the alveolar macrophage can mediate inflammation. These products include the prostaglandins; thromboxane; 12-L-hydroxy-5,8,10-heptadecatrienoic acid (HHT); the hydroxyeicosatetraenoic acids (HETEs); prostacyclin (Samuelsson *et al.*, 1978); and the leukotrienes, which include SRS-A (Samuelsson *et al.*, 1980). Stimuli that increase arachidonic acid oxygenation by the alveolar macrophage include phagocytosis of zymosan and latex, Fc receptor activation, endotoxin, and natural products from cotton bract (Humes *et al.*, 1977; Brune *et al.*, 1978; Grimm *et al.*, 1978; Kurland and Bockman, 1978; Fowler *et al.*, 1981).

Several arachidonic acid metabolites have been shown to cause bronchoconstriction by direct action on airway smooth muscle. Among these are prostaglandin $F_{2\alpha}GA$, thromboxane A_2, and prostaglandin G_2. They may also enhance histamine release and leukotriene production from airway mast cells thus amplifying and prolonging bronchoconstriction.

Other alveolar macrophage secretory products that amplify the inflammatory response are the neutrophile chemotactic factors. At least three are released by the alveolar macrophage under the appropriate stimuli. Two of these are HHT and the hydroperoxyeicosatetraenoic acids. The third is an incompletely characterized lipid (Valone *et al.*, 1980; Hunninghake *et al.*, 1980).

Macrophages release a variety of lysosomal enzymes into the surrounding medium during phagocytosis. These include lysozyme and acid hydrolases. In addition, stimulated macrophages synthesize and secrete neutral proteinases, which play an important role in chronic inflammatory processes. Among these enzymes are collagenase and elastase; both are capable of destroying the interstitial matrix of the lung, unless inactivated by α_1 antitrypsin or α_2 macroglobulin (Page et al., 1978).

The alveolar macrophage has been postulated to be an effector cell in the etiology of byssinosis (Rylander et al., 1975a,b; Greenblatt and Ziprin, 1979). We have demonstrated that macrophages release prostaglandins when stimulated with extracts of cotton bract (Fowler et al., 1981). Prostaglandins may cause the lung-function changes associated with the Monday symptoms of byssinosis. Macrophages also release thromboxane A_2 and chemotactic factors for neutrophils and monocytes upon treatment with cotton-bract extracts (R. Ziprin, S. R. Fowler, and G. A. Greenblatt, unpublished data). Thus, alveolar macrophages may account for two pathophysiological changes associated with byssinosis: the bronchoconstriction responsible for the decline in FEV and the suspected influx of neutrophils into the respiratory air space.

Mast cell granules contain four classes of mediators: (1) vasoactive and smooth muscle-reactive mediators, including histamine and serotonin; (2) chemotactic mediators, including eosinophil chemotactic factor of anaphylaxis, eosinophilotactic oligopeptides, and high molecular weight, neutrophil chemotactic factor; (3) structural proteoglycans, including heparin, chondroitin sulfate, and dermatan sulfate; and (4) enzymes, including chymase, arylsulfatase, N-acetyl-β-glucosaminidase, kallikrein, and β glucuronidase. Several more mediators are generated by Fc receptor stimulation. Among these are leukotrienes, other arachidonic acid metabolites, and platelet-activating factor (Wasserman, 1979). Platelet-activating factor may be the most potent constrictor yet known. The cellular toxicology of mast cells, in relation to byssinosis, is largely unexplored.

III.   CAUSATIVE AGENT RESEARCH

Most investigators have described the causative agent as a water-soluble agent capable of passing through a 0.22-μm filter (Hamilton et al., 1973). In addition to the compounds discussed previously, several candidates have appeared in the literature and include a fraction obtained by Buck and Bouhuys (1980, 1981) and two products described by Russell et al. (1980) and Rohrbach et al. (1980). The consensus is that the causative agent is a component of bract.

Rylander has set forth logical criteria for defining a causal agent of byssinosis. Paraphrased, they are (1) a dose–response relationship is essential; (2) experimental manipulation of exposure levels should be possible; (3) the toxicity of the suspect agent must be known; and (4) quantitative analysis of the suspect agent should be possible. These requirements are reasonable, but criteria 3 and 4 appear to preclude a search for the new. It is also quite probable that byssinosis is not a result of one agent but of the interaction of more than one agent on a complex physiological system that varies somewhat from individual to individual. The variations include such factors as genetics, smoking, and additional underlying cryptic disease.

Our efforts have been directed toward the development of cellular bioassay systems for isolating and identifying those substances in cotton dusts that by either direct or amplified action might account for acute byssinosis. These substances act on alveolar macrophages, which then produce a variety of biologically active mediators that result in the acute symptoms. The role for alveolar macrophages is one point of agreement among ourselves, Rylander, and others who have studied byssinosis.

In our experiments, we have worked primarily with those substances that are extracted from cotton bract with water and will partition from water into diethyl ether (EECB). Originally, the lacinilenes, which move from water into ether, were of interest to us. As our experience with these extracts increased, it became apparent that many substances in aqueous extracts of bract would move into ether. We have detected qualitative differences in the ether solubles and related these to varietal differences among the bract samples extracted. For example, lacinilene C 7-methyl ether is not present in the bract of *G. arborium* cv. Nanking (G. A. Greenblatt, unpublished data). We observed an inhibitory effect of the ether solubles obtained from aqueous solutions of bract on alveolar macrophage chemiluminescence (Greenblatt and Ziprin, 1979) and decided to investigate the effects of these extracts on other alveolar macrophage activities.

We have demonstrated that rabbit alveolar macrophages produce prostaglandins $F_{2\alpha}$GA and E when treated with EECB (Fowler *et al.*, 1981). Prostaglandin release increased with increasing concentrations of extract. Not all cultivars tested were capable of causing prostaglandin release. EECB from field-dried bract of Rogers' Glandless 6 cultivar was much less effective in this respect than the glanded cultivars tested. EECB from green bracts of glandless cultivars also have been shown to be less effective in this assay than EECB from green bracts of glanded cultivars (Ziprin *et al.*, 1982).

In our original experiments on prostaglandin production, we did not attempt to identify or exclude any component of aqueous extract of cotton bract that moved into ether. Questions were raised as to the possibility of our EECB containing endotoxin (R. Rylander, personal communication). We have now demonstrated by three methods that endotoxin is not the active component of these extracts. We partitioned aqueous solutions of endotoxin with ether and assayed the ether fractions for endotoxin by the CFW-mouse, endotoxin-induced histamine hypersensitivity assay (Bergman *et al.*, 1977). Endotoxin could not be detected in the extracts by this system, which is sensitive to 5 ng.

Samples of the green bracts tested in the prostaglandin assay were measured for endotoxin content by the *Limulus* lysate assay and found to contain approximately equal amounts of endotoxin (J. J. Fischer, personal communication). The differences observed in ability to stimulate macrophages to produce prostaglandin cannot therefore be explained by endotoxin content variation among the different cultivars; another substance, probably of plant origin, must be the active agent.

Finally, we have determined that alveolar macrophages from C3H/ HeJ mice produce prostaglandin when exposed to EECB. This mouse strain has a genetic defect that causes it to be nonresponsive to endotoxin (Rosenstreich *et al.*, 1977; Rosenstreich and Vogel, 1980). Cells from these mice do not have the membrane receptor that mediates responses to endotoxin. This system provides conclusive evidence that endotoxin is not the active principle in the EECB (R. L. Ziprin, S. R. Fowler, and G. A. Greenblatt, unpublished data); therefore, we conclude that the activity observed must be caused by components of plant origin.

We have attempted to conduct experiments in conjunction with the Scott and White Clinic and Hospital, Division of Pulmonary Disease, Temple, Texas, to verify that human alveolar macrophages will produce prostaglandin upon exposure to our extracts. However, time, distance, and the general difficulties of conducting human experimentation have caused us to abandon the effort. Nevertheless, it is important to verify this possibility by experimentation. If we could isolate specific active substances from EECB that induce prostaglandin production by human alveolar macrophages, our research could then proceed to an investigation of the prevalence of the suspect agent in working mills.

Greenblatt (1978) has reported previously that there are striking differences among the abilities of aqueous extracts of glandless and glanded bracts to cause histamine release from porcine whole blood. The glandless cultivars, especially Rogers' Gl-6, were markedly less active in their histamine-releasing activity than the glanded cultivars studied. To look for additional differences in biological activity between glanded and glandless cultivars, we have sent a variety of bract samples to M. S. Rohrbach. Rohrbach has observed wide fluctuations in the abilities of aqueous extracts of different bracts to cause release of 5-

hydroxytryptamine (serotonin, 5-HT) from platelets (M. S. Rohrbach, personal communication). The agent that causes 5-HT release is believed to be a tannin with a molecular weight greater than 13,000. The smooth muscle contracting agent already mentioned has a molecular weight of less than 500. Rohrbach and Russell have not, as of this writing, had an opportunity to determine whether there are cultivar-dependent differences in the extractable quantities of their smooth muscle contracting agent, also.

Rohrbach and associates used a guinea pig model in which aerosolized aqueous extracts were administered by endotracheal tubes and compliance, resistance, and respiratory rate were determined. Neither their tannin fraction nor their low molecular weight fraction (in combination or alone) provoked the same effects on the three parameters as the unfractioned extract. This observation suggests a multideterminant etiology for byssinosis (M. S. Rohrbach, personal communication).

Buck and Bouhuys (1980, 1981) have partially isolated a bronchoconstrictive agent that, when inhaled by human volunteers, causes a drop in FEV. They have been able to remove some the active agent from glanded bract by mechanically "deglanding" the bract prior to preparing extracts; however, they can not detect a difference when the subjects are exposed to extracts of the glanded and glandless bracts we have used in our experiments (personal communication). It is certainly not a coincidence that we have measured differences attributable to the glanded characteristic in two systems: prostaglandin production and histamine release. Buck and Bouhuys (1981) has seen a third difference with human volunteers exposed to mechanically deglanded bract. Perhaps we are looking at water-soluble gland components that have not yet been described. In any case, we are now preparing extracts using Buck's methods in an attempt to correlate our macrophage responses with the human responses reported by her.

The airway constrictor agent in Buck's preparations has been partially characterized. The constrictor agent has a molecular weight below 1000. This substance has been examined by Ainsworth in collaboration with Buck and they have found that it will cause smooth muscle contraction and that this *in vitro* activity is blocked by methysergide. The airway constrictor agent is therefore similar, and perhaps identical, to the smooth muscle constrictor described by Russel and associates. We believe that these observations may represent the first instance in byssinosis research of independent laboratories independently making the same observation. Furthermore, this may also be the first instance in which an *in vitro* bioassay has been linked to an *in vivo* human response.

Byssinosis causative agent research has been complicated by the use of a wide variety of *in vitro* assays that have not been linked in a meaningful way to the human *in vivo* response. The sole exception known to us is the above-mentioned work by Ainsworth and Buck. Causative agent research has also been complicated by the wide variety of cotton dusts and plant parts used by various investigators. It is known that there are variations in biological activity of extracts from different bract samples.

Dr. Bob Jacobs of Cotton Incorporated has recently convened a workshop attended by most of the active causative agent researchers. At this meeting Ainsworth proposed, and it was agreed, that Cotton Incorporated would collect a pool of cotton dust that will serve as a standard starting material. This dust will be evaluated for *in vivo* constrictor activity by Buck and then distributed to others for *in vitro* studies. Additionally, we agreed on a standard initial extraction procedure. We urge anyone contemplating research in byssinosis to contact the active investigators and to contact Dr. Bob Jacobs of Cotton Incorporated for samples of the standard dust and details of the standard extraction procedure.

IV. CONCLUSION

A number of substances and hypotheses have been proposed to explain the cause of byssinosis. None has been studied thoroughly. Little, cohesive, causative agent research aimed at realistically understanding byssinosis is being conducted. Those who have studied the role of endotoxin in byssinosis have some evidence that endotoxins may be important. They can invoke a variety of well-understood mechanisms to show how endotoxin may contribute to byssinosis. The ubiquitous distribution of endotoxin and the difficulties in ascertaining that various preparations are endotoxin-free have made it difficult for investigators to implicate alternative agents. The number of substances suggested as causative agents and the variety of their chemical nature suggest to us that byssinosis has a multideterminant origin.

Investigators who do not have ready access to human volunteers can never prove unequivocally that a given substance is byssinogenic. Furthermore, investigators who do have such access must be certain that administered agents are endotoxin free before they can eliminate endotoxin as a source of confusion. Attempts at isolating active agents using human volunteers and FEV measurements as the basic bioassay are exceedingly expensive and technically difficult; therefore, fundamental information must be obtained through animal and cellular investigations. Agents identified as having byssinogenic potential by laboratory studies must then be tested in human volunteers but only concomitantly with or subsequent to quantitative studies showing their presence in mill areas where byssinosis is prevalent. Most causative agent research is still at the early laboratory stage.

The difficult task is to identify unequivocally plant products that can trigger the physiological responses we call byssinosis. Ultimately, byssinosis may be eliminated by breeding cotton varieties free of the offending substance(s).

ACKNOWLEDGMENTS

This work was supported in part by Cotton Incorporated, The Cotton Foundation, and the Natural Fibers and Food Protein Commission of Texas. We acknowledge with thanks the photographs provided by Dr. John Halloin and by Burlington Industries, Inc.

REFERENCES

Ainsworth, S. K., and Neuman, R. E. (1977). Biological activity of cotton dust and cotton plant extracts detected by chemotaxis and by leukocyte histamine release assay. *Proc.—Spec. Sess. Cotton Dust Res. Beltwide Cotton Prod. Res. Conf., 1977* pp. 76–79.

Ainsworth, S. K., and Neuman, R. E. (1980). Chemotactic activity of cotton mill dust extracts: Polypeptides as mediators the acute byssinotic reaction. *Proc.—Spec. Sess. Cotton Dust Res. Beltwide Cotton Prod. Res. Conf., 4th, 1980* pp. 14–18.

Ainsworth, S. K., McCormick, J. P., and Neuman, R. E. (1979a). Chemotaxis and histamine releasing properties of compounds in cotton mill dust. *Proc.—Spec. Sess. Cotton Dust Res. Beltwide Cotton Prod. Res. Conf., 3rd, 1979* pp. 21–29.

Ainsworth, S. K., Neuman, R. E., and Harley, R. A. (1979b). Histamine release from platelets for assay of byssinogenic substances in cotton mill dust and related materials. *Br. J. Ind. Med.* **36,** 35–42.

Ainsworth, S. K., Mundie, T. G., and Pilia, P. A. (1981). Serum levels of immunoglobulin and complement. An investigation of immunological mechanisms in byssinosis. *Proc.—Spec. Sess. Cotton Dust Res. Conf. Beltwide Cotton Prod. Res. Conf., 5th, 1981* pp. 7–8.

Anonymous (1945). Special Bulletin No. 18. Labor Stand. Div. U. S. Dept of Labor.

Anonymous (1974). Criteria for a recommended standard . . . occupational exposure to cotton dust. *HEW Publ. (NIOSH) (U. S.)* **75–118,** 22–24.

Anonymous (1978). Occupational exposure to cotton dust. *Fed. Regist.* **43**(123), 27350–27463.

Asboe-Hansen, G., and Glick, D. (1958). Influence of various agents on mast cells isolated from rat peritoneal fluid. *Proc. Soc. Exp. Biol. Med.* **98,** 458–461.

Ballard, P. L., Mason, R. J., and Douglas, W. H. J. (1978) Glucocorticoid binding by isolated lung cells. *Endocrinology* **102,** 1570–1575.

Barlowe, R. G. (1980). World cotton situation and outlook. *Proc.—Beltwide Cotton Prod. Res. Conf.* pp. 204–209.

Batawi, M. A., Shash, E. I., and El-Din, S. (1962). An epidemiological study on etiological factors in byssinosis. *Int. Arch. Gewerbepathol. Gewerhehyg.* **19,** 393–402.

Battigelli, M. C., Craven, P. L., and Fischer, J. J. (1977). The role of histamine in byssinosis. *J. Environ. Sci. Health, Part A A12,* 327–339.

Beasley, J. O. (1940). The origin of American tetraploid *Gossypium* species *Am. Nat.* **64,** 285–286.

Bell, A. A., and Stipanovic, R. D. (1978). Biochemistry of disease and pest resistance in cotton. *Mycopathologia* **65,** 91–106.

Bell, A. A., Stipanovic, R. D., Howell, C. R., and Fryxell, P. A. (1975). Antimicrobial terpenoids of *Gossypium:* Hemigossypol, 6-methoxyhemigossypol and 6-deox-yhemigossypol. *Phytochemistry* **14,** 225–231.

Bell, A. A., Stipanovic, R. D., O'Brien, D. H., and Fryxell, P. A. (1978). Sesquiterpenoid aldehyde quinones and derivatives in pigment glands of *Gossypium. Phytochemistry* **17,** 1297–1305.

Bergman, R. K., Milner, K. C., and Munoz, J. J. (1977). New test for endotoxin potency based upon histamine sensitization in mice. *Infect. Immun.* **18,** 352–355.

Berry, G., and Molyneux, M. K. B. (1981). A mortality study of workers in Lancashire cotton mills. *Chest* **79,** 11S–15S.

Berry, G., McKerrow, C. B., Molyneux, M. K. B., Rossiter, C. E., and Tombleson, J. B. L. (1973). A study of the acute and chronic changes in ventilatory capacity of workers in Lancashire cotton mills. *Br. J. Ind. Med.* **30,** 25–36.

Berry, G., Molyneaux, M. K. B., and Tomleson, J. B. L. (1974). Relationships between dust level and byssinosis and bronchitis in Lancashire cotton mills. *Br. J. Ind. Med.* **31,** 18–24.

Bouhuys, A., and Nicholls, P. J. (1967). The effect of cotton dust on respiratory mechanics in man and in guinea pigs. *Inhaled Part. Vap. Proc. Int. Symp., 2nd, 1967* pp. 75–84.

Bouhuys, A. (1974). Byssinosis *In* "Breathing: Physiology, Environment and Lung Disease" (A. Bouhuys ed.), pp. 416–440. Grune & Stratton, New York.

Bouhuys, A. (1979). "Cotton Dust and Lung Disease: Presumptive Criteria for Disability Compensation." A report for the Assistant Secretary of Labor for Policy, Evaluation and Research, pp. 45–51. U.S. Dept. Labor, Washington D.C.

Bouhuys, A., Lindell, S. E., and Lundin, G. (1960). Experimental studies on byssinosis, *Br. Med. J.* **1,** 324–326.

Bouhuys, A., Wolfson, R. L., Horner, D. W., Brain, J. D., and Zuskin, E. (1969). Byssinosis in cotton textile workers—respiratory survey of a mill with rapid labor turnover. *Ann. Intern. Med.* **71,** 257–269.

Bouhuys, A., Schoenberg, J. B., Beck, G. J., and Schilling, R. S. F. (1977). Epidemiology of chronic lung disease in a cotton mill community. *Lung.* **154,** 167–186.

Bouhuys, A., Beck, G. J., and Schoenberg, J. B. (1979). Priorities in prevention of chronic lung disease. *Lung* **156,** 129–148.

Bowling, A. L. (1981). World cotton outlook. *Proc.—Beltwide Cotton Prod.-Mech. Conf. 1981* pp. 2–4.

Braun, D., Jurgiel, J. A., Kaschak, M. C., and Babyak, M. A. (1973). Prevalence of respiratory signs and symptoms among U.S. cotton textile workers. *J. Occup. Med.* **15,** 414–419.

Britten, R. H., Bloomfield, J. J., and Goddard, J. C. (1933). The health of workers in a textile plant. *U. S. Public Health Serv. Bull.* **207.**

Brown, D. F., Wall, J. H., Piccolo, B., and Berni, R. J. (1978). Analysis of weave room dust from southeastern U.S. cotton mills. *Proc.—Spec. Sess. Cotton Dust Res. Beltwide Cotton Prod. Mech. Conf., 2nd, 1978* pp. 122–126.

Brown, T. C. (1981). Evaluating work relatedness of diseases. *Chest* **79,** 127S–128S.

Brune, F., Glatt, M., Kalin, H., and Peskar, B. (1978). Pharmacological control of prostaglandin and thromboxane release from macrophages. *Nature (London)* **274,** 261–263.

Buck, M. G., and Bouhuys, A. (1980). Constriction of airways by cotton bract extracts. *Proc.—Spec. Sess. Cotton Dust Res. Beltwide Cotton Prod. Res. Conf., 4th, 1980* pp. 31–32.

Buck, M. G., and Bouhuys, A. (1981). A purified extract from cotton bracts induces airway constriction in humans. *Chest* **79,** 43S–49S.

Cavagna, G., Foa, V., and Vigliani, E. C. (1969). Effects in man and rabbits of inhalation of cotton dust or extracts of purified endotoxins. *Br. J. Ind. Med.* **26,** 314–321.

Cinkotai, F. F., and Whitaker, C. S. (1978). Airborne bacteria and the prevalence of byssinotic symptoms in 21 cotton spinning mills in Lancashire. *Ann. Occup. Hyg.* **21,** 239–250.

Cinkotai, F. F., Lockwood, M. G., and Rylander R. (1977). Airborne microorganisms and prevalence of byssinotic symptoms in cotton mills. *Am. Ind. Hyg. Assoc. J.* **38,** 554–559.

Corley, T. E. (1966). Basic factors affecting performance of mechanical cotton pickers. *Trans. ASAE* **9,** 326–332.

Davenport, A., and Paton, W. D. M. (1962). The pharmacological activity of extracts of cotton dust. *Br. J. Ind. Med.* **19,** 19–32.

David, J. R. (1975). Macrophage activation by lymphocyte mediators. *Fed. Proc., Fed. Am. Soc. Exp. Biol.* **34,** 1730–1736.

Davies, P., and Bonney, R. J. (1979). Secretory products of mononuclear phagocytes: A brief review. *J. Reticuloendothel. Soc.* **26,** 37–47.

Davies, P., Bonney, R. J., Humes, J. L., and Kuehl, F. A., Jr., (1979). The use of mononuclear phagocytes in cell culture to study mechanisms of inflammation. *In* "Advances in Inflammation Research" (G. Weissman, B. Samuelsson, and R. Paoletti, eds.), pp. 189–195. Raven Press, New York.

De Maria, T. F., and Burrell, R. (1980). Effects of inhaled endotoxin-containing bacteria. *Environ. Res.* **23,** 87–97.

Diaz, P., Jones, D., and Kay, A. B. (1979). Histamine receptors on guinea pig alveolar macrophages—Chemical specificity and the effects of H_1 receptor and H_2 receptor agonists and antagonists. *Clin. Exp. Immunol.* **35,** 462–469.

Douglas, W. H. J., Del Vecchio, P., Teel, R. W., Jones, R. M., and Farrell, P. M. (1976). Culture of type II alveolar lung cells. *In* "Lung Cells in Disease" (A. Bouhuys, ed.), pp. 53–67. North-Holland Publ., Amsterdam.

Edwards, C., Macartney, J., Rooke, G., and Ward, F. (1975). The pathology of the lung in byssinosis. *Thorax* **30,** 612–623.

Edwards, J. (1981). Mechanisms of disease induction. *Chest* **79,** 38S–43S.

Edwards, J. H., and Jones, M. B. (1974). Immunology of byssinosis: A study of the reactions between the isolated byssinosis 'antigen' and human immunoglobulins. *Ann. N.Y. Acad. Sci.* **22,** 59–63.

Elissalde, M. H., Elissalde, G. S., and Ziprin, R. L. (1981). Effects of aqueous extracts of cotton bract on isolated mast cells: A bioassay for byssinosis. *Am. Ind. Hyg. Assoc. J.* **42,** 893–896.

Evans, E., and Nicholls, P. J. (1973). Methods for the gas chromatographic determination of urinary 1,4-methylimidazole acetic acid. *J. Chromatogr.* **82,** 391–397.

Evans, E., and Nicholls, P. J. (1974a). Comparative study of histamine release by cotton dust from the lungs of several species. *Comp. Gen. Pharmacol.* **5,** 87–90.

Evans, E., and Nicholls, P. J. (1974b). Preliminary characterization of the histamine releasing activity of cotton dust. *J. Pharm. Pharmacol.* **26,** 115–116.

Fischer, J. J. (1979). The microbial composition of cotton dusts, raw cotton lint samples, and the air of carding areas in mills. *Proc.—Spec. Sess. Cotton Dust Res. Beltwide Cotton Prod. Res. Conf., 3rd, 1979* pp. 8–10.

Fischer, J. J. (1980). Relation of airborne bacteria to endotoxin levels in cotton mills. *Proc.—Spec. Sess. Cotton Dust Res. Beltwide Cotton Prod. Res. Conf., 4th, 1980* pp. 29–30.

Fowler, S. R., Ziprin, R. L., Elissalde, M. H., and Greenblatt, G. A. (1981). The etiology of byssinosis—Possible role of prostaglandin $F_{2\alpha}GA$ synthesis by alveolar macrophages. *Am. Ind. Hyg. Assoc. J.* **42**, 445–448.

Fox, A. J., Tombleson, J. B. L., Watt, A., and Wilkie, A. G. (1973). A survey of respiratory disease in cotton operatives. Part II. Symptoms, dust estimations, and the effect of smoking habit. *Br. J. Ind. Med.* **30**, 48–53.

Fryxell, P. A. (1969). A classification of *Gossypium* L. (Malvaceae). *Taxon* **18**, 585–591.

Gee, J. B. L., and Smith, G. J. W. (1981). Lung cells and disease. *Basics R. D.* **9**, 1–6.

Greenblatt, G. A. (1977). Histamine release from excised pig lung tissue induced by cotton bracts. *Proc.—Spec. sess. Cotton Dust. Res., Beltwide Cotton Prod. Res. Conf., 1977* pp. 71–72.

Greenblatt, G. A. (1978). Effect of bract and dust extracts on macrophage phagocytosis and histamine release from blood. *Proc. Spec. Sess. Cotton Dust Res., Beltwide Cotton Prod. Mech. Conf., 2nd, 1978* pp. 118–119.

Greenblatt, G. A., and Ziprin, R. L. (1979). Inhibition of luminol-dependent chemiluminescence of alveolar macrophages by possible etiological agents of byssinosis. *Am. Ind. Hyg. Assoc. J.* **40**, 860–865.

Grimm, W., Seitz, M., Kirchner, H., and Gemse, D. (1978). Prostaglandin E synthesis in spleen cell cultures of mice injected with *Corynebacterium parvum*. *Cell. Immunol.* **40**, 419–426.

Gupta, S., and Good, R. A. (1980). Lymphocyte subpopulations and functions in hypersensitivity disorders. *In*, "Cellular, Molecular, and Clinical Aspects of Allergic Disorders" (S. Gupta and R. A. Good, eds.), pp. 87–138. Plenum New York.

Hamilton, J. D., Halprin, G. M., Kilburn, K. H., Merchant, J. A., and Ujda, J. R. (1973). Differential aerosol challenge studies in byssinosis. *Arch. Environ. Health* **26**, 120–124.

Hanna, C. J., Bach, M. K., Pare, P. D., and Schellenberg, R. R. (1981). Slow-reacting substances (Leukotrienes) contract human airway and pulmonary vascular smooth muscle *in vitro*. *Nature (London)* **290**, 343–344.

Haworth, E., and Macdonald, A. D. (1937). On histamine in cotton dust, and in the blood of cotton workers. *J. Hyg.* **37**, 234–242.

Helander, I., Salkinoja-Salonen, M., and Rylander, R. (1980). Chemical structure and inhalation toxicity of lipopolysaccharides from bacteria and cotton. *Infect. Immun.* **29**, 859–862.

Heyden, S., and Pratt, P. (1980). Exposure to cotton dust and respiratory disease. *JAMA, J. Am. Med. Assoc.* **244**, 1797–1798.

Hitchcock, M., Piscitelli, D. M., and Bouhuys, A. (1973). Histamine release from human lung by a component of cotton bracts and by compound 48/80. *Arch. Environ. Health* **26**, 177–182.

Hudson, A. R., Halprin, G. M., Kilburn, K. H., and McKenzie, W. N. (1977). Granulocyte recruitment to airways exposed to endotoxin aerosols. *Am. Rev. Respir. Dis.* **115**, 89–95.

Humes, J., Bonney, R., Pelus, L., Dahlgren, S., Sadowski, S., Kuehl, F., and Davies, P. (1977). Macrophages synthesize and release prostaglandins in response to inflammatory stiumli. *Nature (London)* **269**, 149–151.

Hunninghake, G. W., Gadek, J. E., Kawanami, O., Ferrans, V. J., and Crystal, R. G. (1979). Inflammatory and immune processes in the human lung in health and disease: Evaluation by bronchoalveolar lavage. *Am. J. Pathol.* **97**, 149–206.

Hunninghake, G. W., Gadek, J. E., Fales, H. M., and Crystal, R. G. (1980). Human alveolar macrophage-derived chemotactic factor for neutrophils. *J. Clin. Invest.* **66**, 473–483.

Hutchinson, J. (1959). "The Application of Genetics to Cotton Improvement." Cambridge Univ. Press, London and New York.

Imbus, H. R. (1981). Worker monitoring in byssinosis. *Chest* **79,** 122S–123S.

Johnson, K. J., Chapman, W. E., and Ward, P. A. (1979). Immunopathology of the lung: A review *Am. J. Pathol.* **95,** 795–839.

Jones, R. N., Hughes, J., Hammad, Y. Y., Glindmeyer, H., III, Butcher, B. J., Diem, J. E., and Weill, H. (1981). Respiratory health in cottonseed crushing mills. *Chest* **79,** 30S–33S.

Kilburn, K. H., Lynn, W. S., Tres, L. L., and McKenzie, W. N. (1973). Leukocyte recruitment through airway walls by condensed vegetable tannins and quercetin. *Lab. Invest.* **28,** 55–59.

Kilburn, K. H., McCormick, J. P., Schafer, T. R., Thurston, R. J., and McKenzie, W. N. (1977). Biologically active compounds in cotton dust. *Proc., Spec. Sess. Cotton Dust Res., Beltwide Cotton Prod. Res. Conf., 1977* p. 66.

Kilburn K. H., Lynn, D., McCormick, J. P., and Shaffer, T. (1979). Effects of synthetic lacinilene C-7 methyl ether on hamster airways. *Proc.—Spec. Sess. Cotton Dust Res., Beltwide Cotton Prod. Res. Conf., 3rd, 1979* pp. 19–22.

Kondakis, X. G., and Pournaras, N. (1965). Byssinosis in cotton ginneries in Greece. *Br. J. Ind. Med.* **22,** 291–294.

Kurland, J. I., and Bockman R. (1978). Prostaglandin E production by human blood monocytes and mouse peritoneal macrophages. *J. Exp. Med.* **147,** 952–957.

Kutz, S. A., Olenchock, S. A., Elliot, J. A., Pearson, D. J., and Major, P. C. (1979). Antibody-independent complement activation by cardroom cotton dust. *Environ. Res.* **19,** 405–414.

Kutz, S. A., Mentnech, M. S., Olenchock, S. A., and Major, P. C. (1980). Precipitation of serum proteins by extracts of cotton dust and stems. *Environ. Res.* **22,** 476–484.

Kutz, S. A., Mentnech, M. S., Olenchock, S. A., and Major, P. C. (1981). Immune mechanisms in byssinosis. *Chest* **79,** 56S–58S.

Lagunoff, D. (1976). The mast cell. *In* "Bronchial Asthma: Mechanisms and Therapeutics" (E. Weiss and M. Segal, eds.), pp. 383–407. Little, Brown, Boston, Massachusetts.

Lane, H. C., and Schuster, M. F. (1981). Condensed tannins of cotton leaves. *Phytochemistry* **20,** 425–427.

Lopez-Merino, V., Lombart, R. L., Marco, R. F., Carnicero, A. S., Guillen, F. G., and Bouhuys, A. (1973). Arterial blood gas tension and lung function during acute response to hemp dust. *Am. Rev. Respir. Dis.* **107,** 809–815.

Lynn, W. S., Munoz, S., Campbell, J. A., and Jeffs, P. W. (1974). Chemotaxis and cotton extracts. *Ann. N. Y. Acad. Sci.* **221,** 163–173.

McMichael, S. C. (1960). Combined effects of glandless genes gl_2 and gl_3 on pigment glands in the cotton plant. *Agron. J.* **46,** 385–386.

Mares, T., and Sekul, A. A. (1980). Chromatographic separation of the cotton dust/bract antigen into two components. *Proc.—Spec. Sess. Cotton Dust Res., Beltwide Cotton Prod. Res. Conf., 4th, 1980* pp. 5–6.

Mason, R. J., Stossel, T. P., and Vaughan, M. (1972). Lipids of alveolar macrophages, polymorphonuclear leukocytes, and their phagocytic vesicles. *J. Clin. Invest.* **51,** 2399–2407.

Mason, R. J., Williams, M. C., and Greenleaf, R. D. (1976). Isolation of lung cells. *In* "Lung Cells in Disease" (A. Bouhuys, ed.), pp. 39–52. North-Holland Publ., Amsterdam.

Massoud, A., and Taylor, G. (1964). Byssinosis: Antibody to cotton antigens in normal subjects and in cotton cardroom workers. *Lancet* **2,** 607–610.

Merchant, J. A., and Ortmeyer, C. (1981). Mortality of employees of two cotton mills in North Carolina. *Chest* **79**, 6S–11S.

Merchant, J. A., Kilburn, K. H., O'Fallon, W. M., Hamilton, J. D., and Lumsden, J. C. (1972). Byssinosis and chronic bronchitis among cotton textile workers. *Ann. Intern. Med.* **76**, 423–433.

Merchant, J. A., Lumsden, J. C., Kilburn, K. H., Germino, V. H., Hamilton, J. D., Lynn, W. S., Byrd, H., and Baucom, D. (1973). Preprocessing cotton to prevent byssinosis. *Br. J. Ind. Med.* **30**, 237–247.

Merchant, J. A., Halprin, G. M., Hudson, A. R., Kilburn, K. H., McKenzie, W. M., Jr., Bermanzohn, P., Hurst, D. J., Hamilton, J. D., and Germino, V. H., Jr. (1974). Evaluation before and after exposure—The pattern of physiological response to cotton dust. *Ann. N. Y. Acad. Sci.* **221**, 38–43.

Merchant, J. A., Halprin, G. M., Hudson, A. R., Kilburn, D. H., McKenzie, W. M., Jr., Hurst, D. J., and Bermanzohn, P. (1975). Response to cotton dust. *Arch. Environ. Health* **30**, 222–229.

Mohammed, Y. S., El-Gazzar, R. M., and Adamyova, K. (1971). Byssinosan, an aminopolysaccharide isolated from cotton dust. *Carbohydr. Res.* **20**, 431–435.

Molyneux, M. K. B., and Tombleson, J. B. L. (1970). An epidemiological study of respiratory symptoms in Lancashire mills. *Br. J. Ind. Med.* **27**, 225–234.

Morey, P. R. (1979a). Botanically what is raw cotton dust? *Am. Ind. Hyg. Assoc. J.* **40**, 702–708.

Morey, P. R. (1979b). Morphology and fluorescence of respirable plant part dusts. *Proc.— Spec. Sess. Cotton Dust Res., Beltwide Cotton Prod. Res. Conf., 3rd, 1979* pp. 2–3.

Morey, P. R. (1980). Botanical composition of dusts in the primary and secondary cotton industries. *Proc.—Spec. Sess. Cotton Dust Res, Beltwide Cotton Prod. Res. Conf., 4th, 1980* p. 67.

Morey, P. R., Bethea, R. M., Wakelyn, P. J., Kirk, I. W., and Kopetzky, M. T. (1976a). Botanical trash present in cotton before and after saw-type lint cleaning. *Am. Ind. Hyg. Assoc. J.* **37**, 321–328.

Morey, P. R., Sasser, P. E., Bethea, R. M., and Kopetzky, M. T. (1976b). Variation in trash composition in raw cottons. *Am. Ind. Hyg. Assoc. J.* **37**, 407–412.

Morey, P. R., Fischer, J. J., and Sasser, P. E. (1980). Gram-negative bacterial content of cotton bracts and raw cotton. *Proc.—Spec. Sess. Cotton Dust Res., Beltwide Cotton Prod. Res. Conf., 4th, 1980* pp. 68–69.

Morgan, W. K. C. (1975). Byssinosis and related conditions. *In* "Occupational Lung Disease" (W. K. C. Morgan and A. Seaton, eds.), pp. 274–288. Saunders, Philadelphia, Pennsylvania.

Morrison, D. C., and Betz, S. J. (1977). Chemical and biological properties of a protein-rich fraction of bacterial lipopolysaccharides. II. The *in vitro* rat peritoneal mast cell response. *J. Immunol.* **119**, 1790–1795.

Morrison, D. C., and Ulevitch, J. (1978). The effects of bacterial endotoxins on host mediation systems. *Am J. Pathol.* **93**, 527–617.

Muramoto, H., Sherman, R. S., and Ledbetter, C. (1981). A progress report on the breeding of caducous bract cotton. *Proc.—Cotton Dust Res. Conf. Beltwide Cotton Prod. Res. Conf., 5th, 1981* p. 23.

Myrvik, Q. N., and Kohlweiss, L. (1975). Immunobiology of the alveolar macrophage. *In* "International Academy of Pathology Monograph, No. 16" (J. W. Rebuck, C. W. Berard, and M. R. Abell, eds.), pp. 65–80. Williams & Wilkins, Baltimore, Maryland.

Nicholls, P. J. (1962). Some pharmacological actions of cotton dust and other vegetable dusts. *Br. J. Ind. Med.* **19**, 33–41.

Northrup, S., Presant, L., Kilburn, K. H., McCormick, J., and Pachlatko, P. (1976). Lacinilene C-7 methyl ether, an agent from cotton dust causing leukocyte chemotaxis and histamine release (byssinosis). *Fed. Proc., Fed. Am. Soc. Exp. Biol.* **35,** 632.

Page, R. C., Davies, P., and Allison, A. (1978). The macrophage as a secretory cell. *Int. Rev. Cytol.* **52,** pp. 119–157.

Palmer, A., Finnegan, W., Herwitt, P., Waxweiler, R., and Jones, J. (1978). Byssinosis and chronic respiratory disease in U.S. cotton gins. *J. Occup. Med.* **20,** 96–102.

Paterson, N. A. M., Wasserman, S. I., Said, J W., and Austen, K. I. (1976). Release of chemical mediators from partially purified human lung mast cells. *J. Immunol.* **117,** 1356–1362.

Pernis, B., Vigliani, E. C., Cavagna, C., and Finulli, M. (1961). The role of bacterial endotoxins in occupational diseases caused by inhaling vegetable dusts. *Br. J. Ind. Med.* **8,** 120–129.

Pratt, P. C. (1981). Comparative prevalence and severity of emphysema and bronchitis at autopsy in cotton mill workers vs. controls. *Chest* **79,** 49S–52S.

Ramazzini, B. (1964). Diseases of workers. Translated from Latin text *"De Morbis Artificum* of 1713," by W. C. Wright, p. 257. Hafner, New York.

Roach, S. A., and Schilling, R. S. F. (1960). A clinical environment study of byssinosis in the Lancashire cotton industry. *Br. J. Ind. Med.* **17,** 1–9.

Rohrbach, M. S., Rolstad, R. A., Tracey, P., and Russell, J. A. (1980). Action of cotton bracts tannin on platelets. *Physiologist* **23,** 45.

Rooke, G. B. (1981a). What is byssinosis? A review. *Tex. Res. J.* **51,** 168–173.

Rooke, G. B. (1981b). Compensation for byssinosis in Great Britain. *Chest* **79,** 124S–127S.

Rooke, G. B. (1981c). The pathology of byssinosis. *Chest* **79,** 67S–71S.

Rosenstreich, D. L., and Vogel, S. N. (1980). Central role of macrophages in host response to endotoxin. *In* "Microbiology—1980" (D. Schlessinger, ed.), pp. 11–15. Am. Soc. Microbiol., Washington, D.C.

Rosenstreich, D. L., Glode, L. M., Wahl, L. M., Sandberg, A. C., and Mergenhagen, S. E. (1977). Analysis of the cellular defects of endotoxin-unresponsive C3H/HeJ mice. *In* "Microbiology—1977" (D. Schlessinger, ed.), pp. 314–320. Am. Soc. Microbiol., Washington, D. C.

Russell, J. A., Gilberstadt, M. L., and Rohrbach, M. S. (1980). Constrictor effect of cotton bracts extract on isolated canine trachealis strips. *Am. Rev. Respir. Dis.* **121,** 252.

Rylander, R. (1980). Lung cell reactions and early inflammation: Experimental and epidemiological aspects. *Eur. J. Respir. Dis.* **61,** Suppl. 107, 73–81.

Rylander, R. (1981). Bacterial toxins and etiology of byssinosis. *Chest* **79,** 34S–37S.

Rylander, R., and Haglind, P. (1981). Experimental cardroom tests with reference to airborne lipopolysaccharide. *Proc.—Cotton Dust Res. Conf., Beltwide Cotton Prod. Res. Conf., 5th, 1981* pp. 14–15.

Rylander, R., and Snella, M. C. (1980). Effects of subchronic endotoxin exposure. *Proc.— Spec. Sess. Cotton Dust Res., Beltwide Cotton Prod. Res. Conf., 4th, 1980* pp. 26–27.

Rylander, R., Nordstrand, A., and Snella, M. C. (1975a). Bacterial contamination of organic dusts: Effects on pulmonary cell reactions, *Arch. Environ. Health* **30,** 137–140.

Rylander, R., Snella, M. C., and Garcia, I. (1975b). Pulmonary cell response patterns after exposure to airborne bacteria. *Scand. J. Respir. Dis.* **56,** 195–200.

Samuelsson, B., Goldyne, M., Granström, E., Hamberg, M., Hammarström, S., and Malmsten, C. (1978). Prostaglandins and thromboxanes. *Annu. Rev. Biochem.* **47,** 997–1029.

Samuelsson, B., Borgeat, P., Hammarström, S., and Murphy R. (1980). Luekotrienes: A new group of biologically active compounds. *Adv. Prostaglandin Thromb. Res.* **6,** 1–18.

Sandusky, C. B., Johnson, A. R., and Moran, N. C. (1973). Effect of endotoxin and anaphylatoxin on rat mast cells and chopped rat lung. *Proc. Soc. Exp. Biol. Med.* **143**, 764–768.

Schachter, E. N., Brown, S., Zuskin, E., Beck, G. I.., Buck, M., Kolack, B., and Bouhuys, A. (1981). The effect of mediator modifying drugs in cotton bract-induced bronchospasm. *Chest* **78**, 73S–77S.

Schilling, R. (1981). Worldwide problems of byssinosis. *Chest* **79**, 3S–5S.

Schilling, R. S. F., Hughes, J. P. W., and Dingwall-Fordyce, I. (1955a). Disagreement between observers in an epidemiological study of respiratory disease. *Br. Med. J.* **1**, 65–68.

Schilling, R. S. F., Hughes, J. P. W., Dingwall-Fordyce, I., and Gilson, J. C. (1955b). Epidemiological study of byssinosis among Lancashire cotton workers. *Br. J. Ind. Med.* **12**, 217–227.

Snell, J. D. (1966). Effects of inhaled endotoxin. *J. Lab. Clin. Med.* **67**, 624–632.

Stenson, W. F., and Parker, C. W. (1980). Prostaglandins, macrophages and immunity. *J. Immunol.* **125**, 1–5.

Stipanovic R. D., Wakelyn, P. J., and Bell. A. A. (1975). Lacinilene C, a revised structure and lacinilene C 7-methyl ether from *Gossypium* bracts. *Phytochemistry* **14**, 1041–1043.

Stipanovic, R. D., Greenblatt, G. A., Beier, R. C., and Bell, A. A. (1981). 2-Hydroxy-7-methoxycadaline, the precursor of lacinilene C 7-methyl ether in *Gossypium*. *Phytochemistry* **20**, 729–730.

Svenson, J., Strandberg, K., Tuvemo, T., and Hamberg, M. (1975). Thromboxane A_2: Effects on airway and vascular smooth muscle. *Prostaglandins* **14**, 425–436.

Taylor, G., Massoud, A., and Lucas, F. (1971). Studies on the etiology of byssinosis. *Br. J. Ind. Med.* **28**, 143–151.

Territo, M. C. and Golde, D. W. (1979). The function of human alveolar macrophages. *J. Reticuloendothel. Soc.* **25**, 111–120.

Unanue, E. R. (1976). Secretory function of mononuclear phagocytes. *Am. J. Pathol.* **83**, 396–417.

Unanue, E. R. (1978). The regulation of lymphocyte functions by the macrophage. *Immunol Rev.* **40**, 227–255.

Valic, F., and Zuskin, E. (1973). Pharmacological prevention of acute ventilatory capacity reduction in flax dust exposure. *Br. J. Ind. Med.* **30**, 381–384.

Valone, F. H., Franklin, M., Sun, F. F., and Goetzl, E. J. (1980). Alveolar macrophage lipoxygenase products of arachidonic acid: Isolation and recognition as the predominant constituents of the neutrophil chemotactic activity elaborated by alveolar macrophages. *Cell. Immunol.* **54**, 390–401.

Vargaftig, B. B., Lefort, J., Chignard, M., and Benveniste, J. (1980). Platelet-activating factor induces a platelet-dependent bronchoconstriction unrelated to the formation of prostaglandin derivatives. *Eur. J. Pharmacol.* **65**, 185–192.

Wakelyn, P. J., Greenblatt, G. A., Brown, D. F., and Tripp, V. W. (1976). Chemical properties of cotton dust. *Am Ind. Hyg. Assoc. J.* **37**, 22–31.

Walker, R. F., Eidson, G., and Hatcher, J. D. (1974). Influence of cotton dust inhalation on free lung cells in rats and guinea pigs. *In* "Cotton Dust, Proceedings of a Topical Symposium. Sponsored by the American Conference of Governmental Industrial Hygienists, Atlanta, 1974" (H. Ayer, ed.), pp. 87–100. Am. Conf. Govt. Ind. Hyg., Cincinnati, Ohio.

Wasserman, S. I. (1979). The mast cell and the inflammatory response. *In* "The Mast Cell. Its Role in Health and Disease" (J. Pepys and A. M. Edwards, eds.), pp. 9–20. Univ. Park Press, Baltimore, Maryland.

Watkins, G. M. (1981). Cotton diseases—General concepts. *In* "Compendium of Cotton Diseases" (G. N. Watkins, ed.), pp. 2–10. Am. Phytophathol. Soc., St. Paul, Minnesota.

Weibel, E. R., Gehr, P., Haies, D., Gil, J., and Bachofen, M. (1976). The cell population of the normal lung. *In* "Lung Cells in Disease" (A. Bouhuys, ed.), pp. 3–16. North-Holland Publ., Amsterdam.

Ziprin, R. L., Fowlers, S. R., Greenblatt, G. A., and Elissalde, M. H. (1981). The etiology of byssinosis—Comparison of prostaglandin $F_{2\alpha}$ synthesis by alveolar macrophages stimulated with extracts from glanded and glandless cotton cultivars. *Am. Ind. Hyg. Assoc. J.* **42**, 876–879.

10

Physiological Effects of Lead Dusts

Samarendra N. Baksi

I. INTRODUCTION

Lead toxicity affects multiple organ systems, and the underlying mechanism of lead's toxicity has not been fully elucidated. It is apparent, however, that exposure to lead provokes a variety of physiological responses in many different organ systems. Research disclosing the effects of lead on hematopoiesis (Sasa *et al.*, 1973; Roels *et al.*, 1976; Piomelli *et al.*, 1975), central and peripheral nervous system (Grandjean, 1978; Seppalainen *et al.*, 1975; Pentschew and Garrow, 1966; Lilis *et al.*, 1977),

281

AIR POLLUTION
ISBN 0-12-483880-4

behavior (Silbergeld and Goldberg, 1973), spermatogenesis (Lancranjan *et al.*, 1975) fetal development (Ferm and Carpenter, 1967; Ferm and Ferm, 1971), cardiovascular responses (Williams *et al.*, 1977; Kopp *et al.*, 1980), and bone and vitamin D metabolism (Anderson and Danychuk, 1977; Sorrel *et al.*, 1977; Baksi and Kenny, 1978a, 1979; Rosen *et al.*, 1980) have added new dimensions to the problem of lead intoxications. Thus, in recent years, much attention has been given to chronic low-level lead exposure (Needleman, 1980) that is itself insufficient to cause recognizable clinical poisoning but may be associated with harmful effects.

Lead's toxicity is influenced by the amount of calcium in the diet (Six and Goyer, 1970), other nutritional factors (Mahaffey, 1974; Mahaffey and Michaelson, 1980; Barton *et al.*, 1978b; Sleet and Soares, 1979; Smith *et al.*, 1978), and a number of physiological or environmental factors such as age (Ziegler *et al.*, 1978; Quarterman and Morrison, 1978), pregnancy and lactation (Hammond, 1969; Momcilovic, 1978), and ultraviolet light (Goyer and Mahaffey, 1972).

This chapter will briefly review the pharmacokinetics and important physiological effects of lead and its mechanism of action.

II. BACKGROUND

Lead is a widely distributed heavy metal, which has been and continues to be a persistent source of toxicological concern. The sources of lead are both natural and man-made. The environmental signficance of lead reflects both its use and, in comparison to most other toxic heavy metals, its abundance. Lead in its various forms has been known and used by man from ancient times. The world production of lead is approximately 4 million tons, which is more than any other toxic heavy metal. In the United States, lead is consumed mainly (1) as lead metal products, such as lead solder, lead pipes, lead sheeting, storage battery components, and ammunition, and (2) in synthesis of organolead fuel additives such as tetraethyl lead. The use of lead as either a fuel additive or paint base is being reduced considerably. In a modern industrial society, lead is present in food, water, air, soil, dust fall, and other materials with which man and other animals come in contact. Lead-using industrial activity and high density automobile traffic with associated lead emissions generate appreciable contamination of the air and soil. Deteriorating urban housing also presents a potentially high source of lead exposure because of lead-based paint once used in these houses. Lead has no known beneficial biological effect and is viewed purely as a deleterious agent. The so-called normal blood level of lead has been a subject of controversy (Patterson, 1965; Goldwater and Hoover, 1967). Recently, Piomelli *et al.* (1980) have reported that blood- and air-lead concentrations in a remote Himalayan population are much lower than in New York City.

III.   PHARMACOKINETICS OF LEAD

A.   Routes of Exposure

There are three major routes of lead exposure within the environment. In the general population the greatest exposure takes place via oral intake (food and drink), although in specific circumstances the inhalation of air contaminated with lead or uptake of dust from industrial processes may represent important exposure routes. Inorganic lead salts do not reaily penetrate the intact skin (Rastogi and Clausen, 1976), but lipid-soluble forms such as tetraethyl lead penetrate significantly.

B.   Absorption

Ingested lead is absorbed primarily in the duodenum, at least in the rat. The total body burden of lead does not seem to affect lead absorption (Conrad and Burton, 1978). In human adults not occupationally exposed to lead, and in children older than 6–8 years, dietary lead probably comprises the major fraction of daily intake, recent estimates being 100–500 μg per day. The intestinal absorption of lead in adults is about 10% (Rabinowitz *et al.*, 1976), whereas normal infants and children 5 years of age or younger absorb approximately 40% (Ziegler *et al.*, 1978). Although only 10% of ingested lead is absorbed in adults, up to 40% of inhaled lead will be taken in by the pulmonary route (Moore *et al.*, 1980), provided the particle size is small (Aranyi *et al.*, 1977). Nutritional factors influence the extent of lead absorption from the gastrointestinal tract. In particular, a low dietary level of calcium (Six and Goyer, 1970) has been shown to increase the body burden of lead in rats. A study by Barton *et al.* (1978a) showed that both lead and calcium compete for similar binding sites on intestinal mucosal proteins, which are important in the absorptive process. However, these authors reported also that dietary calcium had no significant effect on the absorption of lead, but calcium-deprived rats had decreased excretion and thus increased body retention of lead. Recently, Mykkanen and Wasserman (1981) investigated in more detail the mechanism of lead transport by the gastrointestinal tract and, particularly, the similarities or dissimilarities between lead and calcium in this process. The absorption of these metals was determined in 3-week-old white Leghorn cockerels, raised on a commercial diet or special diets, using an *in vivo* ligated loop procedure. It was shown that lead is rapidly taken up by the intestinal tissue and only slowly transferred into the circulation, whereas calcium, also accumulated rapidly by the intestine, is released much faster from the tissue in the serosal direction. Their data imply that, in spite of the similarities in the response of lead and calcium absorptive processes to various treatments, there is not direct interaction between these cations in the intestine of the chick.

Vitamin D administration also enhances lead absorption in rats (Smith *et al.*, 1978; Hart and Smith, 1981). In a more direct study, Barton *et al.*, (1980) demonstrated that the manipulation of dietary vitamin D content had no significance on the absorption of lead from isolated gut loops, and parental vitamin D administration did not affect lead absorption in rachitic animals. In contrast, dietary vitamin D deficiency and repletion resulted in increased absorption in intact animals because of prolonged gastrointestinal transit time. It has also been found that retention of orally administered lead is greater in iron-deficient animals (Six and Goyer, 1972). That lead absorption is affected by the state of iron repletion suggests that parental iron is available to block absorptive sites utilized by lead in the intestinal mucosa (Barton *et al.*, 1978b). Dietary constituents that enhance the solubility of lead, such as ascorbic acid and sulfhydryl group amino acids, increase lead absorption (Conrad and Barton, 1978). Lactose also facilitates the intestinal absorption of lead in weanling rats (Bushnell and DeLuca, 1980).

C. Retention

Once absorbed lead is transported in the blood, approximately 95% binds to erythrocytes. In the blood, lead has a biological half-life of approximately 25 days. There are three discrete lead compartments in the body: (1) circulating blood, (2) relatively labile soft tissue fractions, and (3) an inert but appreciable portion deposited in the skeleton (Fig. 1). Lead retained in the soft tissues has a half-life of a few months, but the half-life in the brain may be somewhat longer (Grandjean, 1978). More than 90% of the body lead burden is accumulated in the skeleton. The level of lead in the human skeleton increases at least until the age of 40 years (Barry and Mossman, 1970). The half-life of bone lead in man is estimated to be about 10 years. Blood sample analysis is most commonly used to determine lead retention in the body.

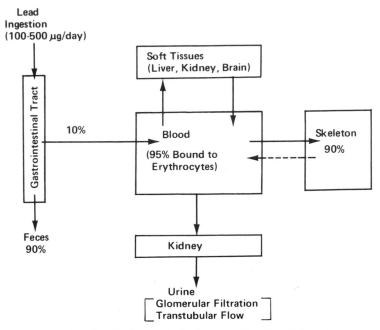

Fig. 1. Absorption, distribution, and elimination of lead in adult humans not occupationally exposed to lead.

D. Excretion

The major route of elimination of ingested lead which has not been absorbed from the gastrointestinal tract is fecal excretion. Primarily, the absorbed fraction of lead is excreted in urine, although lesser amounts are excreted in sweat. Lead is excreted by the kidney in two ways: glomerular filtration and transtubular flow (Goyer and Mahaffey, 1972). A small portion of absorbed lead is excreted in the bile in most animal models studied (Klaassen and Shoeman, 1974). Rates of lead elimination from femurs and whole bodies were apparently less rapid in young than in adult mice (Keller and Doherty, 1980).

IV. PHYSIOLOGICAL RESPONSES TO LEAD OF ORGAN SYSTEMS

Acute cases of lead poisoning are rare in adults. Depending on the severity and duration of exposure, the effect may range from lead colic to encephalopathy and death. Primarily, the more common responses, caused by chronic lead exposure, will be discussed in this section.

A. Hematopoietic System

Anemia from lead poisoning is probably the most well known and best-understood effect of lead in man. A decrease in blood hemoglobin appears at relatively low exposure levels. The anemia caused by lead exposure is attributed to two different actions of lead: lead inhibits heme synthesis, but it also causes increased fragility of the red blood cells and, consequently, an increased rate of destruction. The mechanism for shortened erythrocyte survival is not well understood, but it has been demonstrated that an inhibition of erythrocyte membrane Na^+/K^+-ATPase occurs in people with only moderately elevated lead exposure (Hernberg, 1976). The decreased erythrocyte life span is probably caused by a leak of potassium associated with an inhibition of Na^+/K^+-ATPase, and this leads to an increased mechanical fragility of the cell (Sasa, 1978).

Interference with the heme biosynthetic pathway by lead occurs in two steps: (1) inhibition of the activity of the enzyme aminolevulinic acid dehydralase (ALAD), which mediates the condensation of two units of aminolevulinic acid to form porphobilinogen, and (2) the insertion of iron into protoporphyrin IX to form heme, the final step in heme synthesis (Fig. 2). Most sensitive to a low level of lead is the enzyme ALAD (a zinc-activated enzyme). The mechanism of inhibition probably involves direct competition of lead with zinc for the binding to the sulfhydryl group in close relation to the active site of the enzyme. The inhibition is primarily of the noncompetitive type and can be detected in the erythrocytes at blood lead levels as low as $10-15$ $\mu g/100$ ml (Sasa, 1978). This enzyme inhibition occurs, therefore, in a large proportion of people in industrialized societies. The correlation between blood lead and degree of ALAD inhibition is established so sufficiently that the ALAD activity may be employed as an accurate estimate of blood lead level. There is controversy over the clinical meaning of a moderate inhibition of ALAD by lead, but significant inhibition, to the point that urinary excretion of the substrate aminolevulinic acid is increased, is generally accepted as an indication of significant physiological impairment.

Ferrochelatase, the final mitochondrial enzyme in heme synthesis, is less sensitive to lead than ALAD. When lead exposure is high, its substrate, protoporphyrin IX, accumulates in red blood cells because heme is not found due to inhibition of heme synthesis by lead (Fig. 2). Instead of heme, zinc protoporphyrin (ZPP) is found in large amounts and oc-

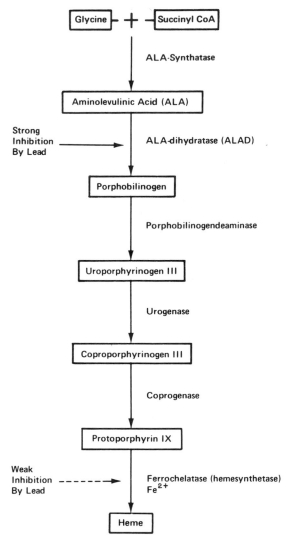

Fig. 2. The effect of lead on heme biosynthesis. Lead strongly inhibits aminolevulinic acid dehydralase (ALAD). The inhibition of ferrochelatase (hemesynthalase) is rather weak. As a result of these inhibitions in heme biosynthesis two intermediates are produced in excess. The first is ALA, which is excreted in the urine; the second is protoporphrin IX, which is found in excess in the erythrocytes.

cupies the molecular site reserved for iron. Because ZPP is table and firmly bound to hemoglobin, the ZPP level in the blood is also an indication of the average toxicity of lead in the bone marrow during the 3-month life span of erythrocytes (Lamola *et al.*, 1975).

Higher concentrations of lead may also interfere with other steps in heme synthesis by depleting cellular protein content by *de novo* induction of the microsomal enzyme heme oxygenase, resulting in increased catabolism of heme (Maines and Kappas, 1977).

B. Nervous System

Next only to the hematopoietic system, the nervous system is most affected by lead exposure. The physiological responses to lead are evident in both the peripheral and central nervous systems.

1. Peripheral

Eariler studies have demonstrated that lead damages peripheral nerves; the pathological change is mainly axonal degeneration (Krigman *et al.*, 1980). The degeneration of spinal nerve roots after lead exposure is controversial (Hyslop and Kraus, 1923). Segmental demyelination is not well established in human lead neuropathy (Behse and Carlsen, 1978) but is common in animals (Fullerton, 1966). Degenerative changes in the sympathetic nerves and ganglia as well as submucosal and myenteric plexuses of the gut after lead exposure have been reported (Cantarow and Trumpa, 1944). Chronic low-level lead exposure in humans also affects nerve conduction velocity (Seppalainen and Hernberg, 1972). Lead exposure, has been suspected as a cause of motor neuron disease (Campbell *et al.*, 1970). The neurophysiological manifestations of lead toxicity are best demonstrated in the peripheral nervous system. The amount of acetylcholine normally released by the action potential at the presynaptic junction is reduced by lead both *in vitro* and *in vivo* (Manalis and Cooper, 1973; Silbergeld *et al.*, 1974). This effect of lead is reversed by the addition of calcium to the system, indicating that lead probably competes with calcium in the presynaptic membrane (Silbergeld and Adler, 1978).

2. Central

The deleterious effects of lead on the central nervous system (CNS) may range from behavioral dysfunctions to encephalopathy, particularly in a developing animal.

An association between lead and at least some form of minimal brain dysfunction or hyperactivity in children has been suggested (Sobotka and Cook, 1974; Lin-Fu, 1972). David *et al.* (1972) reported that concentrations of lead in blood and urine were found to be significantly elevated in the greater proportion of a group of children who had minimal brain dysfunction as compared with those of a control group. In a pilot study, David *et al.* (1976) demonstrated that 50% of the hyperkinetic school children tested in the studies showed marked improvement when lead-chelating medications were used. Whether this subclinical degree of increased blood level induces either transitory or permanent damage of the CNS in young children is not yet established, even though several epidemiological blood lead screening project studies indicate it is a strong possibility.

Lead-induced hyperactivity has been reported in mice (Silbergeld and Goldberg, 1973), and monkeys (Allen *et al.*, 1974). Lead was given in drinking water or food to nursing mothers or directly to pups immediately after birth and before weaning. These animals exhibited significant increase in motor activity and poor learning performance (Brown, 1975). Pharmacologically, these animals responded to certain drugs in a manner comparable to that of hyperactive children (Silbergeld and Goldberg, 1975; LaPorte and Talbott, 1978). In such hyperactive mice, the increased motor activity was suppressed by the administration of amphetamines, methylphenidate, cholinergic agonists, and aminergic antagonists and aggravated by aminergic agonists and the anticholinergic agent atropine (Silbergeld and Goldberg, 1975). This concept of lead-induced hyperactivity is not universally accepted, because several laboratories have not been able to reproduce the behavioral observations reported in the earlier publications, particularly with respect to studies in rats (Sobotka and Cook, 1974; Memo *et al.*, 1980). The behavioral and pharmacological consequences of chronic lead exposure in mice and their similarity to those seen in hyperactive children with increased lead burden prompted relatively extended studies on the neurochemical effects of postnatal lead exposure in animals. Sauerhoff and Michaelson (1973) initially reported that neonatal lead exposure in the rat caused a 20% decrease in dopamine (DA) and no change in norepinephrine (NE) levels in whole brains as compared to controls. Later, the same group of investigators reappraised their previous findings and reported instead no change in either brain content or turnover rate (Golter and Michaelson, 1975). A number of laboratories have reported rather contradictory findings of the effect of chronic lead exposure on brain monoamine levels. Silbergeld and Goldberg (1975) have reported that levels of DA in whole mouse brains after chronic lead exposure remained unchanged. Similar findings were reported in rat brain regions, also (Sobotka and Cook, 1974). Others have found a decrease in

TABLE I

Effect of Chronic Lead Exposure on Steady-State Dopamine Level in Brain

Species	Brain region	Result	Reference
Mouse	Forebrain	No change	Silbergeld and Goldberg (1975)
Mouse	Whole brain	No change	Schumann et al. (1977)
Rat	Whole brain	No change	Golter and Michaelson (1975)
Rat	Cortex, brain stem, hypothalamus, striatum	No change	Grant et al. (1976)
Rat	Whole brain	20% decrease	Sauerhoff and Michaelson (1973)
Rat	Striatum	20% decrease	Jason and Kellogg (1977)
Rat	Midbrain	Increase	Dubas et al. (1978)
Rat	Striatum	17–30% increase	Dubas et al. (1978)
Rat	Hypothalamus	Increase	Dubas et al. (1978)

DA levels in the striata of lead-exposed rats (Jason and Kellog, 1977) or an increase in DA in different brain parts of rats, particularly in the striata (Dubas et al., 1978). Whole brain NE levels in the mouse after lead exposure have been reported to increase by 27% (Silbergeld and Goldberg, 1975) or not change (Schumann et al., 1977). Sobotka and Cook (1974) and Sobotka et al. 1975) reported tha NE levels in rat brain areas after chronic lead exposure were not affected. However, Dubas et al. (1978) reported that chronic lead exposure in rats resulted in a 60–90% increase in midbrain NE levels and a 20–30% decrease in NE levels in the striatum. Goldman et al (1980) have reported recently that chronic lead treatment of rat pups for 21 days significantly increased NE content of the hypothalamus striatum area.

The neurochemical data on the effects of chronic postnatal lead ingestion on the central monoaminergic system is controversial, and the discrepancies among the reports of different laboratories have not been fully resolved. These discrepancies in the results might be due to differences in doses of lead consumed in drinking water and food, duration of exposure, when and how the lead was administered to the animals, the nutritional constituents of the diet, and the age and sex of the animals. The composite data from different laboratories are summarized in Tables I and II.

Another interesting aspect of the effect of chronic lead ingestion on central neurotransmitters, particularly DA and NE, is that DA probably plays an important role in prolactin secretion (McLeod, 1976). Increasing evidence suggests that the prolactin-inhibiting factor in the hypothalamus may be dopamine itself (Gibbs and Neill, 1978). Govoni et al.

TABLE II

Effect of Chronic Lead Exposure on Steady-State Norepinephrine Level in Brain

Species	Brain region	Result	Reference
Mouse	Whole brain	No change	Schumann et al. (1977)
Mouse	Forebrain	27% increase	Silbergeld and Goldberg (1975)
Rat	Whole brain	No change	Sauerhoff and Michaelson (1973)
Rat	Whole brain	13% increase	Golter and Michaelson (1975)
Rat	Cortex, brain stem, hypothalamus, striatum	No change	Grant et al. (1976)
Rat	Brain stem	20% increase	Jason and Kellogg (1977)
Rat	Brain stem	27% increase	Hrdina et al. (1976)
Rat	Midbrain	60–90% increase	Dubas et al. (1978)
Rat	Striatum	20–30% decrease	Dubas et al. (1978)
Rat	Hypothalamus	20% decrease or no change depending on dose	Dubas et al. (1978)

TABLE III

Effect of Lead Acetate or Sodium Acetate in Drinking Water and Normal Calcium Diet for 14 Days on Plasma Prolactin, Calcium, and Bone Lead Content in Adult (60 Days) Male Rats[a]

Group	Plasma Ca (mg/dl)	Bone lead (μg/g dry wt)	Plasma prolactin (ng/ml)
Control, 1.75 mg Ac/ml	9.65 ± 0.12	30 ± 3	39.4 ± 0.98
3.2 mg Pb/ml	9.73 ± 0.1	744 ± 71[b]	63.4 ± 1.6[b]
3.2 mg Pb/ml + bromocryptin (20 mg/kg)	9.88 ± 0.19	800 ± 45[b]	9.6 ± 0.57[b]
Control + bromocryptin	9.78 ± 0.11	28 ± 4	7.3 ± 0.43[b]

[a] Data are means ± SE of 6 rats. Bromocryptin (CB-154; Sandoz Pharmaceuticals) was injected in 75% ethanol for the last 5 days. Prolactin potency was expressed as NIAMDD Rat Prolactin-R.P-1 reference preparation and determined by radioimmunoassay (kit kindly supplied by Dr. A. F. Parlow). (Data from S. N. Baksi and A. D. Kenny, unpublished observations.)

[b] $p < 0.001$ from control.

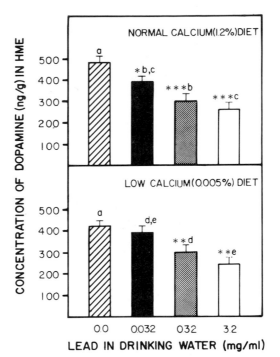

Fig. 3. Concentration of dopamine (DA) in the hypothalamus median eminence (HME) of mature female rats following 3 weeks treatment with lead acetate or sodium acetate control (0.0). The rats were given normal calcium (1.2%) or low calcium (0.005%) diets 1 week prior to lead treatment and continued on the specific diet during the treatment period. The height of each column represents mean concentration of DA, the vertical line represents ±SE of six rats. $*p < 0.05$; $**p < 0.01$; $***p < 0.001$, significantly different from control. Means bearing the same letters are significantly ($p < 0.05$) different. Low dietary calcium alone significantly ($p < 0.05$) reduced DA concentration in HME in lead-free (0.0 Pb/ml) groups. (From Baksi and Hughes, 1982. Reproduced with permission.)

(1978) reported that chronic dietary lead exposure in male rats decreases striatal DA synthesis and increases the serum prolactin level. The authors did not study the DA synthesis activity in the hypothalamus, the area that controls the secretion and release of the pituitary hormone. In a preliminary study, we investigated the effect of lead ingestion on the plasma prolactin level in mature male rats. The results (Table III) indicate that 14 days exposure to lead in drinking water significantly increases the plasma prolactin level as measured by radioimmunoassay. This increase in plasma prolactin is inhibited by bromocryptin, a dopamine agonist. A reduction in plasma prolactin after bromocryptin treatment was observed in both lead-treated and lead-free animals. Recently, we have also investigated the effect of low dietary calcium and lead exposure in adult female rats on hypothalamic and striatal DA and NE contents. The steady-state DA and NE concentrations in the hypothalamus median eminence (HME) area after 4 weeks on two different calcium diets, and the last 3 weeks treatment with lead, are summarized in Figs. 3 and 4. A significant decrease in DA and NE concentrations was seen at all three dose levels of lead treatment in the normal calcium diet group. There was a significant decrease in the DA concentration in the HME in the low dietary calcium group compared to the normal calcium group, even without lead treatment (Fig. 3, upper and lower panels). A similar decrease in NE concentration in the low calcium diet group was also observed (Fig. 4, upper and lower panels). The steady-state DA and NE concentrations in the striatum (STR) after similar treatments are shown in Fig. 5. A significant decrease in the DA concentrations was

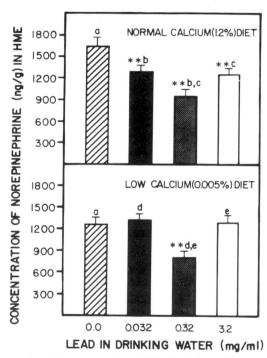

Fig. 4. Concentration of norepinephrine (NE) in the hypothalamus median eminence (HME) of mature female rats following 3 weeks treatment with lead acetate or sodium acetate control (0.0). The other conditions were similar to those in Fig. 3. Low dietary calcium alone significantly ($p < 0.05$) reduced NE concentration in HME in lead-free (0.0 mg Pb/ml) groups. (From Baksi and Hughes, 1982. Reproduced with permission.)

294

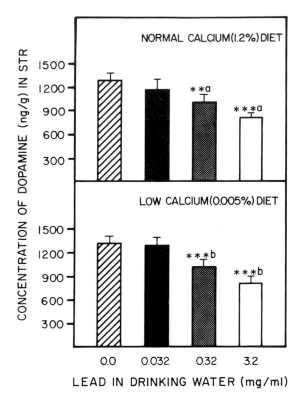

Fig. 5. Concentration of dopamine (DA) in the striatum of mature female rats following 3 weeks treatment with lead acetate or sodium acetate control (0.0). The other conditions were similar to those in Fig. 3. There was no significant difference in DA concentration in striatum (STR) due to low dietary calcium alone in lead-free (0.0 mg Pb/ml) groups. (From Baksi and Hughes, 1982. Reproduced with permission.)

seen in the two higher does of lead-treated (0.32 and 3.2 mg Pb/ml) rats and in both normal and low calcium diets. Unlike DA there was no significant different in the NE concentration in the STR after lead treatment in any group. There was no reduction in DA and NE in the STR due to low dietary calcium intake alone (Baksi and Hughes, 1981, 1982).

Compared to monoaminergic data, the effects of chronic lead exposure on the central cholinergic system are more consistent and have provided extensive information. The effects of chronic lead exposure on the central cholinergic system in rats and mice in terms of steady-state levels in brain regions are summarized in Table IV. The effects of chronic lead exposure on other neurotransmitters such as 5-HT and GABA are also summarized in Table IV.

TABLE IV

Effect of Chronic Lead Exposure on Steady-State Acetylcholine (ACh), Serotonin (5-HT), and Gammaaminobutyric Acid (GABA) Levels in Brain

Component	Species	Brain region	Results	Reference
ACh	Mouse	Forebrain	No change	Carrol et al. (1977); Silbergeld and Goldberg (1975)
ACh	Rat	Cerebellum	No change	Modak et al. (1975)
		Cortex, midbrain, hippocampus, striatum, medulla pons		
ACh	Rat	Cortex	32–48% increase	Hrdina et al. (1976)
5-HT	Rat	Whole brain	No change	Michaelson and Sauerhoff (1974)
5-HT	Rat	Cortex, brain stem, cerebellum	No change	Sobotka et al. (1975); Hrdina et al. (1976)
GABA	Rat	Whole brain	No change	Michaelson and Sauerhoff (1974)
GABA	Rat	Cortex, brain stem	No change	Piepho et al. (1976)
GABA	Rat	Cerebellum	Decrease	Piepho et al. (1976)
GABA	Rat	Cerebellum	Decrease	Silbergeld et al. (1980)

C.   Calcium–Vitamin D Endocrine System

The interaction of lead and calcium metabolism has been known for a long time. It was shown that low dietary calcium enhances lead retention (Lederer and Bing, 1940). This phenomenon was reinvestigated and the findings supplemented by works of Six and Goyer (1970) and Mahaffey *et al.* (1973). These authors have demonstrated that by lowering dietary calcium one could enhance the toxicity of lead. It has been observed in rats that lead ingestion (2 mg/kg/day for 7 days) leads to decreased *active* transport of calcium across the duodenal wall as determined by the *in vitro* everted gut sac technique (Gruden *et al.*, 1974; Gruden, 1975). On the other hand, when total calcium retention of orally administered ^{47}Ca was determined, ingestion of a high dose of lead (20 mg/kg/day for 7 days) caused increased gut absorption of calcium (Gruden and Buben, 1977). Barton *et al.* (1978a) demonstrated that both lead and calcium compete for similar binding sites on intestinal mucosal proteins, which are important in the absorptive process. These authors have also shown that lead is bound mainly to a high molecular weight mucosal protein, whereas calcium is bound primarily to a low molecular protein (calcium binding protein). Vitamin D enhances lead absorption in the distal part of the small intestine of the rat, whereas vitamin D-dependent calcium absorption occurs in the proximal part of the small intestine (Smith *et al.*, 1978). It is possible that vitamin D affects lead absorption. Barton *et al.* (1978a) have shown also that dietary calcium had no significant effect on lead absorption, but calcium-deprived rats had decreased excretion resulting in increased body retention of lead. Prolonged dietary calcium deficiency stimulates parathyroid hormone secretion, and the acute administration of parathyroid hormone in rats has been shown to increase the renal lead accumulation (Mouw *et al.*, 1978a).

The interaction of lead and vitamin D metabolism, which is intimately associated with calcium metabolism, has also been studied. Sorrel *et al.* (1977) first reported that lead-burdened children have low levels of serum 25-hydroxyvitamin D, which is an intermediate metabolite of vitamin D produced in the kidney (Fraser and Kodicek, 1970) and presumed to be the most active metabolite of vitamin D (Haussler *et al.*, 1971; Omdahl *et al.*, 1971). However, our studies in Japanese quail indicated that lead ingestion for 15 days enhances 25-hydroxyl-1-hydroxylase activity in the kidney when dietary calcium intake is low or normal (Baksi and Kenny, 1978a, 1979). Data are summarized in Figs. 6 and 7. We concluded that, probably due to decreased calcium absorption in the presence of lead, kidney enzyme activity was increased. Low dietary

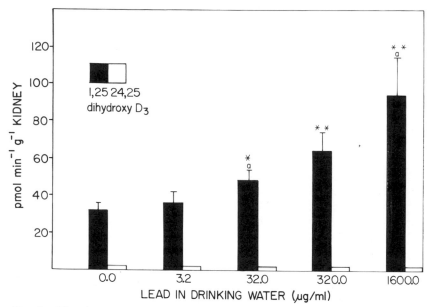

Fig. 6. Vitamin D_3 metabolites (1,25-dihydroxyvitamin D_3 and 24,25-dihydroxyvitamin D_3 production *in vitro* by kidney homogenates prepared from 6-week-old female Japanese female quail placed on normal calcium (2.3–3.3%) diet for 15 days and given drinking water containing lead acetate (3.2, 32.0, 320.0, or 1600.0 (μg Pb/ml) or sodium acetate 1750 μg acetate/ml as control (0.0 mg Pb/ml). Means and SE (vertical brackets) of four birds are given in each group. Significance from the respective control groups is indicated as follows: $^*p < 0.05$, $^{**}p < 0.01$. Means bearing the same superscripts are significantly ($p < 0.05$) different. (From Baksi and Kenny, 1979. Reproduced with permission.)

calcium intake is a potent stimulator for increased 25-hydroxyvitamin D-1-hydroxylase activity in Japanese quail (Baksi and Kenny, 1978b). A careful look at the data of Rosen *et al.* (1980) indicates that lead-burdened children had high plasma parathyroid hormone as well as low serum ionized calcium levels, indicative of calcium deficiency. Since low serum ionized calcium and high serum parathyroid hormone levels have stimulatory effects of the 25-hydroxyvitamin D-1-hydroxylase activity in the kidney, suggestion that lead ions impair renal biosynthesis of 1,25-dihydroxyvitamin D is certainly questionable. We have reported that lead added *in vitro* has no significant effect of the vitamin D hydroxylase in the kidney homogenate (Baksi and Kenny, 1979). Recently, Smith *et al.* (1981) demonstrated that lead ingestion in rats reduced the intestinal calcium transport response. Toraason *et al.* (1981) also showed that maternal lead exposure inhibits intestinal calcium absorption in rat pups.

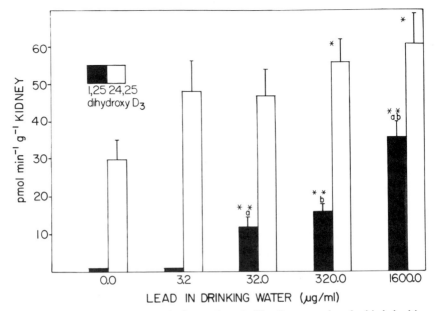

Fig. 7. The conditions were similar to those in Fig. 6, except that the birds had been placed on a low calcium (0.2%) diet for the last 15 days. (From Baksi and Kenny, 1979. Reproduced with permission.)

Acute intravenous administration of lead acetate results in transient hypercalcemia and hyperphosphatemia in rats (Kato *et al.*, 1977). The actual mechanism of induction of hypercalcemia and hyperphosphatemia is a subject of controversy. Whereas Kato *et al.* (1977) suggested that the mechanism results from a direct action of lead on bone mineral, Talmage *et al.* (1978) concluded that hypercalcemia and hyperphosphatemia are caused by the direct interaction of lead with calcium and phosphate in solution in plasma. Peng *et al.* (1979) studied the effect of acute lead-induced increase in serum calcium in the rat and reported no increase in calcitonin secretion as well as no increase in the plasma ionized calcium level.

D. Renal System

The effects of lead on the kidney are manifested in two ways in both adults and children: namely, reduced glomerular function and proximal tubular damage.

In mammals, lead concentrates in the kidney, causing hyper-aminoaciduria, glycosuria, and hyperphosphaturia—all reflecting decreases in the renal, tubular reabsorptive process (Goyer, 1968). Because lead poisoning impairs mitochondrial functions (Alvares, 1978), it is possible that a deficiency of ATP is responsible for the reduction of tubular functions. Lead poisoning in adults is sometimes associated with gout, hyperuricemia, and decreased renal urate clearance. The findings are indicative of either increased tubular reabsorption or decreased tubular secretion of urate. Cases of gout and associated effects occur in conjunction with other renal effects involving a gross reduction in glomerular filtration and progressive renal failure (Lilis et al., 1968; Wedeen et al., 1975). Recently, Batuman et al. (1981) demonstrated that hyperuricemia, renal insufficiency, and nephropathy probably resulted from lead poisoning in gout patients.

Acute administration of lead in dogs increases the urinary excretion of sodium, potassium, calcium, and water, despite a constant glomerular filtration rate. Similar changes, as well as an increase in plasma renin activity, are seen in rats (Mouw et al., 1978b). The level of lead exposure that causes chronic lead nephropathy with associated glomerular and vascular changes in man is not known. In rats, however, Fowler et al. (1980) observed that chronic lead exposure at 5 ppm for 9 months in males and 25 ppm in females produced nuclear inclusion bodies and increased numbers of iron-positive granules within renal proximal tubular cells. Chronic exposure to lead results in the appearance of intranuclear inclusion bodies in renal proximal tubular cells (Goyer, 1971). Chang et al. (1980) have shown that the organolead compound, tetraethyl lead, a gasoline additive, is just as nephrotoxic as inorganic lead compounds. The pathological changes are most prominent in the proximal tubules, producing various morphological and degenerative changes in the tubular epithelial cells and their organelles.

Hong et al. (1980) studied renal function in men occupationally exposed to lead and found that the glucose absorptive capacity was slightly decreased in workers exposed to lead. The glomerular filtration rate was also decreased in four of six patients. The reduced glucose reabsorptive capacity was disproportionately greater than one would expect from the reduced glomerular filtration rate alone.

E. Cardiovascular System

Relationships between cardiovascular diseases and elevated blood lead concentration have been suggested (Beevers et al., 1976). Electrocardiographic abnormalities observed in lead-treated animals and in lead-poisoned humans have been attributed to a disturbance in the function of the autonomic nervous system (Hejtmancik and Williams, 1979;

Webb *et al.*, 1981). It has been postulated that lead plays a role in human hypertension also (Beevers *et al.*, 1976; Morgan *et al.*, 1966). Mouw *et al* (1978b) demonstrated that acute intravenous administration of lead in anesthetized dogs causes an increase in plasma renin activity. In a later study, Goldman *et al.* (1981) demonstrated that lead may increase renin secretion in animals otherwise unstimulated to secrete, but the major mechanism for the short-term rise in plasma renin activity other than lead is the elimination of hepatic removal of renin. Lead also prevents angiotensin II from rising proportionately with plasma renin activity, presumably by inhibiting angiotensin-converting enzyme. From the Goldman *et al.* (1981) studies, it is apparent that lead-induced hypertension may not be related to the renin–angiotensin system.

Williams *et al.* (1977) have demonstrated that rat pups exposed to lead during the neonatal period and subsequently raised on a lead-free diet for 4 months show significantly more cardiac arrythmias in response to norepinephrine. In a later study, Hejtmancik and Williams (1979) showed that, whereas bilateral vagotomy or atropine pretreatment decreased the frequency of cardiac arrythmias, norepinephrine still caused significantly more extrasystoles in lead-exposed than in control rats. Isolated perfused hearts from lead-exposed animals exhibited more irregularities in rhythm after norepinephrine than hearts from control animals, indicating that some direct cardiac effects are also involved.

Kopp *et al.* (1980) demonstrated significant metabolic changes in cardiac function after chronic low-level lead feeding in rats.

F. Immune System

Another interesting aspect of lead toxicity is manifested by the immune system of the host. Prolonged exposure to lead has been shown to suppress the immune system and increase susceptibility to infection which, in a normal condition, the host is able to withstand (Hemphill *et al.*, 1971). Selye *et al.* (1966) demonstrated that a single normally well-tolerated intravenous injection of lead acetate increases the sensitivity of the rat to the endotoxins of various gram-negative bacteria about 100,000 times above normal. Cook *et al.* (1975) also reported that intravenous administration of an acute dose of lead acetate enhanced the susceptibility of rats to challenge with *E. coli* approximately 1000-fold. These authors also observed equivalent vulnerability of lead-treated rats to *E. coli* that had been killed. They concluded that the observed toxicity is probably due to the endotoxin content of the bacteria. An equal dose of gram-positive bacteria *Staphylococci epidermidis* failed to kill lead-intoxicated rats. Trejo *et al.* (1972) demonstrated that, whereas lead acetate induces endotoxin hyperreactivity by impairing the phagocytic as well as endotoxin-detoxifying properties of the macrophages, hepatic parenchymal cell dysfunction may also be a contributing factor. Decreased antibody formation after prolonged lead exposure in mice (Koller and Kovacic, 1974; Koller et al., 1976) and rabbit (Koller, 1973) have been reported. Gainer (1973) induced splenomegaly in lead-treated male mice after injection of Rauscher leukemia virus. Enlarged spleens contained high titers of virus. Untreated virus control mice resisted injection in that they did not develop splenomegaly, and virus was not recovered from their spleen. Gainer (1974) also reported that lead aggravates viral disease, most likely in part through reduced interferon synthesis. However, lead does not inhibit interferon action. Vergic and Mare (1974) studied the effect of lead poisoning in chickens on interferon and antibody production and reported that subclinical lead doses did not affect interferon induction in response to Statolon and New Castle Disease Virus-B$_1$. But interferon concentrations and duration in serum was markedly decreased in chickens that received 320 mg/kg lead orally for 35 days in aqueous solution. However, long-term lead exposure had no marked effect on antibody production to New Castle Disease virus in chickens.

G. Other

Lead exposure in humans has been suspected to affect spermatogenesis (Lancranjan et al., 1975). Lead salts are known to cross the placental barrier (Carpenter, 1974). Developmental malformations resulting from lead administration in animals have been reported (Ferm and Carpenter, 1967; Hilderbrand et al., 1973). Lead-exposed women residing in the Missouri "lead belt" have been reported to have increased incidence of premature deliveries and premature rupture of the membranes as compared to a control group (Fahim et al., 1975).

V. MECHANISM OF ACTION OF LEAD TOXICITY

The mechanism of lead on neurotoxic effects may be described at the molecular and membrane level. Lead binds with the sulfhydryl group or competitively replaced divalent ions such as calcium. Kinetics of dopamine release and reuptake are also affected by lead; lead in the presence of calcium leads to an increased release and decreased reuptake of dopamine in synaptosome preparations from the neostriatum (Silbergeld and Adler, 1978). However, these observations were not corroborated in a recent study using *in vitro* and *in vivo* lead exposure and kinetic analysis (Ramsay *et al.*, 1980). A study by Nathanson and Bloom (1975) reported that lead inhibits adenylate cyclase activity. The implications of this finding are important because the second messenger system of the molecular mechanism of hormone action depends on adenylate cyclase and its substrate, cyclic adenosine 3′,5′-monophosphate. Lead impairs the mixed-function oxidase system of the liver endoplasmic reticulum in rats and, thus, could affect the organ's ability to detoxify drugs and other foreign substances (Alvares *et al.*, 1972). Lead inhibition of the enzyme aminolevulinic acid dehydratase causes increases activity of the GABA system (Muller and Snyder, 1977). Silbergeld and Adler (1978) suggested that many of the behavioral alterations of low-level lead toxicity may be related to the effect of lead on calcium in the synaptic complex with the resulting inhibition of acetylcholine release and enhancement of dopamine and norepinephrine release. The mechanism of lead encephalopathy may be due to either a direct effect on neurons resulting in neural dysfunction or neural dysfunction resulting from lead-induced changes in brain barrier function.

Niklowitz (1977) proposed an interesting multidisciplinary model for lead toxicity. This model points to the basic subcellular mechanism and primary cellular site of lead toxicity. In particular, the model demonstrates a unique shift of essential trace metals in brain tissue after lead exposure. The trace metal shift inhibits cell membrane ATPase, resulting in the breakdown of cell membrane properties (such as the Na^+/K^+ pump) and leads to intracellular and functional impairments. This model of lead toxicity explains the breakdown of the blood–brain barrier and the neurological sequelae of lead exposure.

The mechanism of action of lead toxicity on the hematopoietic system, particularly on the inhibitory effect on the enzyme aminolevulinic acid dehydratase, involves direct lead competition with zinc for the binding to the sulfhydryl group in close relation to the active site of the enzyme.

On the calcium-vitamin D endocrine system, lead probably acts indirectly by inhibitng calcium absorption from gut. It is possible, however, that lead directly inhibits the kidney enzyme, 1α-vitamin D hydroxylase, which produces the active metabolite, 1,25-dihydroxy-vitamin D.

VI. SUMMARY AND CONCLUSIONS

The harmful effects of lead in living organisms are demonstrated in several organ systems. The most prominent effects are on the hematopoietic system, where lead is known to interfere with heme synthesis by inhibiting the enzyme aminolevulinic acid dehydratase (ALAD), which mediates the condensation of aminolevulinic acids to form porphobilinogen, an intermediate product of heme synthesis. Lead is also known to inhibit the enzyme ferrochelatase (heme synthetase), which is involved in the final step of heme synthesis. Lead also causes increased fragility of the red blood cells and an increased rate of destruction.

The effect of lead on the nervous system is also very prominent, although animal experimentation data on neurotransmitter metabolism (particularly the monoaminergic system) are inconsistent and often contradictory. Lead probably decreases dopamine and norepinephrine steady-state concentrations in the hypothalamus and striatum of the rodent. On the other hand, the effects on the cholinergic system are more reproducible and consistent. There is evidence also that lead toxicity results in behavioral disorders in animal models, but its extrapolation to behavioral disorders in children is not well established. Recent findings on lead poisoning and behavioral problems in children indicate that more studies are needed in this area.

The effect of lead on the calcium–vitamin D endocrine system has also been reported. Dietary calcium and lead toxicity have a close relationship. Low dietary calcium enhances the body burden of lead. The effect of ingested lead on vitamin D hydroxylases may be an indirect one, acting through decreased calcium absorption from the gut.

Renal effects of lead are manifested in two ways in both adults and children; namely, reduced glomerular function and proximal tubular damage. Some gout may be associated with lead poisoning in man.

The cardiovascular effects of lead poisoning may be associated with hypertension and some form of disturbed cardiac function. Lead-induced hypertension may not be related to the renin–angiotensin system.

Prolonged exposure to lead suppresses the immune system and increases the susceptibility to infection in animals. Lead induces endotoxin hyperactivity and reduces antibody formation and interferon synthesis.

ACKNOWLEDGMENTS

The writing of this chapter and some of the data reported herein were partially supported by NIH Grant HL 16240, awarded to Maysie J. Hughes, Ph.D.

REFERENCES

Allen, J. R., McWey, P. J., and Suomi, S. J. (1974). Pathological and behavioral effects of lead intoxication in the infant rhesus monkey. *Environ. Health Perspect.* **7,** 239–246.
Alvares, A. P. (1978). Interactions between environmental chemicals and drug biotransformation in man. *Clin. Pharmacokinet.* **3,** 462–477.

Alvares, A. P., Cohn, S., and Kappas, A. (1972). Lead and methyl mercury: effects of acute exposure on cytochrome P-450 and mixed function oxidase system in the liver. *J. Exp. Med.* **135**, 1406–1409.

Anderson, C., and Danychuk, K. D. (1977). The effect of chronic low level lead intoxication on the haversian remodeling system in dogs. *Lab. Invest.* **37**, 466–469.

Aranyi, C., Andres, S., Ehrlich, R., Fenters, J. D., Gardner, D. E., and Waters, M. D. (1977). Cytotoxicity to alveolar macrophages of metal oxides absorbed on fly ash. *In* "Pulmonary Macrophage and Epithelial Cells." (C. L. Sanders, R. P. Schneider, G. E. Dagle, and H. A. Ragan, eds.), pp. 58–65. Technical Information Center, Energy Research and Development Administration, Washington, D.C.

Baksi, S. N., and Hughes, M. J. (1981). Regional alterations of brain catecholamines by lead ingestion in rats: influence of dietary calcium. *Fed. Proc., Fed. Am. Soc. Exp. Biol.* **40**, 258 (Abstr.).

Baksi, S. N., and Hughes, M. J. (1982). Regional alterations of brain catecholamines by lead ingestion in adult rats: influence of dietary calcium. *Arch. Toxicol.* **50**, 11–18.

Baksi, S. N., and Kenny, A. D. (1978a). Effect of lead ingestion on vitamin D_3 metabolism in Japanese quail. *Res. Commun. Chem. Pathol. Pharmacol.* **21**, 375–378.

Baksi, S. N., and Kenny, A. D. (1978b). Vitamin D metabolism in Japanese quail: gonadal hormones and dietary calcium effects. *Am. J. Physiol.* **234**, E622–E628.

Baksi, S. N., and Kenny, A. D. (1979). Vitamin D metabolism in Japanese quail: effects of lead exposure and dietary calcium. *Toxicol. Appl. Pharmacol.* **51**, 489–495.

Barry, P. S. I., and Mossman, D. B. (1970). Lead concentrations of human tissues. *Br. J. Ind. Med.* **27**, 339–351.

Barton, J. C., Conrad, M. E., Nuby, S., and Harrison, L. (1978a). Effect of iron on the absorption and retention of lead. *J. Lab. Clin. Med.* **92**, 536–547.

Barton, J. C., Conrad, M. E., Harrison, L., and Nuby, S. (1978b). Effects of calcium on the absorption and retention of lead. *J. Lab. Clin. Med.* **91**, 366–376.

Barton, J. C., Conrad, M. E., Harrison, L., and Nuby, S. (1980). Effects of vitamin D on the absorption and retention of lead. *Am. J. Physiol.* **238**, G124–G130.

Batuman, V., Maesaka, J. K., Haddad, B., Tepper, E., Landy, E., and Wedeen, R. P. (1981). The role of lead in gout nephropathy. *New Eng. J. Med.* **304**, 520–523.

Beevers, D. G., Erskine, E., Robertson, M., Beattie, A. D., Goldbert, A., Campbell, B. C., Moore, M. R., and Hawthorne, V. M. (1976). Blood lead and hypertension. *Lancet* **2**, 1–3.

Behse, F., and Carlsen, F. (1978). Histology and ultrastructure of alterations in lead neuropathy. *Muscle and Nerve* **1**, 368–378.

Blunt, J. W., Tanaka, Y., and DeLuca, H. F. (1968). The biological activity of 25-hydroxychole-calciferol, a metabolite of vitamin D_3. *Proc. Natl. Acad. Sci. U.S.A.* **61**, 1503–1506.

Brown, D. R. (1975). Neonatal lead exposure in the rat: decreased learning as a function of age and blood lead concentrations. *Toxicol. Appl. Pharmacol.* **32**, 628–637.

Bushnell, P. J., and DeLuca, H. F. (1980). Lactose facilitates the intestinal absorption of lead in weanling rats. *Science* **211**, 61–63.

Campbell, A. M. G., Williams, E. R., and Barltrop, D. (1970). Motor neurone disease and exposure to lead. *J. Neurol. Neurosurg. Psychiat.* **33**, 877–885.

Cantarow, A., and Trumpa, M. (1944). "Lead Poisoning," p. 264. Williams & Wilkins, Baltimore, Maryland.

Carpenter, S. J. (1974). Placental permeability of lead. *Environ. Health Perspect.* **7**, 129–131.

Carrol, P. T., Silbergeld, E. K., and Goldberg, A. M. (1977). Alteration of central cholinergic function by chronic lead acetate exposure. *Biochem. Pharmacol.* **26**, 397–402.

Chang, L. W., Wade, P. R., Reuhl, K. R., and Olson, M. J. (1980). Ultrastructural changes in renal proximal tubules after tetraethyllead intoxication. *Environ. Res.* **23,** 208–223.

Conrad, M. E., and Barton, J. C. (1978). Factors affecting the absorption and excretion of lead in the rat. *Gastroenterology* **74,** 731–740.

Cook, J. A., Hoffmann, E. O., and DiLuzio, N. R. (1975). Influence of lead and cadmium on the susceptibility of rats to bacterial challenge. *Proc. Soc. Exp. Biol. Med.* **150,** 741–747.

David, O. J., Clark, J., and Voeller, K. (1972). Lead and hyperactivity. *Lancet* **1,** 900–903.

David, O. J., Hoffman, S. P., Sverd, J., Clark, J., and Voeller, K. (1976). Lead and hyperactivity. Behavioral response to chelation: a pilot study. *Am. J. Psychiat.* **133,** 1155–1158.

Dubas, T. C., Stevenson, A., Singhal, R. L., and Hrdina, P. D. (1978). Regional alterations of brain biogenic amines in young rats following chronic lead exposure. *Toxicology* **9,** 185–190.

Fahim, M. S., Fahim, Z., and Hall, D. G. (1976). Effects of subtoxic lead levels on pregnant women in the State of Missouri. *Res. Comm. Chem. Pathol. Pharmacol.* **13,** 309–331.

Ferm, V. H., and Carpenter, S. J. (1967). Developmental malformations resulting from the administration of lead salt. *Exp. Mol. Pathol.* **7,** 208–213.

Ferm, V. H., and Ferm, D. W. (1971). The specificity of the teratogenic effects of lead in the golden hamster. *Life Sci.* **10,** 35–39.

Fowler, B. A., Kimmel, C. A., Woods, J. S., McConnell, E. E., and Grant, L. D. (1980). Chronic low-level lead toxicity in the rat. III. An integrated assessment of long-term toxicity with special reference to the kidney. *Toxicol. Appl. Pharmacol.* **56,** 59–77.

Fraser, D. R., and Kodicek, E. (1970). Unique biosynthesis by kidney of a biologically active vitamin D metabolite. *Nature (London)* **228,** 764–766.

Fullerton, P. M. (1966). Chronic peripheral neuropathy produced by lead poisoning in guinea pigs. *J. Neuropath. Exp. Neurol.* **25,** 214–236.

Gainer, J. H. (1973). Activation of the Rauscher Leukemia virus by metals. *J. Natl. Cancer Inst.* **51,** 609–613.

Gainer, J. H. (1974). Lead aggravates viral disease and represses the anti-viral activity of Interferon inducers. *Environ. Health Perspect.* **7,** 113–119.

Gibbs, D. M., and Neill, J. D. (1978). Dopamine levels of hypophysial stalk blood in the rat are sufficient to inhibit prolactin secretion *in vivo. Endocrinology* **102,** 1895–1900.

Goldman, J. M., Vander, A. J., Mouw, D. R., Keiser, J., and Nicholls, M. G. (1981). Multiple short-term effects of lead on the renin–angiotensin system. *J. Lab. Clin. Med.* **97,** 251–263.

Goldman, D., Hejtmancik, M. R. Jr., Williams, B. J., and Ziegler, M. G. (1980). Altered noradrenergic systems in the lead-exposed neonatal rat. *Neurobehav. Toxicol.* **2,** 337–343.

Goldwater, L. J., and Hoover, A. W. (1967). An international study of normal levels of lead in blood and urine. *Arch. Environ. Health* **15,** 132–134.

Golter, M., and Michaelson, I. W. (1975). Growth, behavior and brain catecholamines in lead-exposed neonatal rats: a reappraisal. *Science* **187,** 359–361.

Govoni, S., Montefusco, O., Spano, P. F., and Trabucchi, M. (1978). Effect of chronic lead treatment on brain dopamine synthesis and serum prolactin release in the rat. *Toxicol. Lett.* **2,** 333–337.

Goyer, R. A. (1968). The renal tubule in lead poisoning. I. Mitochondrial swelling and aminoaciduria. *Lab. Invest.* **19,** 71–77.

Goyer, R. A. (1971). Lead and the kidney. *Curr. Top. Pathol.* **55,** 147–176.

Goyer, R. A., and Mahaffey, K. R. (1972). Susceptibility to lead toxicity. *Environ. Health Perspect.* **2,** 73–80.

Grandjean, P. (1978). Widening perspective of lead toxicity. A review of health effects of lead exposure in adults. *Environ. Res.* **17,** 303–321.

Grant, L. D., Breese, G., Howard, J. L., Krigman, M. R., and Mushak, P. (1976). Neurobiology of lead-intoxication in the developing rat. *Fed. Proc., Fed. Am. Soc. Exp. Biol.* **35,** 503 (Abstr.).

Gruder, N. (1975). Lead and active calcium transfer through the intestinal wall in rats. *Toxicology* **5,** 163–166.

Gruder, N., and Buben, M. (1977). Influence of lead on calcium metabolism. *Bull. Environ. Contam. Toxicol.* **18,** 303–307.

Gruder, N., Stantic, M., and Buben, M. (1974). Influence of lead on calcium and strontium transfer through the duodenal wall in rats. *Environ. Res.* **8,** 203–206.

Hammond, P. B. (1969). Lead poisoning: an old problem with a new dimension. *In* "Essays in Toxicology" (F. R. Blood, ed.), Vol. 1, pp. 115–155. Academic Press, New York.

Hammond, P. B. (1977). Exposure of humans to lead. *Ann. Rev. Pharmacol. Toxicol.* **17,** 197–214.

Hart, M. H., and Smith, J. L. (1981). Effect of vitamin D and low dietary calcium on lead uptake and retention in rats. *J. Nutr.* **111,** 694–698.

Haussler, M. R., Boyce, B. W., Littledike, E. T., and Rasmussen, H. (1971). A rapidly acting metabolite of vitamin D_3. *Proc. Natl. Acad. Sci. U.S.A.* **68,** 177–181.

Hejtmancik, M. R., Jr., and Williams, B. J. (1979). Effect of chronic lead exposure on the direct and indirect components of the cardiac response to norepinephrine. *Toxicol. Appl. Pharmacol.* **51,** 239–245.

Hemphil, F. E., Kaeberle, M. L., and Buck, W. B. (1971). Lead suppression of mouse resistance to salmonella typhrmurium. *Science* **172,** 1031–1032.

Hernberg, S. (1976). Biochemical, subclinical and clinical responses to lead and their relation to different exposure levels, as indicated by the concentration of lead in blood. *In* "Effects of Dose-Response Relationships of Toxic Metals" (G. F. Nordberg, ed.), pp. 404–415. Elsevier, Amsterdam.

Hilderbrand, D., Der, R., Griffin, W., and Fahim, M. (1973). Effect of lead acetate on reproduction. *Am. J. Obstet. Gynecol.* **115,** 1058–1065.

Hong, C. D., Hancnson, I. B., Lerner, S., Hammond, P. B., Pesce, A. J., and Pollak, V. E. (1980). Occupational exposure to lead: effects on renal function. *Kidney Int.* **18,** 489–494.

Hrdina, P. D., Peters, D. A. V., and Singhal, R. L. (1976). Effects of chronic exposure to cadmium, lead and mercury on brain biogenic amines in the rat. *Res. Comm. Chem. Pathol. Pharmacol.* **15,** 483–493.

Hyslop, G. H., and Kraus, W. M. (1923). The pathology of motor paralysis by lead. *Arch. Neurol. Psychiat.* **10,** 444–455.

Jason, K., and Kellog, C. K. (1977). Lead effects on behavioral and neurochemical development in rats. *Fed. Proc., Fed. Am. Soc. Exp. Biol.* **36,** 1008 (Abstr.).

Kato, Y., Shoichiro, T., and Ogura, H. (1977). Mechanism of induction of hypercalcemia and hyperphosphatemia by lead acetate in the rat. *Calcif. Tissue Int.* **24,** 41–46.

Keller, C. A., and Doherty, R. A. (1980). Distribution and excretion of lead in young and adult female mice. *Environ. Res.* **21,** 217–228.

Klaassen, C. D., and Shoeman, D. W. (1974). Biliary excretion of lead in rats, rabbits and dogs. *Toxicol. Appl. Pharmacol.* **29,** 434–446.

Koller, L. D. (1973). Immunosuppression produced by lead, cadmium and mercury. *Am. J. Vet. Res.* **34,** 1457–1458.

Koller, L. D., and Kovacic, S. (1974). Decreased antibody formation in mice exposed to lead. *Nature (London)* **250,** 148–150.

Koller, L. D., Exon, J. H., and Roan, J. G. (1976). Humoral antibody responses in mice after single dose exposure to lead or cadmium. *Proc. Soc. Exp. Biol. Med.* **151,** 339–342.

Kopp, S. J., Glonek, T., Erlanger, M., Perry, E. F., Barany, M., and Perry, H. M. (1980). Altered metabolism and function of rat heart folowing chronic low level cadmium/ lead feeding. *J. Mol. Cell. Cardiol.* **12,** 1407–1425.

Krigman, M. R., Bouldin, T. W., and Mushak, P. (1980). Lead. *In* "Experimental and Clinical Neurotoxicology" (P. S. Spencer, ed.), pp. 490–507. Williams & Wilkins, Baltimore, Maryland.

Lamola, A. A., Piomelli, S., Poh-Fitzpatrick, M. B., Yamane, T., and Harber, L. C. (1975). Erythropoietic protoporphyria and lead intoxication, the molecular basis for difference in cutaneous photosensitivity. II. Different binding of erythrocyte protoporphyrin to hemoglobin. *J. Clin. Invest.* **56,** 1528–1535.

Lancranjan, I., Papesca, H., Gonanescu, O., Kepscu, I., and Serbanescu, M. (1975). Reproductive ability of workmen occupationally exposed to lead. *Arch. Environ. Health* **30,** 396–401.

La Porte, R., and Talbott, E. E. (1978). Effects of low levels of lead exposure on cognitive function—a review. *Arch. Environ. Health* **30,** 236–239.

Lederer, L. B., and Bing, F. C. (1940). Effect of calcium and phosphorus on retention of lead by growing organism. *J. Am. Med. Assoc.* **114,** 2457–2461.

Lilis, R., Gavrilescu, N., Nestorescu, B., Dimitriu, C., and Roventa, A. (1968). Nephropathy in chronic lead poisoning. *Br. J. Ind. Med.* **25,** 196–202.

Lilis, R., Fischbein, A., Eisinger, J., Blumberg, W. E., Diamond, S., Anderson, H. A., Rom, W., Rice, C., Sarkozi, L., Kon, S., and Selikoff, I. J. (1977). Prevalence of lead disease among secondary lead smelter workers and biological indicators of lead exposure. *Environ. Res.* **14,** 225–285.

Lin-Fu, J. S. (1972). Undue absorption of lead among children, a new look at an old problem. *New Eng. J. Med.* **286,** 702–710.

McLeod, R. M. (1976). Regulation of prolactin secretion. *In* "Frontiers in Neuroendocrinology" (L. Martini, and W. F. Ganong, eds.), Vol. 4, pp. 169–194. Raven, New York.

Maines, M. D., and Kappas, A. (1977). Metals as regulators of heme metabolism. *Science* **198,** 1215–1221.

Mahaffey, K. R. (1974). Nutritional factors and susceptibility to lead toxicity. *Environ. Health Prespect.* **7,** 107–112.

Mahaffey, K. R., and Michaelson, I. A. (1980). The interaction between lead and nutrition. *In* "Low Level Lead Exposure: The Clinical Implications and Current Research" (H. L. Needleman, ed.), pp. 159–200. Raven New York.

Mahaffey, K. R., Goyer, R. A., and Haseman, J. (1973). Dose-response to lead ingestion in rats on low dietary calcium. *J. Lab Clin. Med.* **82,** 92–100.

Manalis, R. S., and Cooper, G. P. (1973). Presynaptic and postsynaptic effects of lead at the frog neuromuscular function. *Nature (London)* **243,** 354–356.

Memo, M. Lucchi, L., Spano, P. F., and Trabucchi, M. (1980). Lack of correlation between the neurochemical and behavioral effects induced by d-amphetamine in chronically lead-treated rats. *Neuropharmacology* **19,** 795–799.

Michaelson, I. A., and Sauerhoff, M. W. (1974). An improved model of lead-induced brain dysfunction in the suckling rat. *Toxicol. Appl. Pharmacol.* **28,** 88–96.

Modak, A. T., Weintraub, S. T., and Stavinoha, W. B. (1975). Effect of chronic ingestion of lead on the central cholinergic system in rat brain regions. *Toxicol. Appl. Pharmacol.* **34,** 340–347.

Momcilovic, B. (1978). The effect of maternal dose on lead retention in suckling rats. *Arch. Environ. Health.* **33,** 115–117.

Morgan, J. M., Hartley, M. W., and Miller, R. E. (1966). Neuropathy in chronic lead poisoning. *Arch. Intern. Med.* **118,** 17–29.

Moore, M. R., Meredith, P. A., and Goldberg, A. (1980). Lead and heme biosynthesis. *In* "Lead Toxicity" (R. L. Singhal, and J. A. Thomas, eds.), pp. 79–117. Urban & Schwarzenberg, Munich.

Mouw, D. R., Wagner, J. G., Kalitis, K., Vander, A. J., and Mayor, G. H. (1978a). The effect of parathyroid hormone on the renal accumulation of lead. *Environ. Res.* **15,** 20–27.

Mouw, D. R., Vander, A. J., Cox, J., and Fleischer, N. (1978b). Acute effects of lead on renal electrolyte excretion and plasma renin activity. *Toxicol. Appl. Pharmacol.* **46,** 435–447.

Muller, W. E., and Snyder, S. H. (1977). Delta aminolevulinic acid: influences on synaptic GABA receptor binding may explain CNS symptoms of porphyria. *Ann. Neurology* **2,** 340–342.

Mykkanen, H. M., and Wasserman, R. H. (1981). Gastrointestinal absorption of lead (^{203}Pb) in chicks: influence of lead, calcium and age. *J. Nutr.* **111,** 1757–1765.

Nathanson, J., and Bloom, F. E. (1975). Lead-induced inhibition of brain adenyl cyclase. *Nature (London)* **255,** 419–420.

Needleman, H. L. (1980). Lead and neuropsychological deficit: finding a threshold. *In* "Low Level Lead Exposure: The Clinical Implications of Current Research" (H. L. Needleman, ed.), pp. 43–51. Raven, New York.

Niklowitz, W. J. (1977). Subcellular mechanisms in lead toxicity: significance in childhood encephalopathy, neurological sequelae and late dimentias. *In* "Neurotoxicology" (L. Roizin, H. Shiraki, and H. Grcevic, eds.), Vol. 1, pp. 289–298. Raven, New York.

Omdahl, J., Holick, M., Suda, T., Tanaka, Y., and DeLuca, H. F. (1971). Biological activity of 1,25-dihydroxy-cholecalciferol. *Biochemistry* **10,** 2935–2940.

Patterson, C. C. (1965). Contaminated and natural lead environments of man. *Arch. Environ. Health* **11,** 344–360.

Peng, T. C., Gitelman, H. J., and Garner, S. C. (1979). Acute lead-induced increase in serum calcium in the rat without increased secretion of calcitonin. *Proc. Soc. Exp. Biol. Med.* **160,** 114–117.

Pentschew, A., and Garrow, F. (1966). Lead encephalomyelopathy of the suckling rats and its implication on the porphyrinopathic nervous disease. *Acta Neuropathol.* **6,** 266–278.

Piepho, R. W., Ryan, C. F., and Lacz, J. P. (1976). The effects of chronic lead intoxication and the GABA content of the rat CNS. *Pharmacologist* **18,** 125 (Abstr.).

Piomelli, S., Lamola, A. A., Poh-Fitzpatrick, M. B., Seaman, C., and Harber, L. (1975). Erythropoietic protoporphyria and Pb intoxication: the molecular basis for difference in cutaneous photosensitivity. I. Different rates of diffusion of protoporphyrin from the erythrocytes, both *in vivo* and *in vitro. J. Clin. Invest.* **56,** 1519–1527.

Piomelli, S., Corash, L., Corash, M. B., Seaman, C., Mushak, P., Glover, B., and Padgett, R. (1980). Blood lead concentrations in a remote Himalayan population. *Science* **210,** 1135–1136.

Quarterman, J., and Morrison, E. (1978). The effect of age on the absorption and excretion of lead. *Environ. Res.* **17,** 78–83.

Rabinowitz, M. B., Wetherill, G. W., and Kopple, J. D. (1976). Kinetic analysis of lead metabolism in healthy humans. *J. Clin. Invest.* **58,** 260–270.

Ramsay, P. B., Krigman, M. R., and Morell, P. (1980). Developmental studies of the uptake of choline, GABA and dopamine by crude synaptosomal preparations after *in vivo* or *in vitro* lead treatment. *Brain Res.* **187**, 383–402.

Rastogi, S. C., and Clausen, J. (1976). Absorption of lead through the skin. *Toxicology* **6**, 371–376.

Richter, G. W. (1976). Evolution of cytoplasmic fibrillar bodies induced by lead in rat and mouse kidneys: relation to clusters of ferritin. *Am. J. Pathol.* **83**, 135–148.

Roels, H., Buchel, J. P., Lauwerys, R., Hubermont, G., Bruaux, P., Claeys-Thoreau, F., LaFontaine, A., and Vanoverschelde, J. (1976). Impact of air pollution by lead on the heme biosynthetic pathway in school-age children. *Arch. Environ. Health* **31**, 310–316.

Rosen, J. F., Chesney, R. W., Hamstra, A., DeLuca, H. F., and Mahaffey, K. R. (1980). Reduction in 1,25-dihydroxyvitamin D in children with increased lead absorption. *New Eng. J. Med.* **302**, 1128–1131.

Sasa, A. (1978). Toxic effects of lead, with particular reference to porphyrin and heme metabolism. *Handb. Exp. Pharmacol.* **44**, 333–371.

Sasa, S., Granick, J. L., Granick, S., Kappas, A., and Levere, R. (1973). Studies in lead poisoning. I. Microanalysis of erythrocyte protoporpyrin levels by spectrofluorometry in the detection of chronic lead intoxication in the subclinical range. *Biochem. Med.* **8**, 135–148.

Sauerhoff, M. W., and Michaelson, I. A. (1973). Hyperactivity and brain catecholamines in lead-exposed developing rats. *Science* **182**, 1022–1024.

Schumann, A. M., Dewey, W. L., Borzelle, J. F., and Alphin, R. S. (1977). Effects of lead acetate on central catecholamine function in postnatal mouse. *Fed. Proc., Fed. Am. Soc. Exp. Biol.* **36**: 405 (Abstr.).

Selye, H., Tuchweber, B., and Bertok, L. (1966). Effect of lead acetate on the susceptibility of rats to bacterial endotoxins. *J. Bacteriol.* **91**, 884–890.

Seppalainen, A., and Hernberg, S. (1972). A sensitive technique for detecting subclinical lead neuropathy. *Br. J. Ind. Med.* **29**, 443–449.

Seppalainen, A. M., Tota, S., Hernberg, S., and Kock, B. (1975). Subclinical neuropathy at safe levels of lead exposure. *Arch. Environ. Health* **30**, 180–183.

Silbergeld, E. K., and Adler, H. S. (1978). Subcellular mechanism of lead neurotoxicity. *Brain Res.* **148**, 451–467.

Silbergeld, E. K., and Chisolm, J. J. (1976). Lead poisoning: altered urinary catecholamine metabolites as indicators of intoxication in mice and children. *Science* **192**, 153–154.

Silbergeld, E. K., and Goldberg, A. M. (1973). A lead-induced behavior disorder. *Life Sci.* **13**, 1275–1283.

Silbergeld, E. K., and Goldberg, A. M. (1975). Pharmacological and neurochemical investigations of lead-induced hyperactivity. *Neuropharmacology* **14**, 431–444.

Silbergeld, E. K., Fales, J. T., and Goldberg, A. M. (1974). Lead evidence for a prejuctional effect on neuromuscular function. *Nature (London)* **247**, 49–50.

Silbergeld, E. K., Hruska, R. E., Miller, L. P., and Eng, N. (1980). Effects of lead *in vivo* and *in vitro* on GABAergic neurochemistry. *J. Neurochem.* **34**, 1712–1718.

Six, K. M., and Goyer, R. A. (1970). Experimental enhancement of lead toxicity by low dietary calcium. *J. Lab. Clin. Med.* **76**, 933–942.

Six, K. M., and Goyer, R. A. (1972). The influence of iron deficiency on tissue content and toxicity of ingested lead in the rat. *J. Lab. Clin. Med.* **79**, 128–136.

Sleet, R. B., and Soares, J. H., Jr. (1979). Some effects of vitamin E deficiency on hepatic xanthine dehydrogenase activity, lead and alpha tocopherol concentrations in tissues on lead-dosed Mallard ducks. *Toxicol. Appl. Phrmacol.* **47**, 71–78.

Smith, C. M., DeLuca, H. F., Tanaka, Y., and Mahaffey, K. R. (1978). Stimulation of lead absorption by vitamin D administration. *J. Nutr.* **108,** 843–847.

Smith, C. M., DeLuca, H. F., Tanaka, Y., and Mahaffey, K. R. (1981). Effect of lead ingestion on functions of vitamin D and its metabolites. *J. Nutr.* **111,** 1321–1329.

Sobotka, T. J., and Cook, M. P. (1974). Postnatal lead acetate exposure in rats: possible relationship to minimal brain dysfunction. *Am. J. Ment. Defic.* **79,** 5–9.

Sobotka, T. J., Brodie, R. E., and Cook, M. P. (1975). Psychophysiological effects of early lead exposure. *Toxicology* **5,** 175–191.

Sorrel, M., Rosen, J. F., and Roginsky, M. (1977). Interactions of lead, calcium, vitamin D and nutrition in lead-burdened children. *Arch. Environ. Health* **32,** 160–164.

Talmage, R. V., Vander Wiel, C. J., and Norimatsu, H. (1978). A reevaluation of the cause of acute hypercalcemia following intravenous administration of lead acetate. *Calcif. Tissue Int.* **26,** 149–153.

Toraason, M. A., Barbe, J. S., and Knecht, E. A. (1981). Maternal lead exposure inhibits intestinal calcium absorption in rat pups. *Toxicol. Appl. Pharmacol.* **60,** 62–65.

Trejo, R. A., DiLuzio, N. R., Loose, L. D., and Hoffman, E. (1972). Reticuloendothelial and hepatic functional alterations following lead acetate administration. *Exp. Mol. Pathol.* **17,** 145–158.

Vergic, V. E., and Mare, C. J. (1974). Lead poisoning in chickens and the effect of lead on Interferon and antibody production. *Can. J. Comp. Med.* **38,** 328–335.

Webb, R. C., Winquist, R. J., Victery, W., and Vander, A. J. (1981). *In vivo* and *in vitro* effects of lead on vascular reactivity in rats. *Am. J. Physiol.* **241,** H211–H216.

Wedeen, R. P., Maeka, J. K., Weiner, B., Lipat, G. A., Lyons, M. M., Vitale, L. F., and Joselow, M. M. (1975). Occupational lead nephropathy. *Am. J. Med.* **59,** 630–641.

Williams, B. J., Griffith, W. H. III, Albrecht, C. M., Pirch, J. H., and Hejtmancik, M. R. (1977). Effects of chronic lead treatment on some cardiovascular responses to norepinephrine in the rat. *Toxicol. Appl. Pharmacol.* **40,** 407–413.

Ziegler, E. E., Edwards, B. B., Jensen, R. L., Mahaffey, K. R., and Fomon, S. J. (1978). Absorption and retention of lead by infants. *Pediat. Res.* **12,** 29–34.

11

Work at High Altitude in Dusty Environments

Robert F. Grover

I. THE HYPOXIA OF HIGH ALTITUDE

When man ascends to high altitude, the most significant consequence is a decrease in his capacity for muscular work. Let us explore the physiological mechanisms responsible for this phenomenon. To begin with, what do we mean by "high altitude?" The earth is surrounded by a blanket of air which extends for thousands of feet above the surface of the earth. Because this air has weight, it exerts pressure, just as the water in the ocean exerts pressure that becomes greater with increasing depth. Thus, when you are at sea level, i.e., at the bottom of the blanket of air, the atmosphere exerts a pressure of 1 ton per square foot. As measured

AIR POLLUTION

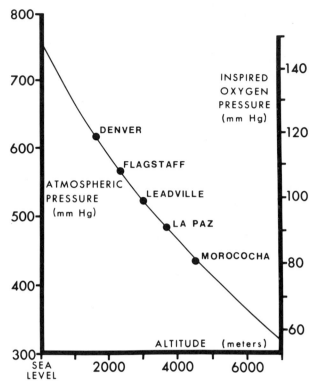

Fig. 1. With ascent from sea level to high altitude, total atmospheric pressure decreases. As a consequence, inspired oxygen pressure also falls. The magnitude of this
atmospheric hypoxia is indicated for Denver, Colorado; Flagstaff, Arizona; Leadville,
Colorado; La Paz, Bolivia; and Morococha, Peru. (From Grover, 1979a, by permission of
Westview Press.)

by a barometer, this pressure will support a column of mercury (Hg) 760
mm high. When we ascend to altitudes above sea level, we move up
through the atmospheric blanket. As a consequence, less of the atmosphere is above us to exert pressure, i.e., the total atmospheric pressure
decreases. Hence, ascent to high altitude means exposure to a reduction
in atmospheric pressure (Fig. 1) (Dill and Evans, 1970).

At all altitudes, the atmosphere consists of 21% oxygen and 79% nitrogen. Consequently, the partial pressure of oxygen is equal to 21% of the
total atmospheric pressure. It follows that if the atmospheric pressure is
decreased, then the partial pressure of oxygen must also be decreased.
What is the magnitude of this atmospheric hypoxia? We express this as
the partial pressure of oxygen inspired into the lung where the air
becomes heated to 37° and saturated with water vapor, which exerts its

own pressure at 47 mm Hg. Because this water vapor pressure contributes to the total atmospheric pressure, it must be subtracted before we can calculate the partial pressure of the remaining gases. When this is done, the inspired oxygen pressure is 150 mm Hg at sea level, 100 mm Hg at 3000 m altitude, and 75 mm Hg at 5500 m altitude (Fig. 1).

II. REDUCED WORK CAPACITY

How much does this atmospheric hypoxia reduce work capacity? Exercise physiologists measure aerobic working capacity as maximum oxygen uptake ($V_{O2_{max}}$). This is done in the laboratory, by having the subject exercise on either a treadmill or a bicycle ergometer. For each work load, oxygen uptake is measured. As the work load is increased progressively, oxygen uptake rises, but as the point of exhaustion is approached, a further increment in work load produces no further rise in oxygen uptake. This plateau is defined as $V_{O2_{max}}$ (Astrand and Rodahl, 1970). Many investigators have examined the influence of altitude on $V_{O2_{max}}$. These collected observations indicate that from sea level to 1500 m elevation, there is no measurable reduction in $V_{O2_{max}}$. With ascent to greater altitudes, however, there is a linear decrease in $V_{O2_{max}}$, averaging 10% for each 1000 m above 1500 m (Fig. 2) (Buskirk, 1969). In other words, the

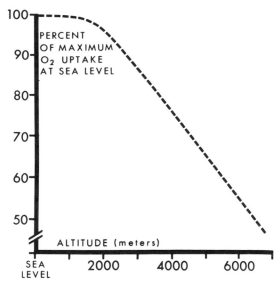

Fig. 2. Maximum oxygen uptake is reduced at high altitude. From sea level to 1500 m, there is no consistent decrease; however, above this threshold, there is a progressive reduction of 10% per 1000 m. (From Grover, 1979a, by permission of Westview Press.)

threshhold for this phenomenon is approximately at the elevation of Denver, Colorado. On the average, when people ascend another 1500 m (to 3000 m), $V_{O2_{max}}$ is reduced 15%; at 4500 m, $V_{O2_{max}}$ is reduced 30%. Again, these are average figures for groups of people. Within a group there will be individual variability, the decrease in $V_{O2_{max}}$ being greater in some people than in others. This individual variability is characteristic of all biological systems (Grover, 1978).

With the exception of athletes, most people (e.g. carpenters) function at work loads far below their aerobic working capacity. How, then, do we extrapolate the reduction in $V_{O2_{max}}$ at high altitude to the effect of altitude on people performing routine activities? When a group of physically trained men were taken to 4300 m altitude, their $V_{O2_{max}}$ was reduced 27% as predicted, with no improvement during the 2 weeks of their sojourn at high altitude. Initially, their endurance at 75% of $V_{O2_{max}}$ was also reduced. Over the next 2 weeks at high altitude, however, their endurance capacity at the same work load increased by 50% over its initial value. Hence, endurance at submaximal exercise improved markedly even though there was no increase in $V_{O2_{max}}$ (Maher *et al.*, 1974). This important observation tells us that measurements of $V_{O2_{max}}$ overestimate the impact of high altitude on submaximal work performance. More studies of submaximal endurance at high altitude are needed to quantitate these relationships.

III.   THE OXYGEN TRANSPORT SYSTEM

Now let us examine the physiological mechanisms by which atmospheric hypoxia reduces physical working capacity. The performance of sustained muscular work requires that the working muscles receive a continuous supply of the oxygen needed for the metabolic processes that generate energy. This supply of oxygen is provided by the so-called oxygen transport system. Oxygen in the atmosphere is brought into the lungs by the act of breathing, i.e., pulmonary ventilation. From the distal airways in the lung, called alveoli, oxygen diffuses across the thin alveolar-capillary membrane into the blood. Within the blood, oxygen is bound chemically to a complex protein called hemoglobin that is contained within the red blood cells. This oxygenated blood is then pumped by the heart into the arteries, which distribute it to all tissues in the body including the working muscles. The arteries then subdivide into capillaries where the oxygen is released from the hemoglobin and diffuses into the surrounding tissue. Hence, the structural components of the oxygen transport system are the lungs, the blood, and the cardiovascular system;

the functional components are pulmonary ventilation, blood oxygenation, and circulation.

A. Pulmonary Ventilation

How do the components of the oxygen transport system respond to the atmospheric hypoxia of high altitude? We know that acute hypoxia stimulates ventilation. This is brought about by the carotid bodies on the walls of the carotid arteries, which carry blood to the brain. These chemoreceptors are sensitive to changes in arterial oxygen tension. Thus, with acute hypoxia, i.e., a lowering of the partial pressure of oxygen in the inspired air, there is a parallel reduction in the partial pressure of oxygen in the blood within the pulmonary capillaries. This lowering of arterial oxygen tension stimulates the carotid bodies, which

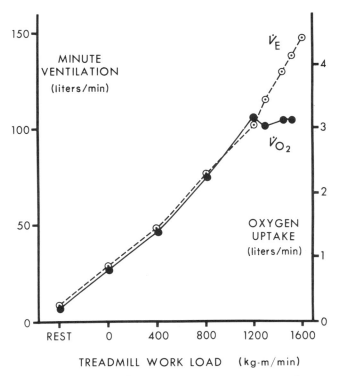

Fig. 3. Pulmonary ventilation ($\dot{V}_E$) does not limit oxygen uptake ($\dot{V}_{O_2}$) at moderately high altitude. With a subject exercising on a treadmill at 3100 m altitude, as the work load (inclination) is increased progressively, both $\dot{V}_{O_2}$ and $\dot{V}_E$ rise. Once maximum $\dot{V}_{O_2}$ is reached, a further increase in work load results in no further rise in $\dot{V}_{O_2}$ even though $\dot{V}_E$ continues to increase by 50%.

send neural impulses to the respiratory center within the brain. The respiratory center then increases its output to the respiratory muscles, including the diaphragm, resulting in deeper breaths of greater frequency (Weil *et al.*, 1970).

When people ascend to high altitude, ventilation increases. Although it is generally assumed that this increase in ventilation results from stimulation of the carotid bodies as a consequence of atmospheric hypoxia, the true mechanisms for the hyperventilation observed at high altitude are not well understood. Rather remarkably, the increase in ventilation at high altitude is just sufficient to offset the decrease in atmospheric pressure (Grover, 1965). Because air is compressible, decreasing atmospheric pressure means that the air is less compressed, i.e., it is less dense. Consequently, a given volume of air will contain fewer molecules of oxygen at high altitude than at sea level. It follows that to bring the required number of molecules of oxygen into the lung, larger volumes of air must be moved, i.e., ventilation must be increased. Fortunately, this does occur, but we do not understand exactly how (Reeves *et al.*, 1976). Does the need to increase ventilation limit $V_{O_{2max}}$? The answer is no, at least at altitudes in the 3000–4500-m range. We have demonstrated that during the incremental work tests employed to measure $V_{O_{2max}}$, as the work load is increased beyond $V_{O_{2max}}$, ventilation continues to increase, but does not result in any further increase in oxygen uptake (Fig. 3). From this, we conclude that pulmonary ventilation is not the limiting factor in oxygen uptake at altitudes up to 15,000 ft.

B. Blood Oxygenation

Once ventilation has brought oxygen into the alveoli of the lung, the next step in oxygen transport is the diffusion of the oxygen across the alveolar capillary membrane and into the red cells within the blood flowing through the pulmonary capillaries. The diffusing capacity (transfer factor) of the lung is defined as the quantity of oxygen transferred from air to blood per minute for each mm Hg difference in oxygen pressure between alveolar air and mixed venous blood. When the oxygen pressure difference is high, large quantities of oxygen are transferred rapidly from air to blood. However, with ascent to high altitude and the consequent lowering of the alveolar oxygen tension, the "driving pressure" becomes smaller and, consequently, the rate of oxygen transfer becomes slower. At rest or during submaximal exercise when the oxygen demands are moderate, the diffusing capacity of the lung poses no problem, even at fiarly high altitudes. Under conditions of maximal exercise, however, when the demands for oxygen are very high and the red cells are exposed to the alveolar gas for only a very short period because of the high blood velocity during exercise, then it is possible theoretically to have a diffusion limitation in oxygen uptake. In practice, however, this becomes a significant factor only at extremely high altitudes (Johnson, 1977; West, 1980).

1. Arterial Desaturation

Once oxygen has diffused across the alveolar capillary membrane to reach the red blood cells in the pulmonary capillaries, it is then bound chemically to hemoglobin. The relationship between the oxygen tension and the quantity of oxygen bound to hemoglobin is described by the hemoglobin–oxygen dissociation curve (Fig. 4). Because of the sigmoid shape of this curve, as oxygen tension is decreased, oxygen saturation falls slowly along the relatively flat portion of the curve. When the steeper portion of the curve is reached, further reductions in oxygen tension produce large decreases in saturation.

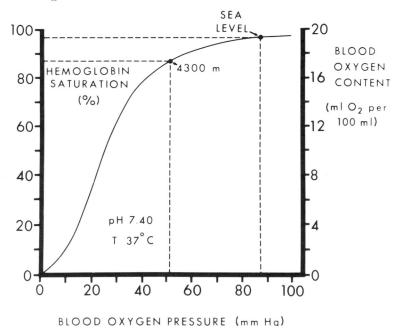

Fig. 4. The hemoglobin–oxygen dissociation curve for human blood, showing the sigmoid relationship between oxygen pressure, the resulting saturation of hemoglobin, and the corresponding blood oxygen content when the hemoglobin concentration is 15 gm/dl. With ascent from sea level to 4300 m altitude, the large reduction in oxygen pressure results in a relatively small decrease in saturation. (From Grover, 1979a, by permission of Westview Press.)

To appreciate the effect of high altitude on arterial blood oxygenation, we must first define its effect on arterial oxygen tension. Earlier, we demonstrated that a decrease in total atmospheric pressure results in a lowering of inspired oxygen tension within the larger airways of the lung. Because this pressure is higher than the oxygen tension of venous blood entering the lung, it provides the "driving pressure" for diffusion of oxygen from air to blood. As diffusion proceeds, the oxygen tension in the venous blood rises, while the oxygen tension in the air falls. Diffusion continues until equilibrium is reached, at which time blood oxygen tension is virtually equal to alveolar oxygen tension (Fig. 5). At sea level, this equilibrium point is at about 90 mm Hg; however, because this equilibrium point varies somewhat from one region of the lung to another, the average blood oxygen tension leaving the lung (i.e., the arterial oxygen tension) is approximately 85 mm Hg. With ascent to high altitude, the arterial oxygen tension falls, approximately, to 60 mm Hg

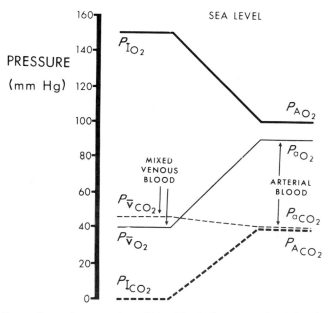

Fig. 5. Gas exchange between air and blood in the lung at sea level. Inspired air has a high oxygen pressure ($P_{I(O_2)}$), but virtually no carbon dioxide ($P_{I(CO_2)}$). Mixed venous blood entering the lung has a low O_2 pressure ($P_{\bar{V}(O_2)}$) and a relatively high CO_2 pressure ($P_{\bar{V}(CO_2)}$). These pressure differences cause diffusion of O_2 from air to blood until the alveolar O_2 pressure ($P_{A(O_2)}$) reaches equilibrium with the pulmonary capillary (arterial) blood ($P_{a(O_2)}$). Concurrently, CO_2 leaves the blood, until the alveolar CO_2 pressure ($P_{A(CO_2)}$) equilibrates with the arterial CO_2 pressure ($P_{a(CO_2)}$). (From Grover et al., 1979b, by permission of W. B. Saunders.)

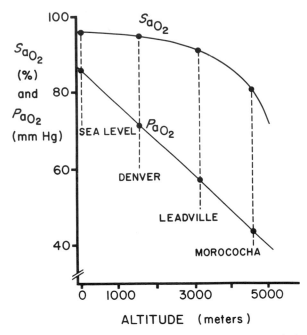

Fig. 6. With ascent from sea level to high altitude, the decrease in inspired oxygen pressure (Fig. 1) produces a linear fall in arterial oxygen pressure ($P_{a(O_2)}$), but a much smaller reduction in saturation ($S_{a(O_2)}$) because of sigmoid shape of the dissociation curve (Fig. 4).

at 3000 m and 50 mm Hg at 4500 m. Now, if we refer to the hemoglobin–oxygen dissociation curve, we find that the arterial oxygen saturations that correspond to these oxygen tensions are 95%, 88%, and 82%, respectively. Put another way, this indicates that whereas ascent to 4500 m altitude lowers arterial oxygen tension from 85 to 50 mm Hg, the effect on arterial oxygen saturation is much less, so that arterial blood is still 82% saturated (Fig. 6).

Because it is the hemoglobin within the red cells that transports oxygen in the blood, the quantity of oxygen contained in a given volume of blood will be determined by the hemoglobin concentration as well as the extent to which that hemoglobin is saturated with oxygen. When 1 g of hemoglobin is saturated completely, it contains 1.34 ml of oxygen. At sea level, the normal hemoglobin concentration is 15 g/100 ml of blood; hence, it has an oxygen-carrying capacity of 20 ml oxygen per 100 ml blood (ml/dl). At 95% saturation, the content would be 19 ml/dl, and at 82% saturation, the content would be 16.6 ml/dl. Obviously, this would reduce oxygen transport and, in fact, this fall in arterial oxygen satura-

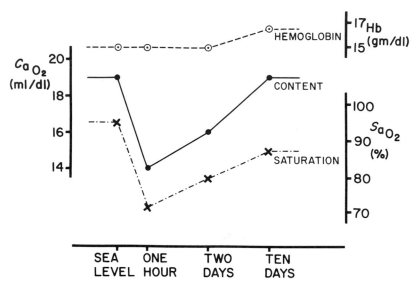

Fig. 7. Blood oxygenation during moderate exercise at high altitude. One hour after ascent to 4300 m, arterial oxygen saturation ($S_{a(O_2)}$) and content ($C_{a(O_2)}$) fall markedly because of relative hypoventilation and poor matching of lung perfusion to ventilation. Within 2 days, saturation and content improve. After 10 days, hemoconcentration raises hemoglobin concentration (Hb), and with further improvement in saturation, arterial oxygen content is fully restored to the original value at sea level. (From Grover *et al.*, 1982, by permission of McGraw-Hill.)

tion (hypoxemia) is a major factor in reducing aerobic working capacity during the first hours and days at high altitude (Fig. 7).

2. *Hemoconcentration*

Compensatory mechanisms now come into play. As stated earlier, pulmonary ventilation increases at high altitude. This hyperventilation eliminates an excessive amount of carbon dioxide from the lungs and, as a consequence, the carbon dioxide tension in the blood falls (hypocapnia). By mechanisms that are not understood, this hypocapnia is associated with the removal of water from the blood plasma, i.e., hemoconcentration. We know that if hypocapnia is prevented in persons exposed to high altitude, hemoconcentration does not occur (Grover *et al.*, 1976b). Removal of water from the plasma increases the concentration of all remaining elements in the blood, including red blood cells. Consequently, the hematocrit, i.e., the concentration of red cells in the blood, rises. The total number of red cells has not changed; they are simply contained in a smaller plasma volume.

3. Arterial Oxygen Content

An increase in hematocrit means that the hemoglobin concentration of the blood has been increased also. For example, ascent to high altitude results often in an increase in hematocrit from 45% up to 51% within the first week, increasing the hemoglobin concentration from 15 to 17 g/100 ml blood. The oxygen-carrying capacity is then increased from 20 to 23 ml/dl. Consequently, even though the saturation has fallen to 82%, the arterial oxygen content has now been increased to 19 ml/dl, i.e., the same value that existed at sea level (Fig. 7). In other words, hemoconcentration counteracts the effect of the reduction in saturation to preserve the arterial oxygen content, even though the individual ascends to 4500 m. This means that blood oxygenation is no longer a factor contributing to the reduction in aerobic working capacity. It should be emphasized that this is a delayed response to high altitude. Upon initial ascent to high altitude, the decrease in arterial oxygen saturation does lower arterial oxygen content. Over the subsequent days at high altitude, however, hemoconcentration occurs and the hematocrit rises, offsetting the effect of desaturation and restoring the arterial oxygen content to normal (Fig. 7).

C. Cardiac Output

1. Heart Rate

The third major component of oxygen transport is the delivery of oxygenated blood to the body by the cardiovascular system. For the body as a whole, this depends upon the cardiac output, which is determined by the quantity of blood ejected from the heart with each beat (i.e., the stroke volume) and the number of beats per minute (i.e., the heart rate). To meet the increased oxygen demands of exercise, cardiac output increases, primarily by an increase in heart rate with little increase in stroke volume. Consequently, maximum cardiac output is determined by maximum heart rate, which is about 200 beats/min in young adults, but decreases progressively with increasing age (Fox et al., 1971). During the initial hours and days following ascent to high altitude, submaximal-exercise heart rate and hence cardiac output increase (Fig. 8). Systemic oxygen transport is defined as the product of arterial oxygen content and cardiac output; therefore, during the early response to high altitude, the decrease in arterial saturation that lowers arterial oxygen content is offset by the increase in cardiac output, and consequently, submaximal oxygen transport is maintained (Fig. 9). This compensation, which requires an increase in heart rate, is limited, however, because maximum heart rate cannot be increased. Consequently, during maximal exertion, even though maximum cardiac output may be as great as it was prior to ascent, the arterial oxygen content remains reduced, which means that maximum oxygen transport is reduced also. This is reflected by the decrease in $V_{O2_{max}}$.

2. Reduced Stroke Volume

As adaptation proceeds over the first week at high altitude, the pattern of oxygen transport changes. As we have stated, hemoconcentration offsets the decrease in saturation and restores arterial oxygen content to normal (Fig. 7); however, as hemoconcentration occurs, there is a decrease in cardiac stroke volume. Furthermore, the initial increase in heart rate is not sustained but returns toward sea-level values. These factors combine to reduce cardiac output to values significantly below those observed at sea level, both at rest and during exercise (Fig. 8)

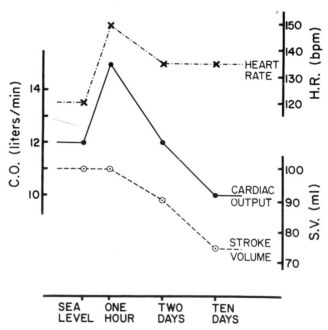

Fig. 8. Time course of adjustment in exercise cardiac output during adaptation to high altitude. Initially, an increase in heart rate (HR) elevates cardiac output (CO). Within 2 days, this tachycardia subsides, in combination with a slight decrease in stroke volume (SV), returning cardiac output to preascent levels. After 10 days, cardiac output has fallen significantly below original sea-level value because of further substantial reduction in stroke volume. (From Grover *et al.*, 1982, by permission of McGraw-Hill.)

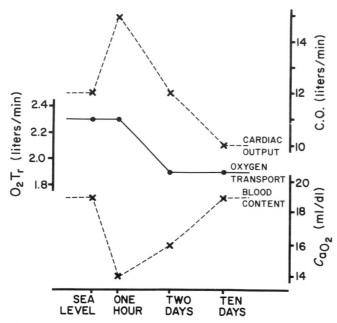

Fig. 9. Oxygen transport (O_2Tr) during submaximal exercise at high altitude. During the first hour at 4300 m, the increase in cardiac output (CO) offsets the decrease in blood oxygen content ($C_{a(O_2)}$) to preserve O_2 transport. After 2 days, however, O_2Tr has been reduced below the original sea-level value because of the fall in cardiac output. After 10 days, even though blood O_2 content has been restored (Fig. 7), a further decline in cardiac output holds O_2 transport at subnormal levels. Greater O_2 extraction is then required to meet O_2 requirements. (From Grover *et al.*, 1982, by permission of McGraw-Hill.)

(Alexander *et al.*, 1967; Vogel *et al.*, 1974b). Thus, after a week or more of adaptation to high altitude, oxygen transport is less than it was at sea level under all conditions. This requires an increase in oxygen extraction from blood in the systemic capillaries to meet the oxygen demands of the tissues (Alexander *et al.*, 1967). During maximal exertion, the smaller cardiac stroke volume, often combined with a lowering of maximal heart rate (Hartley *et al.*, 1974) results in a significant reduction of maximum cardiac output, which now becomes the factor that limits oxygen transport and causes the reduction in V_{O_2max} (Fig. 10) (Vogel *et al.*, 1974b).

Thus, we see that although V_{O_2max} is reduced immediately following ascent to high altitude and remains reduced throughout the sojourn at altitude (Reeves *et al.*, 1967), this is the result of changing aspects of oxygen transport. Initially, maximum oxygen transport is reduced because of a decrease in arterial oxygen content that is not accompanied by a decrease in maximum cardiac output (Stenberg *et al.*, 1966). After

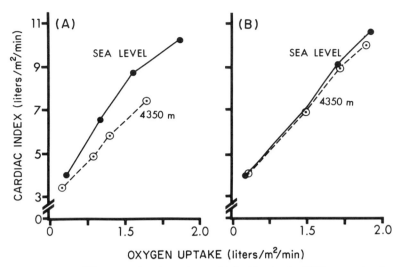

Fig. 10. Racial differences in adaptation to high altitude. At 4350 m, sea-level natives (A) have a decrease in maximum oxygen uptake, as in Fig. 2, caused in part by the reduction in cardiac output, as in Fig. 8. In contrast, when Quechua Indians (B), native to 4350 m in the Andes, descend to sea level, there is no change in either maximum oxygen uptake or cardiac output. Both oxygen uptake and cardiac output (index) have been adjusted for body size and expressed per square meter of body-surface area. (Adapted from Vogel *et al.*, 1974a,b, by permission.)

several days of adaptation, however, oxygen content is restored to normal, but now maximum cardiac output is reduced, and this becomes the factor that limits oxygen transport. [Although this reduction in stroke volume and cardiac output under all conditions has been demonstrated by several investigators, others, who employed rebreathing techniques reported that there is no reduction in either stroke volume or submaximal cardiac output (Pugh, 1964; Saltin *et al.*, 1968) and that maximal cardiac output is reduced only by the reduction in maximum heart rate. Nevertheless, those studies which do show a decrease in stroke volume and a reduction in cardiac output during submaximal exercise are reliable, because they show also the increase in oxygen extraction (Alexander *et al.*, 1967) that would be required to maintain oxygen delivery.]

It is remarkable that cardiac stroke volume should be decreased in healthy young individuals, free of cardiac or coronary artery disease, who ascend to only moderately high altitudes. What is the mechanism responsible for this decrease in stroke volume? It is unlikely that the contractile properties of the myocardium have been impaired by hypoxia, because direct measurements indicate no alteration in myocardial metabolism and no reduction of the oxygen tension of blood draining

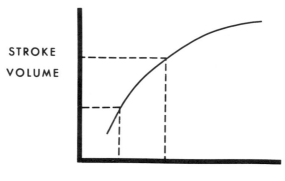

STROKE VOLUME

LV FILLING PRESSURE

Fig. 11. The Frank–Starling function curve for the left ventricle (LV) of the heart. The greater the filling of the ventricle during diastole, the greater the stroke volume during the subsequent systole. Hence, a decrease in diastolic filling reduces stroke volume.

from the myocardium (Grover *et al.*, 1976a), which is a reflection of the tissue oxygen tension. It is, therefore, more likely that the heart is responding to external circulatory influences, such as a reduction in diastolic filling (preload). We have noted that the decrease in stroke volume occurs in association with hemoconcentration. Because hemoconcentration results from removal of water from the plasma, there must be a decrease in plasma volume and, hence, in total blood volume. This could impair the return of venous blood to the heart, particularly during upright exercise. If less blood returns to the heart to fill the ventricle during diastole, there is less blood to eject during the next systole, i.e., stroke volume is decreased (Fig. 11) (Grover *et al.*, 1976b). If this explanation is correct, the heart is behaving in a perfectly normal fashion and is simply pumping less because it receives less. We know that after a return to low altitude, there is a reversal of the changes that occurred during adaptation to high altitude. Water enters the plasma, lowering the hematocrit and thereby reversing the hemoconcentration. Furthermore, stroke volume and cardiac output increase.

IV. LONG-TERM RESIDENTS OF HIGH ALTITUDE

Up to now we have been considering the physiological adjustments that occur in the body during the first days and weeks following ascent to high altitude. Let us now consider the long-term adjustments that occur in inidividuals who reside at high altitude for many years, including people who are born at high altitude. Populations living at altitudes above 10,000 ft are found in only a limited number of locations around the world. Naturally, these are mountainous areas, including the Andes of South America, the Himalaya of central Asia, the Caucasus of southern Russia, the mountains of Ethiopia in northern Africa, and, of course, the Rocky Mountains of the western United States (Grover, 1974). Among the populations living in these various regions, there are important racial differences in their adaptation to high altitude, although most are remarkably well adapted to chronic hypoxia. Both clinical observations and physiological studies indicate, however, that the residents least well adapted to high altitude are those of the Rocky Mountains. Perhaps this is because they are relative newcomers, having lived at high altitude for no more than four generations, with the majority having no more than one or two generations of exposure. In contrast, most other mountain populations have been at high altitude for tens of thousands of years.

Nevertheless, we are most interested in those persons born in the United States whose occupations prompt them to move to high altitude, often to remain for many years. In the state of Colorado, we find this situation in the city of Leadville whose elevation is 3100 m. The industrial base of this community is the nearby Climax molybdenum mine located at 3400 m. Various aspects of adaptation to high altitude in this population have been studied extensively and compared with those of sojourners spending only a few weeks at the same altitude.

Polycythemia

When sufficient to lower arterial oxygen saturation, the atmospheric hypoxia of high altitude leads to an increase in the total number of circulating red blood cells (secondary polycythemia) (Weil *et al.*, 1968). A sustained decrease in saturation causes an increased production of erythropoietin from the kidney, which in turn stimulates the bone marrow to produce more red blood cells. However, since red-cell production is a relatively slow process, it takes several months of sustained hypoxemia to increase total red-cell mass. This in itself tends to increase total blood volume. In addition, the initial shrinkage of plasma volume is not sustained; it returns gradually to the preascent volume. Obviously, this too tends to increase blood volume and as a consequence the individual who has been living at high altitude for many months or several years has a total blood volume larger than that found in sea-level residents (Reynafarje, 1958; Weil *et al.*, 1968). Because both red cell mass and plasma volume have increased, however, the hematocrit and hemoglobin concentrations in Leadville residents are only slightly higher than those in people living at sea level (Weil *et al.*, 1968).

How does this increased blood volume affect the cardiac output of the high altitude resident? If hemoconcentration and a smaller blood volume were the cause of the reduced stroke volume and cardiac output in the sojourner, one could predict that with subsequent expansion of the blood volume, the cardiac output would return to normal. This is, in fact, not the case. Direct measurements indicate cardiac output in long-term residents and natives of Leadville is subnormal by sea-level standards (Harley et al., 1967). Why this should be we do not know; however, we do know that when these people descend to low altitude, the hematocrit falls promptly, indicating further expansion of plasma volume and cardiac output increases. Furthermore, descent to sea level results in an increase in aerobic working capacity of a magnitude comparable to the decrease seen in individuals moving from low to high altitude (Grover et al., 1967). From this observation, we infer that the decrease in aerobic working capacity observed in the sojourner at high altitude persists indefinitely, even for years, and that he never regains the working capacity he had at sea level (except with actual descent to sea level). Hence, in terms of working capacity, the long-term resident of Leadville has no advantage over the newcomer. In contrast, the Indian natives of the high Andes have both an aerobic working capacity far greater than that of the newcomer to 4300 m and cardiac output that is normal by sea-level standards (Vogel et al., 1974a). Furthermore, when the Andean native descends to sea level, he has virtually no increase in cardiac output or in aerobic working capacity (Fig. 10). Obviously, there are important physiological differences between high altitude natives of the Rocky Mountains and those of the Andes, but the basis for these differences is not known.

V. DUST EXPOSURE

In considering further the people who live at high altitude in the Rocky Mountains, we mentioned that most of the men who live in Leadville are employed by the Climax Molybdenum Company. This is a hard-rock mining operation and, consequently, dust exposure is an important consideration. Recall that one basic aspect of adaptation to high altitude is an increase in pulmonary ventilation. At Climax, where the atmospheric pressure is approximately two-thirds of that at sea level, ventilation must be about 50% greater than at sea level under all conditions. This means that for a given concentration of dust in the air, the absolute number of dust particles inhaled per minute will be 50% greater at Climax than at sea level. Obviously, this is a major concern, and the concentration of dust must be reduced to an absolute minimum. In addition, as in all hard-rock mining operations where dust is a consideration, the men exposed to this dust must wear air filters called "respirators." Because any air filter presents some resistance to air flow, what are the implications of wearing air filters during exercise at high altitude?

A. Work of Breathing

To answer this question, we must consider the work of breathing. Air flow into the lung results from expansion of the thorax and descent of the diaphragm, which lowers the air pressure within the lung below that of the ambient air. Conversely, air leaves the lung when the inspiratory muscles are relaxed, permitting the elastic recoil of the thorax and lungs to raise the air pressure within the lung above that of the ambient air. With normal breathing, air moves back and forth, in and out of the lung, the direction of air flow being reversed every 1 to 2 seconds. Although the diameter of the airways does not change with altitude, the density of the air does decrease; consequently, there is less resistance to air flow.

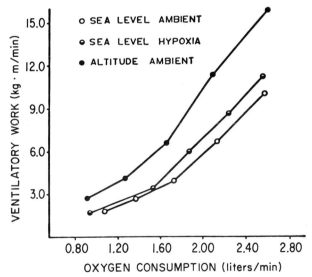

Fig. 12. Ventilatory work at various levels of oxygen consumption during exercise. At sea level, reducing the concentration of oxygen in the inspired air from 21% (ambient) to 15% (hypoxia) without a change in atmospheric pressure produces only a small increase in ventilatory work. With adaptation to the atmospheric hypoxia at 3100 m altitude, however, there is a large increase in ventilatory work. This is caused by the greater volumes of air that must be moved to compensate for the decreased air density, and the greater effort required to overcome the compressibility of this less dense air, particularly at high respiratory frequencies. (From Thoden *et al.*, 1969, by permission.)

Generally, decreased resistance to air flow tends to decrease the work of breathing; however, as altitude increases, greater volumes of air must be moved (50% greater volume at 3400 m elevation). Thus each inspiration requires a greater expansion of the chest, which tends to increase the work of breathing (Thoden *et al.,* 1969).

Another factor not often considered is that less dense air is more compressible. As we stated, air moves in and out of the lung in a cyclic fashion. As the chest expands, the air within the lung will expand before the pressure falls enough to cause air to flow in through the nose. Conversely, with relaxation of the chest, the air within the lung will be compressed before the pressure is elevated enough to force air out of the lung. At low respiratory rates, this effect is minimal; however, when exercise increases ventilation and respiratory frequency exceeds 30 breaths per minute, the air within the lung is expanded and compressed alternately every second. To overcome this compressibility of the air at high altitude, the work of breathing must increase considerably (Fig. 12) (Thoden *et al.,* 1969). Placing an air filter over the nose and mouth can only compound the difficulty of breathing. As a consequence, a given air filter, which may be quite tolerable near sea level, may become virtually intolerable when worn during exercise at high altitude.

B. Respiratory Air Filters

Recommending the use of air filters by men working in dusty environments at high altitude raises very serious practical problems not easily resolved. Is there any objective way of deciding whether a given individual can tolerate an air filter? Unfortunately, relying on the individual's opinion or impression is not entirely satisfactory. To help resolve this problem, a new technique has been introduced. It depends on the use of an ear oximeter to measure arterial saturation at high altitude while the individual is exercising on a bicycle ergometer. Because a healthy person with normal lungs can perform moderate exercise at 3400 m without experiencing a decrease in arterial saturation, the bicycler is tested twice, first not wearing and then wearing the air filter. If his arterial saturation remains normals during both tests, it is assumed the air filter will cause him no undue distress. On the other hand, if arterial saturation decreases during the second test, it is assumed the bicycler cannot tolerate an air filter. In several hundred tests of this kind, the ear oximeter has proved to be a useful index of tolerance to air filters at high altitude (J. Smith, personal communication).

We may summarize the special considerations of work at high altitudes in dusty environments as follows. Ascent to high altitude reduces aerobic working capacity, and the reduced capacity reflects a decrease in cardiac output and, hence, in oxygen transport under all conditions. Impaired performance persists as long as the individual remains at high altitude, but is reversed completely upon descent to low altitude. This means that an individual must work at a slower pace; hence, he must have more time to complete a given task, or others must share his workload. Either way, more man-hours are required, and the economic consequences are obvious. If the task is mining or road construction, either of which involves exposure to dust, the workers may have to wear air filters for their own protection. Such air filters pose special problems at high altitude, because the atmosphere is less dense. Although air-flow resistance is reduced, greater volumes of air must be moved, and the greater compressibility of the less dense air increases the work of breathing substantially at higher respiratory frequencies.

REFERENCES

Alexander, J. K., Hartley, L. H., Modelski, M., and Grover, R. F. (1967). Reduction of stroke volume during exercise in man following ascent to 3,100 meter altitude. *J. Appl. Physiol.* **23,** 849–858.

Astrand, P. O. and Rodahl, D. (1970). Physical work capacity. "Textbook of Work Physiology," pp. 281–286. McGraw-Hill, New York.

Buskirk, E. R. (1969). Decrease in physical work capacity at high altitude. *In* "Biomedicine Problems of High Terrestrial Elevations" (A. H. Hegnauer, ed.), pp. 204–222. U. S. Army Res. Inst. Environ. Med., Natick, Massachusetts.

Dill, D. B., and Evans, D. S. (1970) Report barometric pressure! *J. Appl. Physiol.* **29,** 914–916.

Fox, S. M., Naughton, J. P., and Haskell, W. L. (1971). Physical activity and the prevention of coronary heart disease. *Ann. Clin. Res.* **3,** 404–432.

Grover, R. F. (1965). Effects of hypoxia on ventilation and cardiac output. *Ann. N. Y. Acad, Sci.* **121,** 662–673.

Grover, R. F. (1974). Man living at high altitudes. *In* "Arctic and Alpine Environments" (J. D. Ives and R. G. Barry, eds.), pp. 817–830. Methuen, London.

Grover, R. F. (1978). Adaptation to high altitude. *In* "Environmental Stress: Individual Human Adaptations" (L. J. Folinsbee, J. A. Wagner, J. F. Borgia, B. L. Drinkwater, J. A. Gliner, and J. F. Bedi, eds.), pp. 325–334. Academic Press, New York.

Grover, R. F. (1979a). High altitude physiology. *In* "High Altitude Geoecology" (P. J. Webber, ed.), pp. 127–138. Westview Press, Boulder, Colorado.

Grover, R. F. (1979b). Performance at altitude. *In* "Sports Medicine and Physiology" (R. H. Strauss, ed.), pp. 327–343. Saunders, Philadelphia.

Grover, R. F., Reeves, J. T., Grover, E. B., and Leathers, J. E. (1967). Muscular exercise in young men native to 3100 meters altitude. *J. Appl. Physiol.* **22,** 555–564.

Grover, R. F., Lufschanowski, R., and Alexander, J. K. (1976a). Alterations in the coronary circulation of man following ascent to 3100 m altitude. *J. Appl. Physiol.* **41,** 832–838.

Grover, R. F., Reeves, J. T., Maher, J. T., McCullough, R. E., Cruz, J. C., Denniston, J. C., and Cymerman, A. (1976b). Maintained stroke volume but impaired arterial oxygenation in man at high altitude with supplemental CO_2. *Circ. Res.* **38,** 391–396.

Grover, R. F., Reeves, J. T., Rowell, L. B., Saltzman, H. A., and Blount, S. G., Jr. (1982). The influence of environmental factors on the cardiovascular system. *In* "The Heart, Arteries, and Veins" (J. W. Hurst, R. B. Logue, C. E. Rackley, R. C. Schlant, E. H. Sonnenblick, A. G. Wallace, and N. K. Wenger, eds.), pp. 1654–1670. McGraw-Hill, New York.

Hartley, L. H., Alexander, J. K., Modelski, M., and Grover, R. F. (1967). Subnormal cardiac output at rest and during exercise in residents at 3,100 meter altitude. *J. Appl. Physiol.* **23,** 839–848.

Hartley, L. H., Vogel, J. A., and Cruz, J. C. (1974). Reduction of maximal exercise heart rate at altitude and its reversal with atropine. *J. Appl. Physiol.* **36,** 363–365.

Johnson, R. L., Jr. (1977). Oxygen transport. *In* "Clinical Cardiology" (J. T. Willerson and C. A. Sanders, eds.), pp. 74–84. Grune & Stratton, New York.

Maher, J. T., Jones, L. G., and Hartley, L. H. (1974). Effects of high-altitude exposure on submaximal endurance capacity of men. *J. Appl. Physiol.* **37,** 895–898.

Pugh, L. G. C. E. (1964). Cardiac output in muscular exercise at 5,800 meters (19,000 ft.). *J. Appl. Physiol.* **19,** 441–447.

Reeves, J. T., Grover, R. F., and Cohn, J. E. (1967). Regulation of ventilation during exercise at 10,200 ft. in athletes born at low altitude. *J. Appl. Physiol.* **22,** 546–554.

Reynafarje, C. (1958). The influence of high altitude on erythropoietic activity. *In* "Homeostatic Mechanisms," pp. 132–146. Brookhaven Natl. Lab., Upton, New York.

Saltin, B., Grover, R. F., Blomqvist, C. G., Hartley, L. H., and Johnson, R. L., Jr. (1968). Maximal oxygen uptake and cardiac output after two weeks at 4300 meters. *J. Appl. Physiol.* **25,** 400–409.

Stenberg, J., Ekblom, B., and Messin, R. (1966). Hemodynamic response to work at simulated altitude, 4000 m. *J. Appl. Physiol.* **21,** 1589–1594.

Thoden, J. S., Dempsey, J. A., Reddan, W. G., Birnbaum, M. L., Forster, H. V., Grover, R. F., and Rankin, J. (1969). Ventilatory work during steady-state response to exercise. *Fed. Proc., Fed. Am. Soc. Exp. Biol.* **28,** 1316–1321.

Vogel, J. A., Hartley, L. H., Cruz, J. C., and Hogan, R. P. (1974a). Cardiac output during exercise in sea level residents at sea level and high altitude. *J. Appl. Physiol.* **36,** 169–172.

Vogel, J. A., Hartley, L. H., and Cruz, J. C. (1974b). Cardiac output during exercise in altitude natives at sea level and high altitude. *J. Appl. Physiol.* **36,** 173–176.

Weil, J. V., Jamieson, G., Brown, D. W., and Grover, R. F. (1968). The red cell mass-arterial oxygen relationship in normal men. *J. Clin. Invest.* **47,** 1627–1639.

Weil, J. V., Byrne-Quinn, E., Sodal, I. E., Friesen, W. O., Underhill, B., Filley, G. F., and Grover, R. F. (1970). Hypoxic ventilatory drive in normal man. *J. Clin. Invest.* **49,** 1061–1071.

West, J. B. (1980). Adaptation to extreme altitudes. *In* "Physiology: Aging, Heat and Altitude" (S. M. Horvath and M. K. Yousef, eds.), pp. 269–279. Elsevier/North-Holland, Amsterdam.

Index